多媒体通信技术基础

The Basic Principles of Multimedia Communications

沈晋原　编著

Jinyuan　Shen

国防工业出版社

National Defense Industry Press

·北京·

Bei jing

图书在版编目（CIP）数据

多媒体通信技术基础/沈晋原编著. —北京：国防工业出版社，2015.5
ISBN 978-7-118-10150-8

Ⅰ.①多… Ⅱ.①沈… Ⅲ.①多媒体通信－通信技术－高等学校－教材 Ⅳ.①TN919.85

中国版本图书馆 CIP 数据核字（2015）第 075253 号

※

*国防工业出版社*出版发行
（北京市海淀区紫竹院南路 23 号 邮政编码 100048）
天利华印刷装订有限公司印刷
新华书店经售

*

开本 787×1092 1/16 印张 22¾ 字数 560 千字
2015 年 5 月第 1 版第 1 次印刷 印数 1—2500 册 定价 58.00 元

（本书如有印装错误，我社负责调换）

国防书店：（010）88540777　　　发行邮购：（010）88540776
发行传真：（010）88540755　　　发行业务：（010）88540717

前　言

　　在国内理工科本科生、研究生教材及参考书中，目前还没有一本现成的与多媒体通信专业相关的中英文双语教材/参考书可用，很多开设双语课程的高等院校只能使用英文原版或原版引进版权参考书作为相关课程的教材。这里存在着如下几个问题：

　　1. 英文原版或原版引进版权参考书价格都相对较高，尤其是英文原版技术书籍，每本书大约要 200 美元左右，这是国内高等院校学生无法承受的。即使是原版引进版权参考书，其价格也要在 100 元人民币以上，这对国内高等院校学生，尤其是本科生无疑也是一个负担。

　　2. 原版引进版权参考书虽然相对于英文原版参考书价格低了很多，虽然是个负担，但学生仍可勉强接受。由于原版引进版权参考书数量和品种极为有限，所以内容上很难恰好符合所开课程的要求。

　　3. 国内学生由于受语言环境限制，尤其是本科生的英文水平仍然普遍有限，阅读英文原版专业书籍在速度上和理解上都存在很大困难。这使教师和学生双方对课程的理解和完成都产生了困难。

　　出于同样上述原因，在理工科专业领域执教、工作的专业技术人员及学习深造的研究生、学者也面临着上述同样的问题。尤其随着我国对外开放日益深入，国内外学术、专业交流日益频繁，更需要专业人员、学者能够尽多尽快掌握和了解中英文在自己专业领域中描述的不同，以便更好、更方便地与国外同行交流。

　　出于上述考虑，本书将是第一本这方面的中英文双语专业课教材或参考书。希望能够成为国内本科生同名课程的一本实用教材和国内研究生、同行们的一本必不可少的、方便、实用的参考书。

　　如上所述，本书的目的之一是让同领域内的专业技术人员、学生、学者，手持一书就可以了解中、英文电气、电子工程领域中同一概念、理念的不同说法，所以本书不着重于词对词，句对句的中英文翻译，而是着重于电子工程领域中同一概念、理念中英文的不同说法。换句话说，读者理解了书中同一概念、理念的中文解释后，可以直接用书中相应段落的英文解释和国外同行交流，而用不着再去翻查字典等辅助工具书籍。这是本书的特点之一。

　　另外，由于本书上述的有意识的特殊写法，使得本书的中英文部分可以相对独立地各成一体。这使得它不仅可以成为多媒体通信技术这一电气、电子工程领域中通信、信息及其他专业本科生高年级的专业课中英文双语课教材，也可以作为还没有能力和条件开此双语课课程的中文单语课教材。但不同于其他单语课教材的是，她含有了相应的英文讲解部分，这使得它可以成为一本从单纯中文课到中英文双语课过渡的一本很有价值的教材，这是本书的另一大特点。

　　全书共分 6 章，第一章介绍了多媒体技术的发展简史，引进了多媒体通信技术中所用的数据压缩、有损和无损编码、信息熵等重要概念；第二章进一步论述了多媒体及其多媒体通信技术的一些基本概念，包括它们的基本特性、应用领域及关键技术；第三章阐述了多媒体通信技术基础理论——数据编码的基本概念及理论，进一步较为详细地介绍了几种常用的重要的无损编码算法；第四章介绍了多媒体通信技术基础理论之一的音频压缩编码的基本理论和方法，包括声音数字化及 PCM 等重要概念和方法；第五章阐述了多媒体通信技术基础理论的重要组成部分，静止图像压缩的基本理论和重要方法，详细介绍了二值图像压缩编码和 JPEG 压缩编码方法；第六章阐述了多媒体通信技术基本理论的另一重要组成部分，视频压缩编码的基本理论和重要方法，引进了运动估计、运动补偿等重要概念，详细介绍了 MPEG 系列压缩标准及压缩算法，同时也介绍了相应的 H.264 系列算法。

在编写过程中，参考了中外大量多媒体、多媒体通信技术方面的相关书籍和电子资料，在此向这些书籍的作者和相关网站表示衷心的感谢。

最后，尽管本人在这个技术领域中在国内外学习、工作多年，比较了解国内外工作市场、同行的基本要求，但毕竟个人能力有限，不足之处一定难免，欢迎各位前辈、同行批评指正。同时希望本书能够起到一个抛砖引玉的作用，能引出更多更好地双语专业教科书和参考书问世。

沈晋原
2014 年 12 月

Preface

Currently it is still very hard to find a suitable dual language textbook or technical reference book in science and technology area in mainland China. The universities/colleges which have the dual language courses have to use either the original technical books from abroad or the ones China bought the copyrights and republished in China as the textbooks. There are several issues existing under this circumstance:

1. Both original technical books from abroad and the ones China bought the copyrights and republished in China are too expensive to the Chinese undergraduate, graduate students, and professionals, especially the original ones which cost around USD$200. Even the republished ones still roughly cost more than RMB￥100 which is still a burden to the students and professionals in China.

2. In addition to the cost, the types and options of the republished technical books are very limited that makes it hard to match the reference book's contents to the course's requirements.

3. Under the current circumstances, the Chinese students, especially the undergraduate students'English abilities are still limited. This makes it hard for them to study the original English technical books. General speaking, for most of them, their reading speed and understanding are still

not that good yet. This makes it hard for both the teachers and the students to complete the course.

Same other issues exist among the university teachers, professionals, scholars, and graduate students. Communications between the Chinese scholars and professionals and the foreign scholars and professionals are getting more and more often and deeper along with the Chinese door opening. Chinese scholars, professionals, and students need quickly to know more about the differences between English and Chinese to describe the same technical ideas in the same areas to efficiently communicate with the foreign professionals who are working in the same professional areas.

This book, will be the first Chinese and English dual language professional textbook/reference book in China. Hope it will be a practical textbook for the university grade four undergraduate students' Multimedia Communication Technology course and very useful reference book for Chinese professionals of these technical areas.

As stated above, one of the major reasons to write this book is to provide the Chinese professionals working in the same technical area a good reference book. With this one book, those professionals would be able to know the different descriptions between English and Chinese in Electrical Engineering areas. After knowing those English descriptions, they can directly use them to communicate with the foreign professionals without checking a technical English-Chinese dictionary again as usual. For this reason, instead of like an English-Chinese dictionary translating the technical words or sentence one by one, the book emphasizing the different descriptions of English and Chinese for the same technical ideas. This is one of the advantages to choose this book.

Such a way the book is written, makes its Chinese language part is independent from its English language part. In another word, besides to be served as a dual language teaching textbook and technical reference book, the book could be used as a pure Chinese textbook as well just ignore the English part. This gives the universities without dual language course a good option to use the book as a transfer book from pure Chinese teaching to dual language teaching. In this case, the biggest advantage compared to using a pure Chinese language textbook is that this book gives a good English explanation to the same technical ideas that provides a very good reference for the teachers and the students. For the same reason, the book is a very good reference book for the professionals working in this electrical engineering areas.

The whole book is organized in 6 chapters. Chapter 1 begins with the simple developing history of the multimedia technologies and then discusses some important ideas used in multimedia communication technologies, such as data compression, lossy and lossless coding, and information entropy. Chapter 2 covers the basic ideas of multimedia and multimedia communication technologies including their basic characteristics, application areas, and key technologies. Chapter 3 is devoted to basic principles of multimedia communication technologies, the basic idea and theory of data coding. Furthermore, covers several most often used lossless coding algorithms in detail. Chapter 4 provides a detailed overview of audio compression coding basic theory and methods, one of the basic principles of multimedia communication technologies. This includes audio signal digitalization and PCM etc important ideas and algorithms. Chapter 5 concentrates on the still image compression basic theory and core algorithms, which is a major part of the multimedia communication technology base principles and methods. It further talks about the bi-level image compressing coding and JPEG compression coding in detail. Chapter 6 is devoted to the video compression coding basic theory and core algorithms, which is another major part of the multimedia communication technology base principles and methods. It introduces the motion estimation, motion compensation, and etc important ideas and gives out the detailed MPEG series compression standards and algorithms. Meanwhile, it introduces the correspondent H.264 series algorithms.

During the writing process, I referred quite a bit local and foreign multimedia and multimedia communication technology technical books and professional articles. Here, I would like to say thanks to those books' authors and those webpages.

Finally, I have been studying and working in this technically area for years locally and abroad, understand the job market requirements and colleagues' basic desire. Even this, my personal ability is still limited. I have tried hard. But there must be still some errors in the book. Any suggestions and advices from either former professionals or colleagues are welcome. At the same time, I hope the book could be worked as an introduction for other better ones. After this one, could see more and better dual language technical textbook/reference books are published.

Jinyuan Shen
December, 2014

目　录

第一章 多媒体发展简史

1.1 数据压缩技术半个世纪发展概述

1.1.1 数据压缩及其重要性

首先，来看一下什么是数据压缩，数据压缩为什么重要。

数据压缩有两大功用。

第一，可以节省存储空间。

第二，可以减少对带宽的占用。例如，数据压缩技术的快速发展使在手机上观看高清视频成为现实。

简单地说，如果没有数据压缩技术，就没法用 WinRAR 为 Email 中的附件瘦身；如果没有数据压缩技术，市场上的数码录音笔就只能记录不到 20min 的语音；如果没有数据压缩技术，从 Internet 上下载一部电影也许要花半年的时间……可是这一切究竟是如何实现的呢？数据压缩技术又是怎样从无到有发展起来的呢？

数据压缩的发展史可以形象的归纳为如下一个环链：

由概率起源起始，到数学游戏的出现，从异族传说的流传，发展到现代的音画时尚，转而到未来的展望。

当我们在 BBS 上用"7456"代表"气死我了"，或是用"B4"代表"Before"的时候，我们至少应该知道，这其实就是一种最简单的数据压缩。

Chapter 1   Multimedia Developing Brief History

1.1   Data compression technology development in the past half century

1.1.1   Data Compression and Its Important Role

First of all, what is data compression and why it's important.

Generally speaking, data compression does two major things.

One of them is to save the storage space.

The other one is to reduce the bandwidth usage. For example, years ago, it's like a dream for us to watch movies by our call phones, the fast developing of the data compression technology has made our dreams become true.

Simply speaking, without data compression technology, we can't use the software, WinRAR to reduce the file size attached in our emails; without data compression technology, audio recording pen in the market may only can take less than 20 minutes recording; without data compression technology, it may take half a year to download a movie from the internet...Then how all off those happened? How the data compression technology progressed?

The data compression technology progressed like a ring as follows:

It started form the data probability, then a mathematical game appeared, followed is a legend, progressed to modern audio and videos, next goes to the future...

In China, people use "7456" to represent Chinese words: "Qi Si Wo Le" means "I" m very mad". Both "7456" and "Qi Si Wo Le" have similar pronunciation. Almost all over the world, people use "B4" to represent "Before". This is actually one of the simplest data compression.

1.1.2　数据压缩与信息论

严格意义上的数据压缩起源于人们对概率的认识。

当人们对文字信息进行编码时，如果为出现概率较高的字母赋予较短的编码，为出现概率较低的字母赋予较长的编码，总的编码长度就能缩短不少，这可谓是数据压缩的写照之一。

另一个典型的数据编码例子是著名的 Morse 电码（由美国的摩尔斯在 1844 年发明的，所以电码符号也称为摩尔斯电码——Morse code），电码 符号由两种基本信号和不同的间隔时间组成：短促的点信号"."，读"的"（Di）；保持一定时间的长信号"—"，读"答"（Da）。

常用的 Morse 编码表列在表 1-1 中。

表 1-1　Morse 编码

Morse 电码	Morse 电码	Morse 电码	Morse 电码
A .-	K -.-	U ..-	0 -----
B -...	L .-..	V ...-	1 .----
C -.-.	M --	W .--	2 ..---
D -..	N -.	X -..-	3 ...--
E .	O ---	Y -.--	4 ....-
F ..-.	P .--.	Z --..	5 .....
G --.	Q --.-		6 -....
H	R .-.		7 --...
I ..	S ...		8 ---..
J .---	T -		9 ----.

信息论之父克劳德·香农（C.E.Shannon）第一次用数学语言阐明了概率与信息冗余度的关系。在 1948 年发表的论文中，香农指出，任何信息都存在冗余，冗余大小与信息中每个符号（数字、字母或单词）的出现概率或者说不确定性有关。香农借鉴了热力学的概念，把信息中排除了冗余后的平均信息量称为"信息熵"，并给出了计算信息熵的数学表达式。

这篇伟大的论文后来被誉为信息论的开山之作，信息熵也奠定了所有数据压缩算法的理论基础。从本质上讲，数据压缩的目的就是找出并消除信息中的冗余，而信息熵及相关的定理恰恰用数学手段精确地描述了信息冗余的程度。利用信息熵公式，人们可以计算出信息编码的极限，即在一定的概率模型下，无损压缩的编码长度不可能小于信息熵公式给出的结果。

1.1.2 Data Compression and Information Theory

Scientifically speaking, data compression actually started from the recognition of data probability.

When we encode the text, if we use shorter codes for the characters appeared more often in the text and longer codes for the characters appeared less often in the text, the total length of encoded codes will be reduced a lot. This is one of the advantages of data compression.

Morse code is another typical data encoding. It is invented by Morse, US in 1944. The code consists of two basic signals that separated by deferent intervals. The short dot signal "." sounds "di" and the long dash signal sounds "da".

The common used Morse codes are showing in Table 1-1.

Table 1-1   The Morse Code

Morse code	Morse code	Morse code	Morse code
A .-	K -.-	U ..-	0 -----
B -...	L .-..	V ...-	1 .----
C -.-.	M --	W .--	2 ..---
D -..	N -.	X -..-	3 ...--
E .	O ---	Y -.--	4 ....-
F ..-.	P .--.	Z --..	5 .....
G --.	Q --.-		6 -....
H	R .-.		7 --...
I ..	S ...		8 ---..
J .---	T -		9 ----.

The creator of information theory, C. E. Shannon first time provided the mathematical relationship between data probability and redundancy in 1948. Simply speaking, Shannon indicated in his thesis that any information has redundancy, the quantity of the redundancy depends on the uncertainties, probabilities of the characters, numeral digits, and words representing the information. Shannon used the similar concept, entropy that was introduced into thermodynamics in nineteenth century in the information theory. He defined the information entropy as the average information we need after taking out of the redundancy. He quantifiably defined the mathematical expression of entropy.

Shannon's article was treated as the number one paper in information theory. Entropy becomes the foundation of almost all data compression algorithms. What do those data compression algorithms do is to reduce the redundancy involved in the given information. The entropy mathematical calculation formulas precisely describe how much redundancy existing in the given information. By applying the formulas, people know the encoding limits. In another words, under certain probability distribution, the length of the lossless code can't be shorter than the result calculated from the entropy formulas.

1.1.3 数据压缩算法概要

有了完备的理论，接下来的事就是要想办法实现具体的算法，并尽量使算法的输出接近信息熵的极限了。当然，大多数工程技术人员都知道，要将一种理论从数学公式发展成实用技术，并不是一件很容易的事。

设计具体的压缩算法的过程可比作像是一场数学游戏。科学家首先要寻找一种能尽量精确地统计或估计信息中符号出现概率的方法，然后还要设计一套用最短的代码描述每个符号的编码规则。

1948 年，香农在提出信息熵理论的同时，也给出了一种简单的编码方法——香农编码。1952 年，范诺又进一步提出了范诺编码。这些早期的编码方法揭示了变长编码的基本规律，也确实可以取得一定的压缩效果，但离真正实用的压缩算法还有一定距离。

第一个实用的编码方法是由哈夫曼提出的。

Huffman 编码效率高，运算速度快，实现方式灵活。今天，在许多知名的压缩工具和压缩算法（如 WinRAR、gzip 和 JPEG）里，都有哈夫曼编码的身影。

如果不是后文将要提到的那两个犹太人，还不知要到什么时候才能用上 WinZIP 这样方便实用的压缩工具呢，在此将其称为异族传说。

1.1.3 Abstract of Data Compression Algorithms

After having the theory shown above, people keep trying to find the better and better algorithms to reach the limit of the entropy to get the shortest codes as possible. Engineers and technical people know that it's not that easy to put the scientific theory into the practice.

The process to find the data compression algorithm is somewhat like a mathematical game. The scientists find a method as precisely as possible to describe the appearance probability of the symbols expressing the information wanted. After this, they need to find a practical encoding algorithm to get as short as possible codes representing the symbols.

Shannon provided a relatively simple encoding algorithm in 1948 when he created the information theory. Based on the algorithm, R. M. Fano gave the Fano encoding algorithm in 1952. Those encoding algorithms provided at an earlier time found the basic rules of variable length encoding.

Later on, D. A. Huffman provided first practical encoding algorithm which is named Huffman encoding.

Compared with other algorithms, Huffman encoding has higher efficiency, faster calculation speed, and more flexibility in usage. It has been applied in many famous data compression tools and algorithms, such as WinRAR, gzip, and JPEG.

We're going to mention two Jewish. It's a kind of lengend. Without those two people, we may have to wait for a long time to be able to use the convenient compression tool, WinZIP.

逆向思维有时是科学和技术领域里出奇制胜的法宝。就在大多数人绞尽脑汁想改进哈夫曼或算术编码，以获得一种兼顾了运行速度和压缩效果的"完美"编码的时候，两位聪明的犹太人 J. Ziv 和 A. 哈夫曼独辟蹊径，完全脱离哈夫曼及算术编码的设计思路，创造出了一系列比哈夫曼编码更有效，比算术编码更快捷的压缩算法。人们通常用这两位犹太人姓氏的缩写命名这个算法，统称为 LZ 系列算法。

客观地说，LZ 系列算法的思路并不新鲜，其中既没有高深的理论背景，也没有复杂的数学公式，它们只是简单地延续了千百年来人们对字典的追崇和喜好，并用一种极为巧妙的方式将字典技术应用于通用数据压缩领域。通俗地说，当你用字典中的页码和行号代替文章中每个单词的时候，你实际上已经掌握了 LZ 系列算法的真谛。这种基于字典模型的思路在表面上虽然和香农、哈夫曼等人开创的统计学方法大相径庭，但在效果上一样可以逼近信息熵的极限。而且，可以从理论上证明，LZ 系列算法在本质上仍然符合信息熵的基本规律。

今天，LZ77、LZ78、LZW 算法以及它们的各种变体几乎垄断了整个通用数据压缩领域，较熟悉的 PKZIP、WinZIP、WinRAR、gzip 等压缩工具以及 ZIP、GIF、PNG 等文件格式都是 LZ 系列算法的受益者。

没有谁能否认两位犹太人对数据压缩技术的贡献。可见在工程技术领域，片面追求理论上的完美往往只会事倍功半，如果能像 Ziv 和 Lempel 那样，换个角度来思考问题，有可能发明另一种新的算法。

Thinking up side down sometimes is very useful in the scientific and engineering area. When many people were trying hard to improve the Huffman and Arithmetic coding and expect to find a relatively perfect coding algorithm, two Jewish, J. Ziv and A. Lempel were different from others. Instead of following others, they were independently seeking in another total different way to find the even more efficient and faster algorithms and they really made it. The codes named after their names, LZ series.

Practically speaking, LZ series don't have that much unique thought. They used the traditional dictionary method which has been used for years. They don't have that much mysterious theories and very complicated mathematics. They simply and in a smart way applied the techniques used in the dictionary to the data compression, coding. Actually, one can put it this way, when you know how to use the page and row index in the dictionary to find the words you want, you basically know the LZ series codes. The LZ series codes are totally different from Shannon and Huffman codes in the way they approach the code length limit given in the information entropy. But it has been theoretically approved that the LZ series still actually comply with the information entropy rules.

Currently, LZ series, LZ77, LZ78, LZW algorithms and the ones kindly derived from them occupied quite high percentage of the general data compression market. The algorithms we familiar with like PKZIP , WinZIP, WinRAR, gzip, ZIP, GIF, and PNG are beneficiaries of LZ series.

Nobody can ignore the contribution the two Jewish did to the data compression technology. What we can learn from this is that when we're working in the science and engineering area, we should sometimes try to change the regular way we're searching on and try something total different and we may be able to get the much better results we expected as Ziv and Lempel did.

LZ 系列算法基本解决了通用数据压缩中兼顾速度与压缩效果的难题。但是，数据压缩领域里还有另一片更为广阔的天地等待着人们去探索。

例如，对于生活中更加常见的图片、图像等，通用压缩算法的优势就不那么明显了。幸运的是，科学家们发现，如果在压缩这一类图像数据时允许改变一些不太重要的像素值，或者说允许损失一些精度（在压缩通用数据时，绝不会容忍任何精度上的损失，但在压缩和显示一幅数码照片时，如果一片树林里某些树叶的颜色稍微变深了一些，看照片的人通常是察觉不到的），就有可能在压缩效果上获得突破性的进展。也许，这和生活中常说的"退一步海阔天空"的道理有异曲同工之妙吧。

LZ series basically sold the problems existed in the general data compression to deal with the calculation speed and compression effect at the same time. But this is the only part of the work needs to be done in the data compression area. There are other issues in the area need to be solved.

In our daily life, we deal with pictures, images, and etc. The LZ series are not good enough to efficiently compress the size of those files. Fortunately, scientists found that if we're allowed to change certain amount of less important pixels, or to loss some accuracy (in compression process of general data, we will never tolerate any accuracy losses, especially for the text data, but in compression and display a digital photos or images, we do. In a piece of the woods we won't notice that some leaves' color are a little bit darker.), we could have a breakthrough progress with the compression effect. This could be compared with want we often said in our life: "take a step back, the sea seems broader and the sky seems brighter ".

1.2 声音编码和视频压缩基本原理一

1.2.1 采样率和采样大小（位/bit）

声音是一种能量波，有频率和振幅的特征，频率对应于时间轴线，振幅对应于电平轴线。连续波是无限光滑的，连续波的弦线可以看成由无数点组成，由于存储空间是相对有限的，在数字编码过程中，必须首先对连续波的弦线进行采样。采样的过程就是抽取有限个点的值，很显然，在 1s 内抽取的点越多，则采样后得到的波形与原波形越接近，为了不失真地恢复原波形，一次完整的振动中，必须有至少 2 个点的采样，人耳能够感觉到的最高频率为 20kHz，因此要满足人耳的听觉要求，声波则需要至少每秒进行 40k 次采样，用 40kHz 表达，这个 40kHz 就是采样率。常见的 CD，采样率为 44.1kHz。光有频率信息是不够的，还必须获得该频率的能量值并量化，用于表示信号强度。量化电平数为 2 的整数次幂，常见的 CD 为 16bit 的采样大小，即 2 的 16 次方。采样大小相对采样率更难理解，因为要显得抽象点，举个简单例子：假设对一个波形进行 8 次采样，采样点对应的能量值分别为 A1～A8，但如果只使用 2bit 的采样大小，结果只能保留 A1～A8 中 4 个点的值，舍弃另外 4 个。如果进行 3bit 的采样大小，则刚好记录下 8 个点的所有信息。采样率和采样大小的值越大，记录的波形越接近原始信号。

上述原理同样适用于视频信号，只是采样率和采样大小不同而已。

1.2　Audio and Video Signals Coding Principles 1

1.2.1　Sampling Rate and Bits

Speech sound is a kind of energy wave. It has frequency and magnitude. Frequency correspondent to the time axis and magnitude correspondent to the voltage axis. Analogue wave is unlimited smooth and it can be treated as containing infinite different points. Since the storage space is always relatively limited, it is necessary to first sampling the analogue signals in digital coding process. The process is actually taking relatively limited sample points. Obviously, the more sample points are taking within one second, the closer the sampled signal to the original analogue signal. To completely recover the original wave, at least two sample points should be taken within one complete vibration. The highest frequency the human's ear can feel is 20kHz. So, to be properly heard by human ear, sound wave shall be sampled at least 40k per second. Or say the sampling frequency shall be at least 40kHz. We call it sampling rate. The audio signals saved in the CD we see daily has 44.1kHz sampling rate. Besides the frequency we talked, we need to quantify the signal's energy representing the strength of the signal. We usually represent the magnitudes of the signal as the integrated power of 2. The quantified signal magnitude saved in our normal audio CD used 16 bits which has $2^{13} \times 2^3 =$ different values. The idea expressed here is a little bit hard to understand compared with the sample rate. Furthermore, we can have a look at a simple example. Say we are quantifying a speech wave with A1~A8, 8 different values. But we are using 2 bits sampling size to represent the values. We have to through away 4 of the 8 different values we got from the quantifying process since the size of 2 bits only can representing 4 different values. If we use 3 bits sampling size, then we can just represent the 8 sample values. Usually, the bigger the sample rate and the size, the closer the recorded discrete signal to the original analogue signal.

The above principle applies to video signal as well.

1.2.2　有损和无损编码

根据采样率和采样大小可以得知，相对自然界的信号，音频编码最多只能做到无限接近，至少目前的技术只能这样了，相对自然界的信号，任何数字音频编码方案都是有损的，因为无法完全还原。在计算机应用中，能够达到最高保真水平是 PCM 编码，被广泛用于素材保存及音乐欣赏，CD、DVD 以及常见的 WAV 文件中均有应用。因此，PCM 约定俗成了无损编码，因为 PCM 代表了数字音频中最佳的保真水准，并不意味着 PCM 就能够确保信号绝对保真，PCM 也只能做到无限接近。人们习惯性地将 MP3 列入有损音频编码范畴，是相对 PCM 编码的。强调编码的相对性的有损和无损，是为了告诉大家，要做到真正的无损是非常困难的，就像用数字去表达圆周率，不管精度多高，也只是无限接近，而不是真正等于圆周率的值。

1.2.2　Lossy and Lossless coding

From sampling rate and size we know that compared with the original analog signal, the coded digital speech signal will ever be exact the same as the original one. It only can approach the original one as close as possible. That's what the current technology can do. Under this consideration, any coded digital signal will be a somehow lossy one. In computer application, the closest coding is the PCM coding. It has been used to keep the original data and high quality music. It has been applied in the CD, DVD, and WAV files we use everyday. For this reason, the PCM coding has been treated as lossless coding. But even though, it actually still can not recover the original signal absolutely without any loss. We can say that it is the one that can recover the original signal with minimum loss. We usually treat MP3 as lossy coding compared with PCM. We spend some time to explain the lossy and lossless coding to let people know that it will be very tough to get the absolutely lossless code. Somewhat like we are using a number to express the circular rate, no matter how many digits we want to use, what we can get is still a closer and closer result and not the exact value of the circular rate.

1.2.3 信号频率与采样率的关系

采样率表示了每秒对原始信号采样的次数，人们常见到的音频文件采样率多为 44.1kHz，这意味着什么呢？假设有 2 段正弦波信号，分别为 20Hz 和 20kHz，长度均为 1s，以对应我们能听到的最低频和最高频，分别对这两段信号进行 40kHz 的采样，我们可以得到一个什么样的结果呢？结果是：20Hz 的信号每次振动被采样了 40k/20=2000 次，而 20k 的信号每次振动只有 2 次采样。显然，在相同的采样率下，记录低频的信息远比高频的详细。这也是为什么有些音响发烧友指责 CD 有数码声不够真实的原因，CD 的 44.1kHz 采样也无法保证高频信号被较好记录。要较好地记录高频信号，看来需要更高的采样率，于是有些朋友在捕捉 CD 音轨的时候使用 48kHz 的采样率，这是不可取的，这其实对音质没有任何好处，对抓轨软件来说，保持和 CD 提供的 44.1kHz 一样的采样率才是最佳音质的保证之一，而不是去提高它。较高的采样率只有相对模拟信号的时候才有用，如果被采样的是数字信号，请不要去尝试提高采样率。

1.2.3 The relationship between signal frequency and sampling rate

The sampling rate represents the pick up times per second to the original analogue signal. The digital audio signals we usually used in our daily life has 44.1kHz sampling rate. What does this actually means? Let us have a look at two different sinusoid signals. One of them has frequency 20Hz and the other has frequency 20kHz. Our watching time frame is one second for both of them. The frequencies are correspondent to our human being's hearing limits. If we sample the two signals using same sampling rate of 40kHz, let's have a look what could we get. For the 20Hz signal, within each wave vibration (period), 40k/20=2000 samples will be taken. For the 20kHz signal, within same each wave vibration (period), 40k/20k=2 samples will be taken. Obviously, with the same sampling rate, much more detailed info will be taken in the low frequency signal sampling process. This one the reasons why some Hi-Fi music lovers complain about the digital audio music CD not keeping the real sound. With the sampling rate 44.1kHz, there is still no high quality digital audio signal guarantee for the high frequency analogue signals. To get a better recording of high frequency signal, looks like the higher sampling rate is needed. This made some people try to use 48kHz, higher sampling rate to catch the CD music track. This is actually wrong since it won't help to the music quality at all! To the software catching CD track, instead of using higher sampling rate, keeping the same sampling rate, 44.1kHz the CD provided is the best option. Higher sampling rate only works when dealing with the analogue signals. If you are sampling the digital signal, don't try to increase the sampling rate.

1.2.4 流特征

随着网络的发展，人们对在线收听音乐提出了要求，因此，也要求音频文件能够一边读一边播放，而不需要把这个文件全部读出然后回放，这样就可以做到不用下载就可以实现收听了。也可以做到一边编码一边播放，正是由于这种特征，才实现了在线的直播，使架设自己的数字广播电台成为了现实。

1.2.4 The Character of Streaming

Along with the fast internet developing, more and more people prefer to listen music on line. This requires that the audio files could be played while downloading. It's not like before, the whole audio file has to be downloaded first and play back. This makes it possible to listen to the music (or watch movies) without downloading and saving the files to one's local disk. This also makes one thing possible which is coding while broadcasting. Because of this character, we can have directly on line broadcasting and have our own digital broadcasting station.

1.2.5 电话是如何发明的

亚历山大·格拉汉姆·贝尔（Alexander Graham Bell，1847—1942）美国发明家和企业家，他发明了世界上第一台可用的电话机，创建了贝尔电话公司，被誉为"电话之父"。

1847 年生于英国苏格兰，他的祖父亲毕生都从事聋哑人的教育事业，由于家庭的影响，他从小就对声学和语言学有浓厚的兴趣。开始，他的兴趣是在研究电报上。有一次，当他在做电报实验时，偶然发现了一块铁片在磁铁前振动会发出微弱声音的现象，而且他还发现这种声音能通过导线传向远方，这给贝尔以很大的启发。他想，如果对着铁片讲话，不也可以引起铁片的振动吗?就是贝尔关于电话的最初构想。

是梅布尔鼓舞他进行了所有那些使人精疲力竭的实验，也是梅布尔使他克服了不时产生的沮丧情绪并研制出当时很了不起的一种工具。它能把人说的话转变为电脉冲，之后又在金属丝的末端使之还原成人说的话。

为了纪念贝尔的功绩，将电学和声学中计量功率或功率密度比值的一种单位命名为"贝尔"。

一些额外的话：由于贝尔 1876 年 3 月 10 日所使用的这部电话机的送话器，在原理上与另一位电话发明家菲利浦·格雷（德国科学家）的发明雷同，因而格雷便向法院提出起诉。一场争夺电话发明权的诉讼案便由此展开，并一直持续了十多年。最后，法院根据贝尔的磁石电话与格雷的液体电话有所不同，而且比格雷早几个小时提交了专利申请等这些因素，做出了现在大家已经知道结果的判决，电话发明权案至此画上句号。

1.2.5　How The Telephone Invented

Alexander Graham Bell (1847-1942), American inventor and entrepreneur invented the first usable telephone in the world. He created the Bell enterprise and is named `the father of telephone`.

Alexander Graham Bell was born in Scotland, Britain. His grandfather had been working on deaf and dumb people education all through his life. Affected by family, Alexander had been very interested at the sound and speech. At the very begin, he was interested at studying telegram. Once when he was doing a telegram test, he incidentally found that there will be a miner sound comes out of a metal sheet when it's vibrating in front of a magnet. Further more, he found the sound could be transferred far away via a metal wire. This made Alexander think. He was thinking that if one face the metal to talk, would the speaking cause the metal sheet to vibrate? That was the Bell telephone's first thinking.

It was Mabel who encouraged him did those exhausting boring tests. It was also Mabel who helped him to overcame the disappointments he had from time to time. Those disappointments stop people who are trying hard to get the expected. At the end of the tests he did, he successfully created the great tool to convert people's talking to electrical pulse at one end of the wire and reconvert it back to people's talking at the other end of the wire.

To memory Alexander Bell's great contribution to human being, people named Bell as a unit used to measure the ratio of power to the power density in electricity and acoustic studying.

Some other stories: Since the transmitter Alexander Bell used in March 10, 1876 is very similar in principle to the one another inventor, Philip Gray (German scientist) created, Gray sued Bell. A legal fighting had been started since then and continued for about 10 years. At last, the court justified Bell won. The court's decision based on the fact that Bell's magnet phone is different from Gray's liquid phone and Bell submitted the patent application a couple of hours earlier than Gray did. That was the period of the legal fighting.

1.2.6　常用声音及图像编码法简介

图 1-1 表示出了基本的声音编码示意图。第一步，正常的声音通过编码器（Encoder）压缩变换成待传输的码字。码字如图所示是一连串的 0、1 码。当然这些 0、1 码是按照一定的编码算法得来的。第二步，这些码字可以通过无线或者通过有线通信传输到目的地或说用户端；这些码字也可以存储到存储介质上，如磁盘、CD、DVD，等等。第三步，传送到用户端或存储到存储介质上的压缩码字通过解码器（Decoder) 解码还原成原始的声音，人耳即可听到。

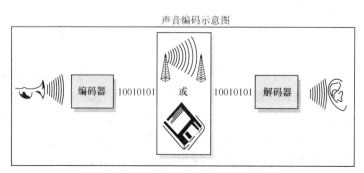

图 1-1　编码、传输、人耳听音的实现

1.2.6 Speech and Image Coding Introduction

Fig. 1-1 shows the basic sound/speech coding procedure. First of all, the normal speech goes through a encoder. The speech signal will be digitalized and compressing encoded into a serious codes. Those codes are continuously either binary digit 0 or 1. They are produced based on certain compression algorithm. Secondly, those code produced in the first step could be either transferred by either wireless or wired communication channel to the destination/user end or saved in the storage media, such as disk, CD, and DVD. Thirdly, at the destination/user end, the received codes or the codes saved on either one of the storage medias will go through the decoder which converts the codes back to the original normal speech/sound people can hear.

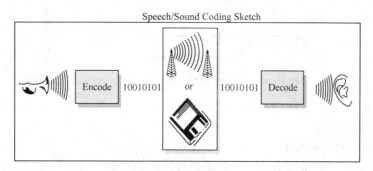

Fig. 1-1　Coding, Transmission, and Hearing

　　人耳能够听到声音的范围是一定的。声压和频率是声音的两个重要指标。人耳所能听到的频率范围是 20Hz～20kHz。声压过低，人耳听不到，而声压过高，人耳会觉得刺耳甚至疼痛，图 1-2 表示出了这种关系。

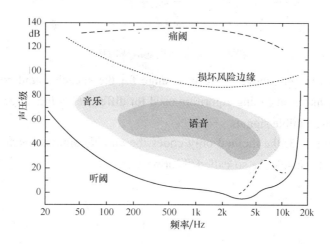

图 1-2　人耳所适应的声压和频率范围

As is often said, our ears have a limit to hear the sound around us. Sound pressure and frequency are the two very important coefficients of sound. The frequency range human being can hear is 20Hz~20kHz. If the sound pressure is too low, we can't hear. On the other side, if the sound pressure is too high, we can feel very uncomforTable and even pain. Figure 1-4 shows the range that is suiTable for human being hearing.

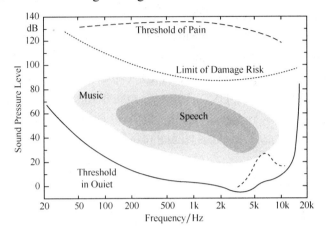

Fig. 1-2 The Hearing Range of Sound Pressure Level and Frequency

根据以上的特点，声音编码是有一定范围的。图 1-3～图 1-6 给出了几种不同编码方法所用的不同编码范围。

1. 只对可闻信号进行编码

如图 1-3 所示，这种编码方式只对听阈曲线以内的可闻信号进行编码，而忽略听阈曲线以外的信号。

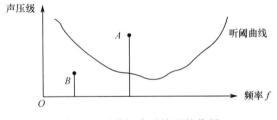

图 1-3　听觉阈度对编码的作用

Based on the characters above, there is a range for the speech/sound coding. Fig ure1-3 to Figure 1-6 gives the different coding algorithm used for different coding range.

1. Only codes the audible signals

As shown in Fig. 1-3, the method only codes the audible signals inside the threshold curve and ignores the signals outside the so-called "hearing area".

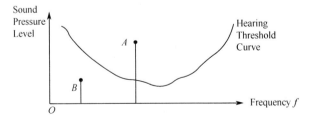

Fig. 1-3　The Hearing Area effects on Coding

12

2．只对幅度强的掩蔽信号进行编码

在这种编码方法中，只对幅度强的掩蔽信号进行编码，而忽略被其掩蔽相对较弱的信号，如图 1-4 所示。

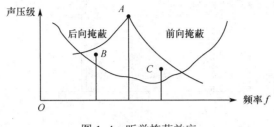

图 1-4　听觉掩蔽效应

2. Only code the signal with bigger amplitude

In this style, only code the stronger signal with bigger magnitude that is called "masker" and ignore the relatively weaker signals that is called "test tone" covered as shown in Fig. 1-4.

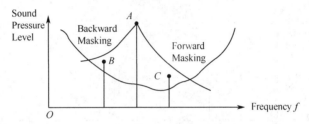

Fig. 1-4　Hearing Masking Effects

3．只对信号与量化噪声的差值进行编码

在这种编码方法中，只对信号与量化噪声的差值进行编码。换句话说，只对信号高于量化噪声的部分进行编码，如图 1-5 所示。

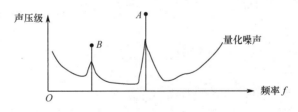

图 1-5　量化噪声对编码的影响

3. Only code the differences between the speech signal and the quantization noise

In this way, only code the differences between the speech signal and the quantization noise. In another ward, only code the portion of the speech signal that bigger than the quantization noise as shown in Fig. 1-5.

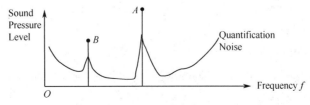

Fig. 1-5　Quantification Noise Effects on Coding

4．通过子带分割来进行优化、编码

在这种编码方法中，是把一个频域内的信号分割成不同的子频带，然后对子频带内的信号进行优化、编码，如图1-6所示。

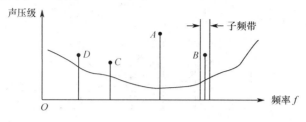

图1-6 子带优化、编码

4. Optimum and coding within the divided sub-bandwidth range

In this method, firstly divide the whole frequency range of the signal into smaller sub-bandwidth range and optimum and coding within the divided sub-bandwidth range as shown Fig. 1-6.

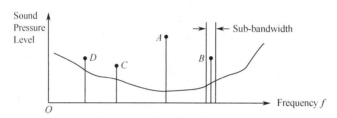

Fig. 1-6 Sub-bandwidth Optimizing and Coding

1.2.7 视频压缩基本原理

视频压缩对时域和空域的冗余采用不同的处理办法。

1．时域冗余信息

使用帧间编码技术可去除时域冗余信息，它包括以下两个部分：

（1）帧间预测

这种方法中，只传送相邻两帧之间变化的部分。

（2）运动补偿

运动补偿是通过先前的局部图像来预测、补偿当前的局部图像，它是减少帧序列冗余信息的有效方法。

1.2.7 Image and video compression basic principles

Image and video compression algorithms use different methods to deal with the spatial and frequency domain redundancy.

1. Process spatial domain redundancy

Apply the coding between frames technology to reduce the spatial domain redundancy. This includes the following two parts.

（1）Prediction between the image frames.

Here only transmit the changed parts between the two consecutive image frames.

（2）Motion compensation.

Motion compensation uses the part of previous image frame to predict and compensate the part of current image frame. It is an efficient way to reduce the redundancy between the frames.

2．空域冗余信息

（1）变换编码。

变换编码将空域信号变换到另一个正交矢量空间，使其相关性下降，数据冗余度减小。

（2）量化编码。

经过变换编码后，产生一批变换系数，对这些系数进行量化，使编码器的输出达到一定的位率。

（3）熵编码。

熵编码是无损编码。它对变换、量化后得到的系数和运动信息，进行进一步的压缩。

2. Frequency domain redundancy

（1）Transform coding.

What the transform coding doing is to convert the spatial signal to another space and reduce the correlations inside the image and data redundancy.

（2）Quantization coding.

Quantization coding is to quantize the coefficients produced from the transform coding to reduce the encoder's output codes length.

（3）Entropy coding.

Entropy coding is one of the lossless coding. It further compress the data after the transform and quantization coding.

前面我们已提到 MPEG4 标准。图 1-7 和图 1-8 是 MPEG-4 的两个实用实例。它们给出了 MPEG-4 中背景全景图和视频对象的合成过程。图 1-7 中是有一个视频对象的情形，图 1-8 中是有多个视频对象的情形。

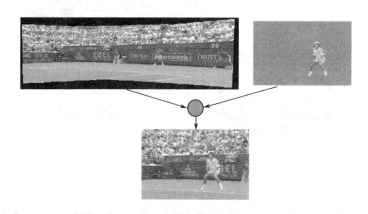

图 1-7　背景全景图+视频对象（VO）＝合成图像　　MPEG-4 应用实例

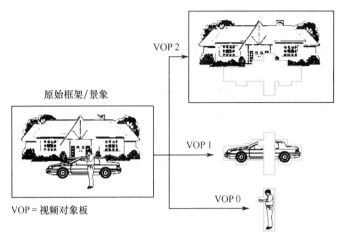

图 1-8　背景全景图+视频对象（VO）＝合成图像　　MPEG-4 应用实例

We have mentioned MPEG-4. Figure 1-7 and Figure 1-8 are two examples of MPEG-4. Both of them show the combined process of background image and video objects. Figure 1-7 has 1 video object and Figure 1-8 has several video objects.

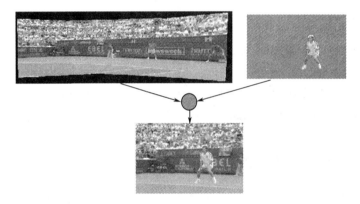

Fig. 1-7　Background image + Video object(VO)=Combined image　　MPEG-4 example

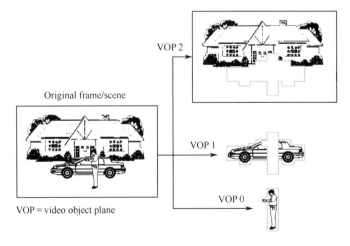

Fig. 1-8　Background image + Video objects(VO)=Combined image　　MPEG-4 example

16

在多媒体通信技术发展过程中有几个很值得一提的事情，ZIP 数据压缩格式的诞生过程是其中之一。

叛逆斗士的胜利——ZIP 格式诞生

在 DOS 年代由于计算机存储介质容量的微小，个人用户对数据压缩软件的渴望是现在的用户无法想象的。例如在 1984 年，个人计算机的标配不过是容量 360kB 的 5.25 寸软盘而已，如果个人能将数据压缩数倍后存储，则不啻于节省了一大笔钱。

于是数据压缩软件就成为了一项必需的工具。1985 年 SEA 公司开发了 MS-DOS 环境下第一个应用 LZW 算法的 ARC 压缩软件，它是当时 MS-DOS 下统治性的压缩软件。从技术角度来说 ARC 确实不错，但使用了专利 LZW 算法的 ARC 当然是标准的商业软件，使用这种软件工作就必须付费。不过当时许多玩家根本买不起 ARC 软件，顺便说一句题外话，那时大多个人计算机玩家基本都没什么富裕的钱，事实上个人计算机本身的发展就是被穷玩家精打细算所推动。不过个人计算机从诞生之日起就充满了叛逆、自由的精神，这也是推动整个个人计算机世界前行的主要动力。此时一个年轻的程序员出现并试图改变压缩世界，这个人叫 Phillip W. Katz（菲利普·卡兹）。

20 世纪七八十年代出售软件的方式和现在截然不同，以 ARC 软件来说，它不仅包括了一份 EXE 可执行文件，还包括它的 C 语言源代码。经常混迹于 BBS 上的菲利普·卡兹同样买不起 ARC，于是他自己将 ARC 的 C 语言源代码进行复制并用汇编语言重写，并将这个压缩工具称作 PKARC，这个程序自然与 ARC 完全兼容，而且由于使用汇编使得速度较 ARC 更快。在当时的计算机世界里这是一种很普遍的现象，并没有程序员认为这种行为不对，甚至只要不与自己冲突，被改写者通常也不在乎。不过这次不太一样，菲利普·卡兹不仅仅是自己和朋友用，而是将这个软件免费向他人开放，大批 ARC 用户自然也就转而使用菲利普·卡兹的软件。

SEA 其实不是什么大企业，它只是一个 3 人起家的小公司，当然无法接受这种毁灭性打击。以现在的眼光看来，最初 SEA 的方式是温和的，它接洽菲利普·卡兹并希望通过授权的方式将 PKARC 纳入旗下，然而并不认为自己有什么过错的菲利普·卡兹一口拒绝，他不想让 PKARC 成为商业软件，他制作这个工具的初衷并不是为了赚钱。最终菲利普·卡兹被 SEA 以侵犯 ARC 压缩格式编码算法的罪名告上了法庭，并输掉了官司。叛逆倔强的卡兹在败诉后依然拒绝将 PKARC 授权给 SEA 公司，而选择了支付法律费用和停止发放 PKARC。

这场官司对菲利普·卡兹的人生观和信念影响巨大，追求自由平等的精神并不意味着盲目和法律对抗，试图劫富济贫的少年侠客行为只能逞一时快意，实质上帮助不了任何人。败诉后菲利普·卡兹决定将 PKARC 完全重写。很显然，这次再也不能去触犯任何编码算法的专利权了，从 3 个基本编码算法来衍生自己的算法是必然的，于是去掉有专利权的 LZW 和 LZ78，剩下的就只有 LZ77。

也许是被激怒后带来了惊人的动力，只用了几周的时间菲利普·卡兹就创造出一个全新的压缩编码算法，该算法完美地结合 LZ77 和哈夫曼编码，也就是后来大名鼎鼎的 PKZIP，而其文件格式扩展名叫作 ".zip"。PKZIP 可将多个文件压缩到一个文件中，无论压缩比、压缩速度都全面超过了商业软件 ARC。菲利普·卡兹将 PKZIP 作为自由软件免费发放，使其如野火般在全美各大 BBS 上蔓延开来，用户以几何级数增长，遭受毁灭性打击的 SEA 公司半年内就无声无息。这段故事最后演变为用自由软件打败商业软件的传奇，菲利普·卡兹更是成为充满幻想的年轻程序员心中十步杀一人的偶像。

然而事情如果仅仅到此为止，那么这也不过是菲利普·卡兹为私人恩怨而快意恩仇的行为，未必能得到后人的真正尊重。不过他做出了一个让所有计算机用户都收益无穷的举动，那就是宣布开放 ZIP 格式，任何人都可以自由使用 ZIP 编码算法而不需要缴纳任何专利费用。这个决定最终改变了压缩的世界，使得通用数据无损压缩领域再无法出现垄断的商业巨鳄，真正意义上帮助了每个需要压缩的计算机用户。凭借这个无私的行为，菲利普·卡兹真正成为他想成为的英雄。

2000 年 4 月 14 日，年仅 37 岁的菲利普·卡兹被人发现倒毙在美国威斯康星州密尔沃基的一家汽车旅馆里，据说死因是慢性酒精中毒引起的并发症。这位天才程序员从未在 ZIP 身上得到半点好处，坚持信念的结果是潦倒的生活。他为世界贡献了一个伟大的免费软件，更为重要的是他缔造了一种大众化的压缩格式，然而他却过早的离开了这个世界。

During the process of multimedia communication development, some things really deserve us to member. The born process of ZIP data compression is one of them.

The victory of a rebel - The born of ZIP

When people were using DOS operating system, the computer storage space is very limited. At the time, people desired to death to have an efficient data compression software to use. For example, in 1984, one of the standard components of a personal computer is a 5.25inch floppy disk with 360kB capacity to save data. Each floppy disk cost money. If one can get the data compressed to several times smaller before the storage, one can absolutely save quite a bit money.

For this reason, data compression software was highly desired. SEA developed the first ARC compression software using LZW algorithm under the MS-DOS environment. It was the predominate software under the MS-DOS environment. Technical speaking, ARC is a good software. But since it used patent algorithm, it is also a standard business software that charges whoever uses it. At the time, not many people who loves computer could afford it. At that time, most the computer lovers don't have that much money for something like floppy disks. But they love the computers. They have to think hard to find the way to get what they need and cost less. It is this hard thinking pushed the computer technology developed. The personal computer could be treated as a rebel compared with the main frames since the first day it was appeared. It's one of the reasons for personal computers developed quickly and shortly after became a pioneer. At this time, one young software developer, Phillip W. Katz dropped in and tried to change compression world.

The style of selling software in 70 to 80's of 20 century is total different from today. Take ARC as an example, it's not only includes a EXE execute file, but also includes the original C language codes. So the price is too high to afford for the BBS players like Phillip W. Katz. What Phillip W. Katz did is to rewrite the ARC original C language codes in assemble language. He called the revised compression tool, PKARC. The program of course is completely compatible with ARC. Besides this, since he used assemble language instead of C language, the compression speed is faster than ARC's. What happened then seems normal in computer world. No software developer thought it was not right. As soon as there is no conflict exists, the original writer doesn't mind either. But this time it's a little bit different. Phillip W. Katz not only used the revised codes himself and his friends, but also opened the codes free to the public and everybody could use it. Many ARC clients turned to use Phillip W. Katz's software.

SEA was not a big firm and actually pretty small with 3 employees. What Phillip W. Katz did was absolutely a disaster to them. We look at it today, SEA was quite friendly at the very beginning. They contacted Phillip W. Katz and tried to give an authorization to him and make Phillip W. Katz to be one of their employees. However, Phillip W. Katz didn't think he did anything wrong and refused the offer from SEA. He simply didn't want the PKARC to become a commercial software. He was not intending to make money on the tool he programmed. Afterwards, Phillip W. Katz was sued by SEA for offending the rights of the ARC compression algorithm and lost. After lost the case, rebel Phillip W. Katz still refused to give SEA the copyright of PKARC. Instead, he took the option paying the lawyer fee and stopping distributing the PKARC.

The case affected Phillip W. Katz a lot and made him realized that seeking freedom doesn't mean to offend the law purposelessly. He also realized that short time against the rich and helping the poor like a young rebel won't last long and won't actually help anyone. Then he decided to completely rewrite the PKARC software. Obviously, he wouldn't want to offence the law and touch any patent of coding algorithms anymore. At the time, there were three matured coding algorithm existed. usually, the easiest way to write a algorithm program is to pick up the exist one and modify or rewrite it. Phillip W. Katz understood this and picked up the LZ77 which was the only didn't have patent among the three exist ones.

The extreme powerful energy may come from the anger. it only took Phillip W. Katz couple of weeks to create the brand new compressing coding algorithm. He perfectly technically combined LZ77 and Huffman coding algorithms and created later famous PKZIP. The extension file name created by this software is ".zip". PKZIP can compress multiple files into one. It is weigh better than the commercial software ARC both in compression ratio and speed. Phillip W. Katz put PKZIP as a free software and open to the public. Quickly, it's like a wild fire spreading over each BBS in North America. Users were in a geogaraphy speed growing up. Under this deadly attack, SEA disappeared within half a year. This story after became a legend of a freedom software defeat a commercial software. Phillip W. Katz became a idol killer of young men who can kill the competitor within ten walking steps.

But the story didn't stop here. The followed computer lovers respected Phillip W. Katz not because he won the fighting for himself but for the open software he provided for those lovers which benefited them a lot. People don't have to worry about the patent anymore and use the ZIP coding algorithm for free. This changed the whole data compression world. No big business monsters can completely control the general data compression industry again. This selfless action really helped every computer user. For this reason, Phillip W. Katz became a real hero as he wanted.

On April 14, 2000, people found Phillip W. Katz dead in a motel located at Milwortz, Wiskonching. That year, he was only 37. People were saying he was sick and dead on alcohol. This genes software developer was not benefited from ZIP at all. He contributed the computer world a great free software and didn't have a good life. More importantly, he created a public compression format. But he himself left this world too early.

1.3　声音编码和视频压缩基本原理二

1.3.1　简化熵的计算公式

考虑用 0 和 1 组成的二进制数码为含有 n 个符号的某条信息编码，假设符号 F_n 在整条信息中重复出现的概率为 P_n，则该符号的熵也即表示该符号所需的位数为

$$E_n = -\log_2(P_n)$$

1.3　Audio and Video Signals Coding Principles 2

1.3.1　Simplified Entropy Calculation Formula

Think about a information consisted of n "0" and "1"s. Suppose the appearance probebility of F_n in this information is P_n. Then the entropy of this symbol, also the necessary bits to represent the symbol, is:

$$E_n = -\log_2(P_n)$$

1.3.2　几种常用编码方法

1. 算术编码

算术编码是由 J. Rissanen 在 1979 年提出的。

算术编码的基本原理为将被编码的信息表示成实数轴上 0 和 1 之间的间隔，信息越长，间隔越小，表示这一间隔所需的二进制位数就越多。

2. 哈夫曼编码原理及计算过程

这需要从变长编码谈起。

变长编码的基本原理为频繁使用的数据用较短的代码代替，较少使用的数据用较长的代码代替，每个数据的代码各不相同。

举个例子：假设一个文件中出现了 8 种符号 S0,S1,S2,S3,S4,S5,S6,S7，那么每种符号要编码，至少需要 3bit。假设编码成 000,001,010, 011,100,101, 110,111（称为码字）。

那么符号序列

$$\text{S0S1S7S0S1S6S2S2S3S4S5S0S0S1}$$

编码后变成

$$000001111000001110010010011100101000000001。$$

共用了 42bit。人们发现 S0，S1，S2 这三个符号出现的频率比较大，其他符号出现的频率比较小，如果采用一种编码方案使得 S0，S1，S2 的码字短，其他符号的码字长，这样就能够减少占用的比数。

例如，采用这样的编码方案：S0 到 S7 的码字分别为 00,10,110,111,0100,0101,0110,0111，那么上述符号序列变成 001001110010011011011011101000101000010，共用了 39bit，尽管有些码字如 S4，S5，S6，S7 变长了（由 3 位变成 4 位），但使用频繁的几个码字如 S0，S1 变短了，所以实现了压缩。

上述的编码是如何得到的呢？随意乱写是不行的。编码必须保证不能出现一个码字和另一个的前几位相同的情况，比如，如果 S0 的码字为 01，S2 的码字为 011，那么当序列中出现 011 时，你不知道是 S0 的码字后面跟了个 1，还是完整的一个 S2 的码字。此处给出的编码能够保证这一点。

1.3.2　Several Normal Used Coding

1. Arithmetic Coding

J. Rissanen proposed the Arithmetic coding in 1979.

The major principle of Arithmetic coding is to represent the input information as the intervals on the real axis between 0 and 1. The longer the information, the smaller the intervals. Correspondently, the more binary bits will be used to representing each interval.

2. Huffman Coding

We need to start from Variable Length Coding.

Variable Length Coding - using shorter code represents the data appears more frequently and using longer code represent the data appears less frequently. Each data has different code.

For example, imagine there are 8 different symbols: S0, S1, S2, S3, S4, S5, S6, S7 in one file. If wants to code the 8 different symbols, minimum 3 bits of binary code will be needed. supposed encoded codes from S0 to S7 look like: 000, 001, 010, 011, 100, 101, 110, 111, then the symbol sequence:

S0S1S7S0S1S6S2S2S3S4S5S0S0S1

after encoding will become:

000001111000001110010010011100101000000001

Total 42 bits are used in this sequence. In the above encoding, the even length codes were used for every symbol. Now, if we pay attention to the sequence, we can find that the symbols S0, S1, and S3 appeared more often than other in the sequence. If we use shorter codes for the more often appeared symbols, S0, S1, and S2 and longer codes for other less often appeared symbols, the total bits used in the encoded codes will be less.

With the thinking above, we can change the encoding plan from the even length codes to the following variable length codes. From S0 to S7, we give codes: 00, 10, 110, 111, 0100，0101, 0110, 0111.

Then the same symbol sequence,

S0S1S7S0S1S6S2S2S3S4S5S0S0S1

after encoding will become:

001001110010011011011011101000101000010

Totally, 39 bits code was used. In this encoding, 4-bit codes were used for S4, S5, S6, and S7 instead of 3-bit ones used in the original encoding. But since the more often appeared symbols S0, S1, and S2 are encoded in 2-bit codes, the total bits used in this one reduced from 42 to 39. The codes are compressed.

How did we get the codes as above? It's not from the random writing. The encoding has to promise that any encoded code will not be the same as the first several bits of any other codes. For example, if the encoded code of S0 is 01 and S1's is 011, then you can't tell whether the bits sequence 011 is the code of S0 followed by a 1 or the complete code of S1. The code we gave above won't let this situation happen.

3. 离散余弦变换

物理意义：将信号从一种表达形式（空间域，即图像的像素值）变成另一种等同的表达形式（频率域，即频率系数），图 1-9 是离散余弦变换的示意图。

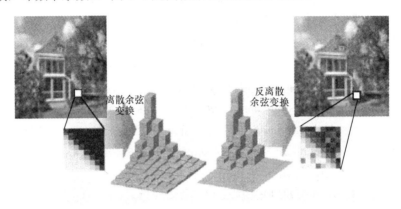

图 1-9　离散余弦变换示意图

3. Discrete Cosine Transform

Meanings: Transform the signals from one representing from (Space domain that is pixel values of the image) to another equivalent representing form (Frequency domain that is frequency coefficients). The process is show in Fig.1-9.

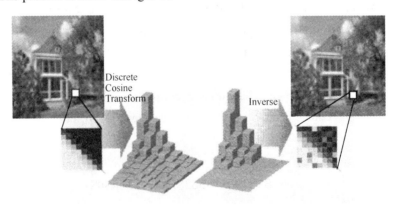

Fig. 1-9　Discrete Cosine Transform Idea

DCT 变换：空间域矩阵（$\boldsymbol{P}$ 矩阵）和频率域矩阵（$\boldsymbol{T}$ 矩阵）

$$T(i,j) = \frac{1}{\sqrt{2N}} C(i)C(j) \sum_{x=0}^{N-1} \sum_{y=0}^{N-1} P(x,y) \cos\left[\frac{(2x+1)i\pi}{2N}\right] \cos\left[\frac{(2y+1)j\pi}{2N}\right]$$

$$C(i), C(j) = \frac{1}{\sqrt{2}}, \text{当} i, j = 0$$

$$C(i), C(j) = 1, \text{当} i, j \neq 0$$

DCT Transform: Space domain array (**P** array) and Frequency domain array (**T** array)

$$T(i,j) = \frac{1}{\sqrt{2N}} C(i)C(j) \sum_{x=0}^{N-1} \sum_{y=0}^{N-1} P(x,y) \cos\left[\frac{(2x+1)i\pi}{2N}\right] \cos\left[\frac{(2y+1)j\pi}{2N}\right]$$

$$C(i), C(j) = \frac{1}{\sqrt{2}}, \text{ when } i, j = 0$$

$$C(i), C(j) = 1, \text{ when } i, j \neq 0$$

离散余弦变换过程

把每一个 8×8 的样本模块变换为 8×8 空间频率系数模块，如图 1-10 所示。

（1）能量倾向性的集中于有限的几个重要系数中；

（2）其他的系数接近于零/不重要。

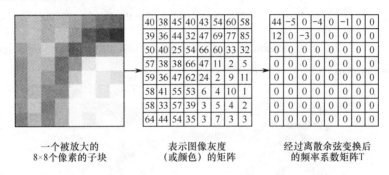

一个被放大的　　　　表示图像灰度　　　经过离散余弦变换后
8×8个像素的子块　　（或颜色）的矩阵　　的频率系数矩阵T

图 1-10　离散余弦变换中的数据块

图 1-11 给出了 DCT 变换后在每个数据块中取不同个频率系数时所得到的图的效果比较。

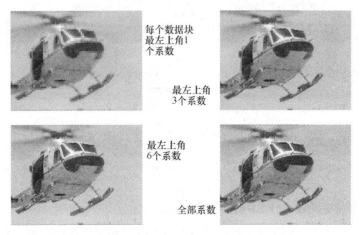

图 1-11　DCT 变换后在每个数据块中取不同个频率系数时所得到的图的效果比较

大部分图像块只包含有数的几个重要系数（通常位于最低"频率"）

Discrete Cosine Transfor

- Transform each 8×8 sample block to a 8×8 spatial frequency coefficients block.
  （1）Energy tends to be concentrated into a few significant coefficients;
  （2）Other coefficients close to zero/insignificant.

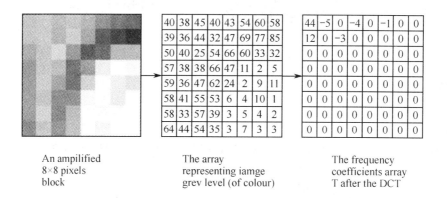

An ampilified
8×8 pixels
block

The array
representing iamge
grev level (of colour)

The frequency
coefficients array
T after the DCT

Fig. 1-10　The Data Block Within the DCT

Fig. 1-11 shows the comparison of different images with different coefficients after the discrete cosine transform.

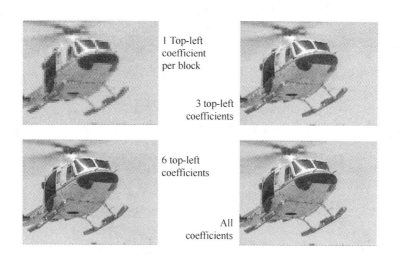

Fig. 1-11　the comparison of different images with different coefficients after the DCT

Most image blocks only contain a few significant coefficients (usally the lowest"frequencies")

4. 运动补偿

运动补偿是当前视频图像压缩技术中最关键的技术之一。

为了消除运动图像的时间冗余,人们普遍采用的办法是从当前帧中减去参考帧(前一帧),得到通常含有较少能量的"残差",从而可以用较低的码率进行编码。解码器可以通过简单的加法完全恢复编码帧。

4. Motion Compensation

Motion compensation is one of the key technologies currently used in video image compression.

To reduce the time redundancy of moving images, commonly used method is to minus the reference frame (the previous frame) from current frame. This way, we could get the "differences" contain much less energy for which we can use lower code rates to finish the encoding. At the other end, decoder could use a simple addition to recover the original frames.

图 1-12～图 1-14 给出了一个乒乓球运动的例子。从这三个系列图中可以看到：

两帧相减会出现什么问题呢？上一帧的地方有乒乓球，下一帧乒乓球的位置发生了移动，两帧直接相减，中间又出现了一个相反的东西。在这种情况下，乒乓球所在位置的编码不仅没少，反而增加了。

这使得人们去想，能不能把上一帧估算一下？乒乓球无非是移动吗，把上一帧乒乓球的位置，根据直线的估算，移动到估算的位置，然后再把这两个相减。相减以后，得到变化的部分就更少，这就叫运动补偿。

运动补偿的基本原理是：当编码器对图像序列中第 N 帧进行处理时，利用运动补偿中的核心技术——运动估值 ME（Motion Estimation），得到第 N 帧的预测帧 N'（N'帧=N–1 帧+位移）。在实际编码传输时，并不传输第 N 帧，而是第 N 帧和其预测帧 N'得差值 Δ。如果运动估计十分有效，Δ 中的概率基本上分布在零附近，从而导致 Δ 比原始图像第 N 帧的能量小得多。编码传输的是（Δ+位移），所需比特数要少得多。这就是运动补偿技术能够去除信源中时间冗余度的本质所在。

图 1-12　运动图像系列：乒乓球运动图像帧 0

图 1-13　运动图像系列：乒乓球运动图像帧 1

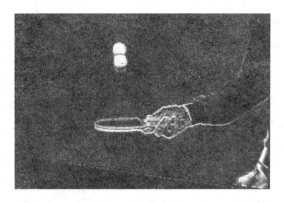

图 1-14　乒乓球运动图像帧差：帧 0 和帧 1

From Fig. 1-12 to Fig. 1-14 shows a Table tennis moving example. From this series of pictures we could see that:

There will be an issue raising up when we directly minus current frame from the previous one. The reason for this is that where the ball was in the previous frame will be moved to another location in the consecutive frame. This way, after the directive subtraction, there will be some operative things left between the ball's current location and the previous one which needs certain bits code to represent it. Then the total bits used in the encoding will be increased instead of decreasing.

The situation made people thinking. Whether it's possible to estimate the ball's previous location based on it's current location? The ball's moving along approximately a straight vertical line in the situation discussed. We know where the ball is, why not estimate it's previous location based on a vertical straight line movement? Then we can get it's estimated previous location which is very close to it's actual previous location for the reason shown above. Afterwards, we use the ball's current location minus it's estimated previous location and get a very small difference which we'll encode it. The encoded code bits will be much less since the difference we got is very small. This is the basic of motion compensation.

In conclusion, the basic principle of motion compensation is that when encoder process the Nth frame of a video images, apply the core technique of motion compensation - ME, Motion Estimation and get the N' frame, the estimation frame of Nth frame (N' frame = $N-1$ frame + motion). In real codes transmission, instead of transmitting the Nth frame, we actually transmit the difference of the Nth frame and it's estimating frame N', Δ. If the estimation is accurate enough, the probability of Δ is distributed around zero that leads the energy inside Δ is much smaller than that inside the Nth frame. What we transmitted is Δ + motion which needs much less bits to represent. This is the base reason why the motion compensation technique could remove the time redundancy inside the information source.

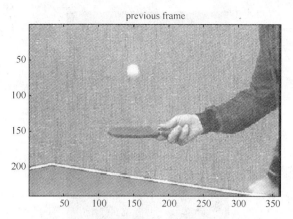

Fig. 1-12　Moving Pic Series: Pingpong Moving Frame 0

Fig. 1-13　Moving Pic Series: Pingpong Moving Frame 1

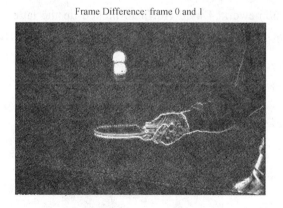

Fig. 1-14　Pingpong Moving difference: Frame 0 & Frame 1

1.3.3　几位多媒体通信发展史中的关键人物

1．戴维·哈夫曼（David Huffman）

戴维·哈夫曼，美国科学家。1950 年，在麻省理工学院信息理论与编码研究生班学习。Robert Fano 教授让学生们自己决定是参加期末考试还是做一个大作业。而哈夫曼选择了后者，原因很简单，因为解决一个大作业可能比期末考试更容易通过。这个大作业促使了哈夫曼算法的诞生。

戴维·哈夫曼

他的算法被广泛地应用于传真机、图像压缩和计算机安全领域。但是哈夫曼却从未为此算法申请过专利或其他相关能够为他带来经济利益的东西。

哈夫曼一生获得了许多荣誉，如：计算机先驱奖、IEEE 的 McDowell 奖、哈明奖章等。1999 年 10 月 7 日逝世。

1.3.3 Several Crucial People in Multimedia Communication History

1. David · Huffman

David Huffman, American scientist.

In 1950, When he was studying at information theory and coding graduate students class in MIT, Professor Robert Fano gave the students in the class an option, that is either taking a final exam or finishing a big homework. David Huffman picked up the latter. The reason was simple. The big homework was easier to do. Just this big homework induced the Huffman coding algorithm later.

David · Huffman

David Huffman's algorithm is broadly used in the fax machine, image compression, and computer security area. But he never applied for any patent or any other things which may benefit himself financially with his this famous algorithm.

David Huffman won many good reputations such as computer pioneer award, IEEE's McDowell award, and Huming award.

David Huffman died in October 7, 1999.

2. 克劳德·香农（Claude Elwood Shannon）

克劳德·香农

何谓信息论?

信息论是一门用数理统计方法来研究信息的度量、传递和变换规律的科学。

信息论之父克劳德·香农（1916 年出生于美国，著名的数学家和物理学家），在 1948 年发表了《通信的数学原理》，成了信息论诞生的标志。

在这篇开山之作里，香农指出文字、电话信号、无线电波、影像等通信交流方式，都能够编码为一种二进制的通用语言——比特（bit）（这也是"比特"一词第一次出现在文章中）。他描绘出的是一幅"数字时代的蓝图"，为以后计算机及数字通信的发展奠定了理论基础。

2001 年 2 月 24 日去世。

2. Claude Elwood Shannon

What is Information Theory?

Information Theory apply the mathematical statistic method to research information's measurement, transmission, and transformation.

Named the father of information, C. E. Shannon was born in United States in 1916 and is a famous mathematician and physicist. He published the paper: "A mathematical theory of communication" in "The bell system technical journal" in 1948. The paper is the landmark of the foundation of information theory.

In this famous paper, C. E. Shannon indicated that the communication contents such as text, telephone signals, wireless telecomm waves, video images, and etc. could be coded as common binary language, bit. That was the first time the bit appeared in this area. What he gave is somewhat like a blue print of digital decade. This set up the theory foundation of the development of computer and digital communication technology.

Claude Elwood Shannon passed away in February 24, 2001.

Claude Elwood Shannon

　　起始用二进制通用语言比特表示信息的实例如下，示于图 1-15 中。此图所示是只有一个烽火台的情况，一个烽火台只有两种状态，点燃和未点燃。所以它只能表示两种不同的信息，如图所示，灭时表示平安无事，没有敌人来犯，点燃时表示有敌人来犯，但这时无法表达更进一步的信息，即当有敌人来犯时，是从哪个方向来犯的。

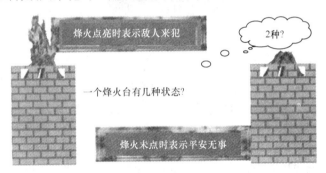

图 1-15　烽火台传递信息

　　用烽火台表示敌人来犯与否的状态。

　　一个烽火台有两种状态，点亮和未点亮。

　　分析：方向数有 4 个，东西南北。如上所述，一个烽火台显然不行，而两个正好，可以表示这四种不同的信息。如图 1-16 所示。注意，这里的前提是已经知道有敌人来犯了，在这里传递这一信息可以另用多种方法，比如没有敌人来犯时可遮挡烽火台，只有敌人来犯时才启用等等。

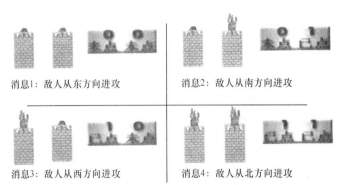

图 1-16　烽火台问题数字化示意图

把烽火台问题进行数字化则只要两位的二进制就可以，具体表示如下。

00 代表敌人从东方向进攻

01 代表敌人从南方向进攻

10 代表敌人从西方向进攻

11 代表敌人从北方向进攻

再进一步看一下信息论。

信息论的研究范围极为广阔，一般分成 3 种类型。

（1）狭义信息论是一门应用数理统计方法来研究信息处理和信息传递的科学。它研究存在于通信和控制系统中普遍存在着的信息传递的共同规律，以及如何提高各信息传输系统的有效性和可靠性的一门通信理论。

（2）一般信息论主要是研究通信问题，但还包括噪声理论、信号滤波与预测、调制与信息处理等问题。

（3）广义信息论不仅包括狭义信息论和一般信息论的问题，而且还包括所有与信息有关的领域，如心理学、语言学、神经心理学、语义学等。

《财富》杂志称信息论是"人类最值得骄傲、最罕有的创造之一，是最深远地影响了人类对世界的看法的理论。"

Original common binary bit representing transmitted information showing below, in Fig. 1-15. The figure shows the situation of one fire beacon tower. Each fire beacon tower only has two different statuses, fired and not fired. This limits that it can only represent two different messages as shown in the figure. When it's not fired means that there is no enemy attacking. When it's fired means there is enemy attack. In this case, it can't further express the more detailed information that from which direction, the enemy is attacking.

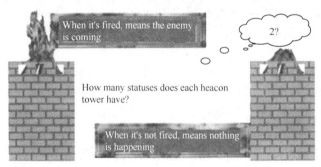

Fig. 1-15　Fire Beacon Tower Passing Messages

Use the fire beacon tower to indicate the status of enemy's invasion.

Each fire beacon tower has two statuses, fired and not fired.

Analyzing: There are four different directions, which are north, south, east, and west. As shown above, one fire beacon tower obviously can't give out those four different messages. But two of them will do as shown in Fig. 1-16. What worth to be mentioned is that here we are supposed to know that there is enemy attacking already. There are different ways to pass this message along, such as blocking the fire beacon tower all the time unless there is enemy attacking.

Transfer the fire beacon tower's representing to binary bit representing, we could get:

00　representing the enemy is attacking from the east;

01　representing the enemy is attacking from the south;

10   representing the enemy is attacking from the west;

11   representing the enemy is attacking from the north.

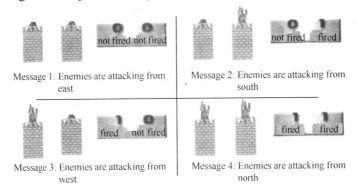

Fig. 1-16   Fire Beacon Question Digitalized Schema

Let us have a further look at the information theory.

The information theory's research range is very broad. It could be classified into three classes.

(1) Narrow sense information theory is a science that apply mathematical and statistic methods to research the information processing and transmitting. It is a kind of communication theory that focuses on the research of common regular pattern of information transmission existed in the communication and control systems. It also focuses on how to improve the efficiency and reliability of different transmission systems.

(2) General sense information theory majorly researches communication issues including noise theory, signal filtering and prediction, modulation and information processing, and etc.

(3) Broaden sense information theory not only includes the contents of narrow sense information theory and general sense information theory, and also includes all other areas which has kind of relationship with information such as, psychology, linguistics, nervous psychology, semantics, and etc.

"Wealth" magazine describes the information theory as one of the most proud of and rarely seen creation of human being. It is one of the most profound and long lasting affected the way people look at the world.

1.4   声音编码和视频压缩发展趋势简述

1.4.1   MPEG 标准

互联网的发展对视频压缩提出了更高的要求。在内容交互、对象编辑、随机存取等新需求的刺激下，ISO 于 1999 年通过了 MPEG-4 标准。MPEG-4 标准拥有更高的压缩比率，支持基于内容的交互操作等先进特性。

互联网上新兴的 DivX 和 XviD 文件格式就是采用 MPEG-4 标准来压缩视频数据的，它们可以用更小的存储空间或通信带宽提供与 DVD 不相上下的高清晰视频，这使人们在互联网上发布或下载数字电影的梦想成为现实。

从信息熵到算术编码，从犹太人到 WinRAR，从 JPEG 到 MP3，数据压缩技术的发展史就像是一个写满了"创新"、"挑战"、"突破"和"变革"的羊皮卷轴。也许，在这里不厌其烦地罗列年代、人物、标准和文献，其目的只是要告诉大家，前人的成果只不过是后人有望超越的目标而已，谁知道在未来的几年里，还会出现几个香农，几个哈夫曼呢？

1.4　The Future of Audio Coding and Video Compression

1.4.1　MPEG Standards

Along with the development of internet, the requirement of video data compression is getting higher. Under the stimulating of content exchanging, objects editing, and random saving and withdrawing, ISO proposed the MPEG-4 standard in 1999. MPEG-4 standard has higher data compression rate. It supports advanced data compression technology such as content based exchange operating.

Currently internet used DivX and XviD format are compressed under the MPEG-4 standard. Those format takes less storage space and smaller communication bandwidth and provides high definition video similar to the DVD's. This makes our dreams watching or quickly downloading movies from the internet become true.

From the entropy to arithmetic coding, from Jewish to WinRAR, and from JPEG to MP3, the history of data compression like a sheep leather roller full of "creating", "challenging", "breaking through", and "changing over". The major purpose here of listing the years, characters, standards, and articles is that all the achievements done by the others are the objects of our current people. Who knows what's going to happen in the near future? There may be a couple of more Shannon and Huffman coming out in our life.

1.4.2　几种发展中的新技术

谈到未来，还可以补充一些与数据压缩技术的发展趋势有关的话题。

分形压缩技术是图像压缩领域近几年来的一个热点。这一技术起源于 B. Mandelbrot 于 1977 年创建的分形几何学。M. Barnsley 在 20 世纪 80 年代后期为分形压缩奠定了理论基础。从 20 世纪 90 年代开始，A. Jacquin 等人陆续提出了许多实验性的分形压缩算法。今天，很多人相信，分形压缩是图像压缩领域里最有潜力的一种技术体系，但也有很多人对此不屑一顾。无论其前景如何，分形压缩技术的研究与发展都提示人们，在经过了几十年的高速发展之后，也许，人们需要一种新的理论，或是几种更有效的数学模型，以支撑和推动数据压缩技术继续向前跃进。

人工智能是另一个可能对数据压缩的未来产生重大影响的关键词。既然香农认为，信息能否被压缩以及能在多大程度上被压缩与信息的不确定性有直接关系，假设人工智能技术在某一天成熟起来，假设计算机可以像人一样根据已知的少量上下文猜测后续的信息，那么，将信息压缩到原大小的万分之一乃至十万分之一，恐怕就不再是天方夜谭了。

1.4.2　Several Developing Technologies

As we are talking about the future, there are several data compression technologies worth to mention here.

Fractal compression is a hot topic in the image compression area in recent years. It starts from the fractal propose by B. Mandelbrot in 1977. M. Bamsley set up the theory foundation for fractal compression at the late 80's of 20th century. A. Jacquin and some others have proposed many experimental fractal compression algorithms since the 90's of 20 century. Nowadays many people believe the fractal compression is the potentially strongest technical system in the image compression area. But some others don't buy it. No matter what kind of future it has, the research and development of fractal compression remind us that after years of high speed developing, we may need some more new theories or several more efficient mathematical models to support and push the data compression technology jumping forward.

Artificial Intelligence (AI) may have significant affection to the future of data compression. As Shannon can indicate in his paper that whether the information can be compressed or not and how much it can be compressed is directly related to the information's uncertainty. If one day the Artificial Intelligence is so developed that the computer could like a human being to guess the whole information based on very limited related contents, to compress the information to 1/10000 or even 1/100000 of the original one's would be possible.

1.4.3 多媒体技术的形成

多媒体技术基本上是随着电视、计算机和通信三大技术的快速发展而发展起来的。图 1-17 简单地展示了多媒体技术的发展过程，由图可见，它是由上述三种技术发展到一定程度的一种必然结果，图 1-18 展示了这种综合发展结果。

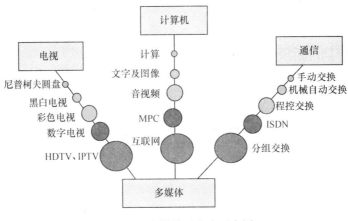

图 1-17　多媒体形成史示意图

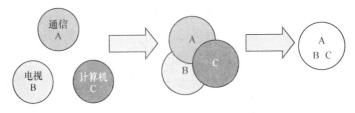

图 1-18　多媒体综合发展图

1.4.3　The Forming of Multimedia Technology

Multimedia technology basically takes form with the quickly development of TV, computer, and communication three major technologies. Figure 1–17 simply shows the taking form process of multimedia technology. From the picture we know that it is the eviTable result of TV, computer, and communication technologies developed to certain stage. Figure 1–18 shows this relationship.

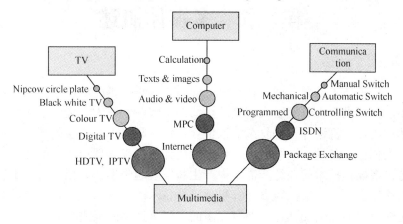

Fig. 1–17   Multimedia Technology Forming Schema

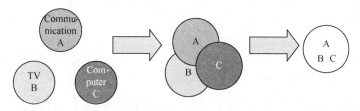

Fig. 1–18   Multimedia Technology Integration Evolution Schema

练　习　题

参照课上所讲到的"数据压缩"与"互联网"的发展简史，通过查找相关资料，写一篇"多媒体通信技术"的小论文。

Exercises

Refer to the developing history of "Data compression" and "Internet", find related information and write a simple thesis of "Multimedia communication technology".

第二章　多媒体概述

近年来，计算机技术迅猛发展，计算机的应用领域越来越广泛，在各行各业所起的作用也越来越重要。计算机不仅改变了人们处理事情的方式，而且大大提高了人们处理事情的效率。人们最初利用计算机进行一些复杂的数值处理计算和简单的文字处理，随着科学技术的进步和发展，人们已不满足于这些应用，希望能用计算机来处理更多的媒体形式，如动画、声音、视频等，多媒体技术就在这一需求中产生和发展起来了。数据压缩技术在这个过程中起了决定性的作用。几十年来，随着计算机软硬件技术的发展以及声音、视频处理包括其核心技术数据压缩技术的逐渐成熟，已经有众多的多媒体产品陆续进入市场，并且已经渗入到计算机应用的各个领域中，深入了大众的生活，如可视电话、电视会议、视频点播、远程教育、多媒体通信等。

Chapter 2　Multimedia Technology

Recently, computer technology developed very fast. Its usage is getting broader and broader. Computer is getting more and more important in every professional area. The computer not only changed the way people dealing with normal routines, and also increased the efficiency of processing those routines. At the very beginning, people used computer to deal with complicated mathematical calculations and simple text processing. Along with the development of science and technology, people are not satisfied anymore. They hope that they can use the computer to deal with more medias such as moving pictures, audio, video, and etc. Under such a requirement, the multimedia technology quickly developed. Data compression technology played a key role in this development. Along with the mature development of computer hardware and audio, video, and its core technology, data compression, many multimedia products have been appeared in the market since decades ago. Those products have been getting into almost every computer applied area and affect people's daily life. Those products include visual telephone, teleconference, video play as required, remote education, multimedia communications, and etc.

2.1　多媒体的发展

从 1949 年 2 月 15 日第一台 ENIAC 计算机问世以来，计算机的应用大大改变了人类的生活，而且人类与计算机的交互方式变成了推动计算机技术发展的一个重要因素。最初，人们使用计算机，需要由专门的操作人员将程序转换成二进制纸带后由计算机读入，这是因为计算机内部是以 0、1 组成的二进制代码进行运算的，而这个过程人类是看不到的，这种操作方式极大地局限了计算机的应用。随着计算机技术的进步，人类可以将自己所编写的高级语言源程序或命令利用键盘输入计算机，由计算机执行，从而实现了人和计算机的直接交互。而随着微电子技术的迅猛发展，促使计算机领域、通信网络领域和电视领域得到了飞速发展，人们对计算机的要求也越来越高，如要求尽量操作简便、交互方式灵活等，多媒体技术就在这一需求中应运而生，而且发展十分迅速。多媒体技术正在一步步地改变着人们的生活。它不仅使计算机的应用更广阔，而且使计算机更接近人的思维习惯，拓宽了人们与外界信息的交流方式。多媒体技术的发展与成熟为计算机应用翻开了新的一页，必将会对计算机业乃至整个社会产生深远影响。

2.1　The Development of Multimedia

Since the invention of first computer, ENIAC in February 15, 1949, the application of computers have changed people's living style. The interactive way between the computer and human being has become an important encourage factor in the computer technology development. At the earlier computer usage stage, a special computer operator was hired to translate the programs to the binary paper tapes read by the computer. This is because the computer uses the binary codes composed with "0" and "1" to calculate and the process is not visible. This extremely limited the computer's applications. Afterwards, along with the development of computer, people could key in the programmed language programs by keyboard. This realized the direct interactivity between the computer and human being. The quickly development of microelectronic technology pushed the fast development in the areas of computer, communication, and television. The demands on computers are getting higher and higher. Those demands include the easier and more flexible to operate and interact. The multimedia technology appeared and developed very fast under those requirements. The multimedia technology is further affecting people's living styles. It makes the computer's application areas getting broader and the computer's running getting closer to human being's thinking. It gives more options to human being to communicate with the natural world. The multimedia technology's mature and fast development will definitely deeply affect computer technology and human being's living styles.

"多媒体"一词最早出现于 20 世纪 80 年代，由于计算机快速、方便的特点使得计算机的应用越来越广泛，人们对人机交互界面的要求也在不断提高，人们不仅希望能够用计算机来做数值计算和文字处理，也希望人机界面和人机交互方式丰富多彩，即计算机不仅具有文字方式，还应具有图形方式、图像方式、声音方式和动画方式等多种方式的结合。

1984 年，美国苹果公司引入了位映射的概念来对图形进行处理，并使用了窗口和图形符号作为用户接口，用鼠标取代了键盘，这一方式极大地改善了人机界面。1985 年，微软公司推出了 Windows 操作系统，它极大地改善了人机交互界面，友好的可视界面方便了人们的使用，并且随着版本的不断更新，使之成为很多家庭和企业运用最广的一种操作系统平台。1986

年，Philips 公司和 SONY 公司联合推出了交互式光盘系统（Compact Disc Interactive，CDI），并公布了 CD-ROM 的文件格式。该系统把各种多媒体信息如声音、文字、计算机程序、图形、动画等以数字化的形式存放在容量为 650MB 节的只读光盘上，极大地方便了人们对多媒体数据的存储、携带。1987 年，美国 RCA 公司推出了交互式数字视频系统（Digital Video Interactive，DVI），并制定了 DVI 标准。它与 CDI 一样是用标准光盘来存储图形、图像、音频等大容量数据的。DVI 系统曾在 1991 年美国计算机大展上荣获"Comdex 91"最佳奖。该技术标准对交互式数字视频技术进行了规范化和标准化。随着多媒体技术的快速发展，在 1989 年，IBM 公司又推出视听连接系统（Audio Visual Connection，AVC），它提供立体声输入输出、全真彩色图像输入输出，以及声音和图像编辑、展示等功能。1990 年，为了适应多媒体技术的发展，使不同厂家生产的产品能互相方便地组成多媒体个人计算机系统，由 Philips、Microsoft 等 14 家厂商组成了多媒体市场协会，制定了一套相适应的多媒体计算机标准 MPCI，随后又制定了 MPC2 和 MPC3 标准。这些标准对多媒体计算机所需的软、硬件规定了相应的标准和量化指标。从 MPCI 到 MPC3，多媒体计算机在高容量的存储器及高质量的视频和音频方向有了飞速发展。MPC 平台标准的特点是兼容性、个性化或家庭化，MPC 的任务是让每个 PC 用户在软、硬件上的投入能够得到延续。1992 年，Microsoft 公司推出了窗口式操作系统——Windows 3.1，成为计算机操作系统发展的一个里程碑。Windows 3.1 是一个多任务的图形化操作环境，使用图形菜单，能够利用鼠标对菜单命令进行操作，极大地简化了操作系统的使用。它综合了原有操作系统的多媒体技术，还增加了多个具有多媒体功能的软件，如媒体播放器、录音机以及一系列支持多媒体事件的技术，使得 Windows 3.1 成为真正的多媒体操作系统。1995 年以后，随着 Windows 95 的广泛使用和国际互联网的兴起，多媒体计算机的用户界面操作变得更加简单，功能更加强大，多媒体技术的发展也得到了更加广泛的发展。而近年来，随着计算机领域、电视领域和通信领域这三大领域的相互融合和相互渗透，多媒体技术的发展更加迅速。多媒体计算机处理器的处理速度在不断升级，操作系统的版本在不断提升，其功能在不断地增加，可视化程度和可操作性也在不断增强。各种高级的外部设备也层出不穷，硬盘存储容量和存储速度在大幅度提高，各种大容量的移动存储设备（如 U 盘、移动硬盘等）在不断升级，各种数字化的音频、视频等外部设备（如刻录机、数码相机等）和与之相匹配的软件也在大量涌现。同时，各种音频、视频压缩方法也极大地提高了多媒体计算机处理音频和视频的能力。此外，网络的出现，尤其是国际互联网的飞速发展促进了多媒体技术网络化的发展。多媒体技术的发展与成熟为计算机应用翻开了新的一页，多媒体技术正在一步步地迈向新的台阶。

The word "Multimedia" appeared in 80's 20th century. Because of the computer's quick and convenient usage, it's expanded into more and more application areas. At the same time, people's requirements regarding to the interactive faces are increasing. People not only want the computer can help with complicated calculations and text processing, but also want the computer has colourful contents of interactive faces and actions. That means the computer not only has text interactive face, but also has photograph, image, audio, and moving picture combined interactive face.

Apple Company proposed using the position reflection to process photograph idea in 1984. It used window and photograph sign as interactive face. It also used mouse to replace the keyboard.

This greatly improved the interactive face. Microsoft proposed the Windows in 1985 which dramatically improved the interactive face again. The friendly graphic interactive face made the computer usage more convenient. Furthermore, along with the improving version appeared, made the Windows become the most applied operating systems of commercial and residential computers. Philip and Sony proposed the Compact Disk Interactive, CDI together in 1986. They proposed the CD-ROM format as well. This format could save many multimedia formats such as, audio, text, computer language program, graphics, and moving pictures into a 650M capacity CD. This made it much easier for people to save and carry multimedia data.

RCA in US proposed Digital Video Interactive, DVI and its related standards in 1987. Similar to CDI, it also uses the standard optical disk to save graphics, images, and audio massive data.

Multimedia – an interactive presentation of speech, audio, video, graphics, and text, has become a major theme in today's information technology that merges the practices of communications, computing, and information processing into an interdisciplinary field. In recent years, there has been a tremendous amount of activity in the area of multimedia communications: applications, middleware, and networking. A variety of techniques from various disciplines such as image and video processing, computer vision, audio and speech processing, statistical pattern recognition, learning theory, and data-based research have been employed.

This volume has the intention of providing a resource that covers introductory, intermediate, and advanced topics in multimedia communications, which can respond to user requirements in terms of mobility, easy of use, flexibility of systems, as well as end-to-end interoperability with specific quality requirements.

2.2 基本概念

为了认识多媒体，了解多媒体，首先要掌握它的基本概念。

2.2.1 媒体

在现代人类社会中存在着很多的信息，信息的表现形式是多种多样的。所谓媒体，就是指人们日常生活中所接触的各种信息的总称，也就是说，媒体是信息的载体。信息的媒体表现形式有很多，这些媒体可以是图形、图像、声音、文字、视频、动画等信息的表示形式，可以是显示器、扬声器、打印机等信息的显示形式，也可以是用于传输信息的光纤、电缆、电磁波等传输形式，还可以是磁带、磁盘、光盘等信息的存储形式。因此广义地说，按照国际电信联盟的定义，可以将媒体分为以下五大类。

（1）感觉媒体：它指的是用户所接触信息的感觉形式，直接作用于人的感官，如视觉、听觉和触觉等。

（2）表示媒体：它指的是信息的表示和表现形式，是为了能更有效地加工、处理和传输感觉媒体而人为研究和构造出来的一种媒体，如图形、图像、声音、文字、动画和视频等。

（3）显示媒体：它指的是显示和获取信息的物理设备，如显示器、打印机、扬声器、键盘和摄像机等。

（4）存储媒体：它指的是存储信息的物理设备，如磁带、磁盘、光盘等。

（5）传输媒体：它指的是传输信息的物理设备，如电缆、光纤、电磁波等。

人们通常所指的媒体是狭义的说法，指的就是表示媒体，因为作为多媒体系统来说，所处理的媒体最主要还是各种各样信息的表示和表现，而且归根结底可分为三种最基本的媒体形式：声音、图像、文字。而用于存储、呈现或传输信息的这三大类媒体狭义来说应该都属于媒介的范畴。

2.2　Basic Concepts

To know the multimedia and better understand it, we have to know its basic concepts first.

2.2.1　Media

There is a lot of different information existing in our current daily life which is showing in kinds of formats. Simply speaking, the media is the general name of all information we have known. In another word, medium is an information vehicle. Media have different formats. Graphics, images, audios, texts, videos, moving pictures, and etc. are media presenting formats. Monitor, speaker, printer, and etc. are media representing formats. Fibre, cables, electromagnetic waves, and etc are media transportation formats. Magnetic tape, computer disk, compact disk, and etc. are media saving formats.

Generally speaking, based on the definition of ITU (International Telecommunications Union), media could be classified into the following five groups.

(1) Perception media (P):

It refers to the format that people receive the information and act directly with human being's organs such as eyes, ears, and skins.

(2) Representation media (R):

It refers to the media's presenting and representation formats. It is made by human being to better deal with, process, and transit perception media. It includes graphic, image, audio, text, moving picture, and video.

(3) Presentation media (Pre-M):

It refers to the physical equipment to present and catch the information such as monitor, printer, speaker, keyboard, video camera, and etc.

(4) Storage media (Sto-M):

It refers to the saving physical equipment such as magnetic tape, disks, compact disk, and etc.

(5) Transportation media:

It refers to the information transportation equipment such as electrical/electronic cable, fibre, magnetic wave, and etc.

The media we normally talked about are the narrow ranged media idea which refers to the representation media. The reason for this is that the major media the multimedia system processing are still the presenting and representation of different kinds of information which could be concluded into three basic media formats: audio, image, and text. The other three formats:

presentation, storage, and transportation would be narrowly classified into physical media.

2.2.2 多媒体

"多媒体"（Mutimedia）是最近几年来才流行的一个名词，它是一个组合词，该词由 "multiple" 和 "medium" 的复数形式 "media" 组合而成。如果单从字面上来理解，"multiple" 是 "多重、复合" 的意思，而 "media" 是 "介质、中间" 的意思，因此该组合词就可解释为多种媒体的综合，即用 "多媒体" 来表示包含文字信息、图形信息、图像信息、声音信息、动画信息和视频信息等不同信息类型的一种综合体。不过这种解释不太准确，它没有体现出该词的内涵，所以另一种解释就是：多媒体的含义应将 "多"、"媒"、"体" 三个词的含义都体现出来，即 "多" 是多种媒体表现，多种设备、多种学科；"媒" 是媒介；"体" 是综合、集成一体化。而且每个时代对 "多媒体" 的理解也不同，如在 20 世纪 90 年代以前，人们对 "多媒体" 的解释为信息的表现媒体，如电影、电视、录像、影碟机、录音机、收音机等；而进入 21 世纪后，人们对 "多媒体" 的理解已更加深入，"多媒体" 不仅应含有信息的表现媒体，还应该包括那些能对信息进行传输、存储的媒介，如电磁波、光纤、多媒体技术与应用磁盘、光盘等，而且应该是多种媒体的有机统一的综合。

从以上所述可看出现代多媒体与传统媒体有着很重要区别：传统媒体所处理的基本上是模拟信号，而现代多媒体所处理的信息是数字信号；现代多媒体具有丰富的人机交互功能；传统媒体提供给人们的信息量往往是有限的，而现代多媒体平台提供的信息容量几乎是无限的，且存储成本较低，获取方式方便快捷；传统媒体在信息的传播中不能根据每个人的个性化需要进行信息传输，而现代多媒体可以针对每个人的不同需求进行定制传播，从而最大化利用了信息，充分满足人们各自的不同要求。

2.2.2 Multimedia

"Multimedia" has appeared as a modern word in recent years. It is a combined word and composed by multiple and media two words. Simply speaking, it means multiple media and refers to a composed medium including text info, graphic info, image info, audio info, moving picture info, and video info. More precisely speaking, multimedia should show the meaning multiple and media integrated together. That means it should imply multiple media representation, multiple equipment, and multiple technical subjects are integrated together. People's understanding to the multimedia is different from time to time. For example, before 90's, 20th century, people understand multimedia as information representing formats such as movie, TV, video, DVD, recorder, radio, and etc. Started and after 21st century, people understand the multimedia further deeper. It not only includes the information representation formats, but also includes the information transportation and saving formats. It should be multiple media integration.

As we know, there is big difference between the traditional media and modern media. Traditional media basically deal with analogue signals and modern media majorly deal with digital signals. The modern has plenty of interacting functions between human being and the computer. The traditional media provides limited information and the modern media provides almost unlimited information. Additional, the modern media's saving cost is much lower and accessing is much more flexible. The traditional media can't transmit the information based on individual needs and the modern media can. This way, the modern media completely transport the information to meet the individual needs.

2.2.3 多媒体技术

多媒体的实现离不开计算机，正是由于计算机的处理能力，才使多媒体能够进行显示、存储等处理。从某种意义上来说，多媒体技术就是指利用计算机对文字、图像、图形、动画、音频、视频等多种信息进行数字化处理的综合技术。其中包括信息的采集、获取、压缩/解压缩、编辑、显示、存储、传输等。

多媒体技术是多种技术的综合，如计算机技术、通信技术、电视技术等。多媒体技术的发展和这些技术的飞速发展紧密相连。大规模集成电路的飞速发展促进了计算机的系统处理能力，大容量存储设备的出现促进了多媒体存储技术的发展。多媒体技术处理的数据对象主要有文本、图形、图像、动画、音频、视频等6种。

1. 文本

文本是以文字及各种专用符号表达信息的形式，它是我们现实生活中使用最多的一种信息存储和传递方式，是计算机文字处理的基础。人们习惯用文本来表达信息，进行相互交流，它主要用于对知识的描述性表示，如阐述概念、定义、原理和问题以及显示标题、菜单等内容。文本的多样化体现在文字的变化上，即文字的格式、文字的定位、文字的字体及文字的大小等的变化。通常采用文字编辑软件（记事本、写字板、Microsoft Word 等应用软件）生成文本文件，或者使用图像处理软件形成图形方式的文字。

2. 图形

图形是指由点、线、面到三维空间的黑白或彩色几何图形。图形的表示不直接描述构成它的每一点，而是描述产生这些点的过程和方法，可用矢量来表示，因此也称为矢量图。它具有体积小、数据表示精确、处理过程中可以分别控制图中的各个部分等特点。矢量图形的主要缺点是处理起来比较复杂，所花费的时间较长。因此，图形这种形式主要用于线型图画、工程制图及美术字等。可采用算法语言或某些图形处理软件生成矢量化图形，如图形处理软件 AutoCAD。

3. 图像

图像是多媒体中最重要的信息表现形式之一，是检验多媒体软件视觉效果的关键因素。图像也称为位图图像，其最基本的单元是像素。所谓像素是指图像中的点，而这些点是用将图像按行和列进行数字化后所得到的一个数字化值来表示的，位图图像是所有视觉表示方法的基础。位图中的位用来表示图中每个像素的颜色和亮度。位图图像具有表现细腻、层次和色彩丰富、包含大量细节等特点，而且位图显示比图形显示快。图像的一个重要因素是分辨率。图像分辨率以水平和垂直像素点来表示，通常一幅图像的像素点越多，则其图像质量越好，越接近自然状况，不过需要的存储空间也越大。图像的另一个重要因素就是图像的灰度值，也就是其颜色色彩的多少，它采用二进制位来表示。例如，单色图像的灰度用 1 位二进制码来表示亮或暗；如果用 4 位二进制码来表示的话，就可以表示 2 的 4 次方即 16 种颜色；如果用 8 位则可以表示 2 的 8 次方即 256 种颜色；若为 24 位则可以表示的颜色数目可达 2 的 24 次方即 1677 万种。不过颜色的数目越多，其数据量越大，所需要的存储空间也随之越大。

一幅图像的大小可以采用以下方式计算：

图像数据量大小（字节）＝高×宽×灰度位数/8 其中，高是指垂直方向的像素值，宽是指水平方向的像素值。如一幅分辨率为 600×800 的图像，如果每个像素点采用 8 位二进制数来表示颜色数目，则其所需的空间大约为 480000B；如将同一图像采用 1024×768 的分辨率，则其所需的空间则上升为 786432B，存储空间比原来大了约 1.6 倍。

由上可见，位图所要求的存储空间很大，因此，在实际图像的存储和传输时，需要对图像进行压缩。图像的存储格式主要的有 gif、bmp、png、tif、jpg 等。对图像文件可进行改变图像尺寸、对图像进行编辑修改、调节调色板等处理，可用软件技术减少图像灰度，用较少的颜色描绘图像，以减少图像的数据量。

图形和图像在用户看来没什么区别，而对多媒体制作者来说则完全不同。对同样一幅图，在上面画一个圆，若采用图形媒体元素，其数据记录的信息为圆心坐标点、半径值及颜色编码；若采用图像媒体元素，其数据记录的信息为在哪些坐标位置上有什么颜色的像素点。二者比较之，图形数据信息处理起来更灵活，而图像数据则与实际更加接近。随着计算机技术的飞速发展，二者之间的差别越来越小，通常先用矢量图形创建复杂的图，然后转换成位图来进行各种处理。

可采用数码相机、录像机、扫描仪等进行图像的获取而生成图像，也可采用图形图像处理软件（Photoshop、CorelDraw、Freehand 等）来制作和编辑图像。

4. 动画

动画是快速播放一系列连续运动变化的图形图像而利用人的视觉暂留特性来获得的，可包括画面的缩放、旋转、变换、淡入淡出等特殊效果。合理使用动画，可以极大地丰富多媒体作品的视觉效果。

与视频不同，动画采用的是计算机产生出来的图像或图形，而视频采用直接采集的真实图像。视频影像要想输入计算机进行处理，就必须实现模拟信号向数字信号的转换。动画不仅包括矢量动画，也包括帧动画，不仅有二维动画还有三维动画等多种形式。在各种媒体对象中，动画所要求的硬件环境是最高的。它不仅需要高速的 CPU，还需要较大的内存。在动画的播放过程中，为了保证其播放效果，要求视频的帧速应该不小于 15 帧/s，如果小于该值，画面就会出现抖动。电影采用的帧速是 24 帧/s，NTSC 制式电视的帧速是 30 帧/s，PAL 制式电视的帧速是 25 帧/s。

运动图像每秒钟的数据量是帧速乘以单帧数据量。若一幅图像的数据量为 2MB，其帧速为 24 帧/s，则一秒钟的数据量为 48MB。可见运动图像的数据量比静态图像的更大，因此数据压缩是必须的。

计算机设计动画分为矢量动画和帧动画，也有二维动画和三维动画之分，生成方法有所不同。矢量动画是对物体的每一个部分进行分别设计，赋予每个部分一些特征，然后用这些部分构成完整的画面，常用软件有 3dsMAX 等；而帧动画则是由一幅幅位图组成的连续画面，要分别设计每屏所显示的画面，帧动画与传统动画的原理一致，常用软件有 Flash 等。

5. 音频

音频信号可提供其他任何媒体不能取代的效果，它能够烘托气氛，增加活力。声音是音频信号的一种形式，是人们用来传递信息、交流感情最简捷的方式之一。除了波形声音之外，音频信号还包括语音、音乐等形式。语音是对讲话声音的一次抽象。语音也可以表示为波形声音，但波形声音表示不出语言、语音学的内涵。

声音是一种机械波，通常用一种模拟的连续波形来表示，为了能在计算机中处理声音，必须将模拟的声音信号转换成数字信号即声音数字化。影响数字声音波形质量的有三个主要因素，它们为采样频率、采样精度和通道数。计算机对音频的处理有声音的采集、声音的数字化、声音的压缩及声音的播放等。如上所述，影响数字音频文件的质量的主要因素有三个：第一个是采样频率，采样频率越高，则声音质量越好；第二个是采样精度，采用二进制位数来表示，位数越多，则声音质量越好；第三个是声道数，一般有单声道和立体声道两种。

音频文件有多种格式，常采用的有波形音频文件（wav 格式）、数字音频文件（mid 格式）和 mp3 压缩格式等。可采用声卡、录音机等进行声音的采集而获得音频信号，也可利用声音处理软件（UleadMediaStudio、SoundForge、CoolEdit、WaveEdit 等）进行声音的编辑。

6. 视频

视频信号是动态的图像，是将一组图像按照时间的先后顺序进行连续显示而形成的一种信号，它非常类似于电影和电视，既有声音又有图像，在多媒体中充当着重要的角色。不过它和电视视频不同，电视视频是模拟信号，而计算机视频是数字信号，虽然这两种视频正在融合，如高清晰度数字电视（HDTV），但两者之间仍有差距。具有代表性的视频格式有 avi 格式的电影文件、压缩格式的 mpg 视频文件等。可用录像机、摄像机等采集影像而形成视频信号，也可采用视频处理软件（UleadMediaStudio、AdobePremiere 等）来对影像进行编辑。

这些媒体元素比起单一的文本元素而言，具有数据量大、处理复杂、传输时间长等特点，因此需要相关多媒体关键技术的支持。

2.2.3 Multimedia Technology

There will be no multimedia technology without computer. Just because of the data processing ability of computer, multimedia technology could deal with the data display, storage, and etc. processes. In some way, the multimedia technology is to use the computer digitally processing the diverse information format such as text, graph, image, cartoon, audio, video, and etc. The processing includes the information collection, acquisition, data compression/decompression, data edition, display, storage, transportation, and etc.

Multimedia put diverse technologies together and formed an integrated technology. Those diverse technologies include computer technology, communication technology, TV technology, and etc. The development of multimedia technology is closely related to the fast development of those diverse technologies. The fast developed integrate circuit pushed the computer's system processing ability. The appearance of high capacity storage equipment pushed the development of multimedia storage technology. Multimedia technology majorly processes text, graphic, image, cartoon, audio, and video.

1. Text

Text uses words and other special signs to represent the information. It's the most likely used information storage and transportation format in our daily life. It's the base of computer processing words. We have got used to use text to express information and communicate. It is majorly used to represent the knowledge descriptively such as express the abstract, definition, principle, and display the title, manual, and etc. The different style of the text shows in the changes of the text format, position, font, size, and etc. People usually apply text processing software such as Notepad,

WordPad, Microsoft word and etc to produce text files. People also apply image processing software to produce graphic format texts.

2. Graphic

Graphic means the geometric pictures from the simple dot, line, and plane to complicated 3D object. The representation of the graphic doesn't show every point in the picture and does show the process and the method how those points produced. It could be presented by vectors. So the graphic is called vector diagram as well. It has good characters like small occupy space, precise data representation, and separate parts control during the processing. The bad part of it is that the process is complicated and takes longer time. It is majorly applied in liner diagrams, engineering drawings, and article words. It could be produced by algorithm languages or image processing software such as AutoCAD.

3. Image

Image is one of the most important information formats in multimedia. It is the key factor to check the multimedia software's vision quality. Image is also called spot image. Its most basic unit is pixel. It shows as the dot inside the image. It is represented as a digital valve after digitalizing the image both horizontally and vertically. Spot image is the foundation of all other visual representations. The spot in the image shows the colour and brightness of the corresponding pixels. The advantages of the spot images include presenting lively, rich layers and colours, and consisting of details. Its display speed is faster than the graphic's.

One important factor of the image is its resolution. It is represented by the horizontal and vertical pixels. Usually, the more pixels the image has, the better quality the image is. The image is closer to the natural. But it will need more space for the storage.

Another important factor of the image is its grey value. It's the factor showing how much colour the image has. It is represented in binary digits. For example, the grey value of the black and white image is represented by one binary digit in either light or dark. If use 4 binary digits to represent a kind of colour, we could get power 4 of 2, 16 different colours included. If use 8 binary digits, we could get power 8 of 2, 256 different colours included. While with 24 binary digits, would give us power 24 of 2, 16770000 different colours. On the other hand, the more colours we get, the more quantity data we have and the more storage space we'll need.

The data size of one image could be calculated as follows: The data size of an image (bytes) = Height×Width×Grey digits/8. Inside the formula, Height is stand for the amount of vertical pixels and Width is stand for the amount of horizontal pixels. Take one 600×800 resolution image as an example, if each pixel's colour represented by 8 binary digits, the storage space the image needs is 480000 bytes. If we increase the same image's resolution to 1024×768, its storage space will be increased to 786432. It is roughly 1.6 times bigger.

From above we know that the spot image takes a lot storage spaces. So, in the real world, when store and transport images, we need to compress them first. The major image storage formats include gif, bmp, png, tif, jpg, and etc. We could process the image as changing the image size, editing or modifying the image, and adjusting the colour of the image. We could use software to reduce the grey levels of the image and decrease the colour describing the image in order to save the image storage space.

Graphic and image look no big difference to the viewers and big difference to the multimedia products developers. For the same picture, draw a circle on it. If we use graphic format, the data recorded will be the coordinate of the center of the circle, radium, and the colour code. If we use image format, the data recorded will be the pixels along the circle with desired grey levels. Compare the above two of them, the graphic format is more flexible and the image format is much closer to the real. Along the quick development of the computer technology, the differences between the two are getting smaller and smaller. What people usually do is to use the graphic format to create the complicated pictures and process it after using the image format.

We could use digital cameras, camcorders, scanner, and etc to get the images. We also could use the relative software, such as Photoshop, CorelDraw, Freehand, and etc. to produce and edit the images we want.

4. Cartoon

Cartoon is achieved based on the human being's natural vision delay character by quickly broadcasting a series of changing graphics or images. It could include compress and expand, turn, transform, zoom in and out, and other particular effect on the pictures. Reasonable using cartoon could greatly improve the vision quality of the multimedia products.

Different from video, cartoon takes either images or graphics produced by computer. Video takes the images created from the real world. In order to process the videos with computer, one needs to transfer the analogue signals to digital ones. Cartoon includes vector ones and frame ones. It also includes two-dimension ones and three-dimension ones. Among the multimedia information formats, the cartoon's hardware environment requirement is the highest. It not only needs the high speed of the CPU, and also needs bigger memory. In the process of broadcasting, to keep the broadcasting quality, the frame speed shall not be slower than 15 frame per second. If the speed is slower than this, the images showing will be shaking. The movie's frame speed is 24 frames/second. The NTSC TV's frame speed is 30 frames/second and PAL's is 25.

The data quantity of a moving image is the product of frame speed and the single frame's data. Suppose a single frame's data is 2M bytes, its frame speed is 24 frames/second, the data quantity of this moving image within one second is 48M bytes. Obviously, the data quantity of a moving image is even bigger than a still one's. Data compression is absolutely necessary.

Computer design cartoon can be classified into vector cartoon and frame cartoon or two-dimension cartoon and three-dimension cartoon. Vector cartoon is to design each individual part of the object and give them the necessary characters. Then put the designed individual part together and get the whole picture. The often used software is 3dsMAX and etc. On the other hand, frame cartoon is to design each frame of the whole cartoon. This is the same as the traditional cartoon principle. The often used software is Flash and etc.

5. Audio

Audio could provide the special effect on the multimedia product that other formats couldn't. It could create the atmosphere and add vividness to the product. Sound is one of the simplest formats of audio for people to transfer information and communicate. Besides the sound wave, audio includes speech, music, and etc as well. Speech is abstracted from the sound. Speech could also be represented as wave signals. At the same time, wave signal can't express the inside details of languages and phonetics.

Sound is a mechanic wave and often presented as an analogue continuous wave. To process the sound with computer, one needs to covert the analogue signal to digital signal. It is also called digitalizing. There are three major factors that can affect the quality of the digital sound wave and they are sampling frequency, sampling accuracy, and channel quantities. Computer sound wave processing includes sound wave collection, digitalizing, compressing, broadcasting, and etc. As indicated above, there are three major factors affect the digital sound quality. One of them is the sampling frequency. The higher the sampling frequency, the better the digital sound quality is. Another one of them is the sampling accuracy. If presented in binary digits, the more digits used, the better digital sound wave quality we could get. The last one is the channel quantities. There are two common channels. One of them is single channel and the other is stereo.

Sound files have many different formats. The common ones include wave sound file (wav format), digital sound file (mid format), and mp3 compress format. We could use audio card or audio recorder to get the sound signals. We also could use sound processing software such as UleadMediaStudio, SoundForge, CoolEdit, WaveEdit, and etc. to edit the sound files.

6. Video

Video signal is dynamic image. It's a kind of signal that continuously broadcasts a group of images in time sequence. It's very similar to the well known movie and TV. It has both sound and image. It plays a important role in the multimedia products. But it's different from the TV video signals which are analogue signals. Computer video is digital signal. The two are combining together such as high definition TV (HDTV). But they are still different from each other. The typical video formats are avi movie file, compressed format mpg, and etc. We could get the video signals by using VCR, camcorder, and etc. We also could use video processing software such as UleadMediaStudio, AdobePremiere, and etc. to edit the video files.

Compared with the pure text, the other multimedia formats mentioned above have the characters of occupying bigger data quantity, processing complicated, taking longer time to transmit, and etc. They need more multimedia key technologies' support.

2.3　多媒体技术的基本特性

多媒体技术是以计算机为中心，对多种媒体元素进行各种数字化处理，如对各种媒体元素进行编辑、压缩、存储、传输等处理。多媒体技术具有许多特性，如多样性、交互性、集成性、非线性和实时性等。

1. 多样性

通信技术及计算机技术的发展，使人们进行处理的媒体信息更加多样化，不再局限于文本一种方式，而是文本、声音、图形、图像、动画、影像和视频等多种形式的结合。由于多样化主要用于计算机的信息输入和信息输出上，因此，可以使计算机处理的范围和空间扩大，大大丰富信息的表现力和增加信息的表现效果，而且也使计算机与用户之间的关系更密切，使用户更容易操作和控制计算机，从而可以更全面、更准确地接受信息。

2. 交互性

交互性是指用户与计算机之间传递信息的方式。用户与计算机之间不仅传递简单的数据信息，而且也要传递各种多媒体信息，如文字、图像、图形、动画、视频等。交互功能不仅使人们从计算机中获取到了更多的信息，而且使人们在处理多媒体的过程中更加具有主动性和可控制性，与计算机的交互也更加亲切友好。交互可以增加用户对信息的注意力和理解力，延长信息在大脑中的存留时间，而且交互可以提高人对信息表现形式的选择和控制能力，同时也提高信息表现形式与人的逻辑和创造能力结合的程度。借助交互过程，用户可以获得更多的信息，而且可以对信息的运行过程进行控制，如用户可以利用检索系统找出自己想读的书籍、想看的电视节目，可以在浏览时跳过自己不感兴趣的内容等。在多媒体远程计算机辅助教学系统中，学习者可以主动学习自己所感兴趣的内容，而且可以通过网络向讲授者提出问题和上交作业，充分体现学习者的主动性和自觉性。因此，可以说人机交互不仅仅是一个人机界面的问题，它与人类的智能活动有密切的关系。

3. 集成性

集成性是指能够集成处理多种信息的能力。多媒体技术包含了计算机领域、电视领域及通信领域内较新的硬件技术和软件技术，并将不同功能、性质的设备和媒体处理软件集成为一体，以计算机为中心来综合处理各种信息。以前，以上形成多媒体技术的各个单个技术都是单独使用的，为了更加有效地组织和表现信息，就必须将这多种技术进行综合而形成多媒体技术。多媒体的集成性主要表现在两个方面：一个是多媒体信息的集成，即可以同时地获取信息、统一地表示信息、组织信息和存储信息；另一个是处理这些媒体信息的设备和系统的集成。换句话说，多媒体系统具有能够处理各种媒体信息的高速及并行处理体系、大容量的存储设备、各种各样的输入输出设备等。

4. 非线性

多媒体技术的非线性特点将改变人们传统的顺序性读写模式。传统的顺序性读写模式大都采用从上到下、从左到右的方式，循序渐进地获取知识。而在多媒体技术中采用超文本链接的方法，它把内容以一种更灵活、更具变化的方式呈现给用户，而用户也可以按照自己的目的、需要和兴趣来使用信息，可以任意读写其中的图、文、声等媒体信息，并可以重新组织信息，如增加、删除或修改节点等，重新建立新的链接。

5. 实时性

当用户给出操作命令时，相应的多媒体信息都能够得到实时控制。总的来说，多样性、交互性和集成性是多媒体技术最重要的三个特性。

2.3　Basic Characterristics of Multimedia Technology

The multimedia technology process the different formats of the multimedia products based on the computer technology. Those processing includes edit, compress, storage, transmission, and etc. It has many characters such as multiplicity, interactivity, integration, nonlinear, real time, and etc.

1. Multiplicity

The development of communication and computer technology has made the processed multimedia formats get more and more. The formats are not limited to the text only, but also include audio, graphic, image, cartoon, TV and movie, video, an etc. This kind of multiplicity majorly used in the computer inputs and outputs. Because of this, it made the computer's processing range becomes wider. This improves the information representation ability and effects. It also made the user and the computer being getting closer. This way, it'll be easier for user to manipulate and control computer. In turn, user could more broadly and precisely acquire the necessary information.

2. Interactivity

Interactivity means the communication format between user and the computer. The user and the computer not only exchange the simple data information, but also exchange diversity multimedia information such as text, image, graphic, cartoon, video, and etc. The interactivity not only makes users get more information from the computer, but also gives the user more domains and controlling during the processing and makes the interface more friendly. Because of the interactivity, user could pay more attention to the information and better understand it. User also could keep the information in head for a longer time and improve the information representation formats pick up and control ability. Meanwhile, this improves the combination of information representation formats and user's logical and control ability. By interactivity, user could get more information and control the information processing procedure. For example, user could index the books or TV programs one wanted and skip the contents not interested. In multimedia remote computer aided teaching system, learner could freely pick up the materials interested, via the network asking questions and submit homework. It completely shows the learner's self-awareness and initiative. From this we can see that the interactivity is not only a simple interface, but also has a close relationship with human being's intelligent activities.

3. Integration

The integration here means the ability that could deal with multiple kinds of information at the same time. Multimedia technology combines the new hardware and software technologies together in the computer, communication, and TV field. It takes computer as a core and integrates the different functional and characteristic equipment and software to process kinds of information. Before, each technology formed the multimedia technology is working separately. To more efficiently organize and present the information, it is necessary to combine the separated technologies together and form the multimedia technology. The integration of multimedia is shown in two ways. One of them is the multiple information integration that is get the different information at the same time and present, organize, and transmit the information in an integrated one way. The other one is the integration of the equipment and systems which process the information. In another word, the multimedia system has the high speed and parallel processing structure to deal with different media information. It has high capacity storage equipment, sorts of input and output equipment, and etc.

4. Non-linearity

The non-linearity of the multimedia technology is changing people's traditional sequence reading and writing way. The traditional reading and writing way usually takes from top to bottom and from left to right format to gain the knowledge needed. In multimedia technology, it takes link format to show the content to the readers in a more flexible and exchangeable way. Meanwhile, the readers could chose the information based on their own intent, need, and interest. For example, they could purposely read and write the images, texts, sounds, and etc. media information and even reorganize the information such as add, delete, and edit the link points or recreate the new links.

5. Real-time

When user gives the operating command, the corresponding multimedia information could be controlled in a real time.

General speaking, multiplicity, interactivity, and integration are the most important characters of multimedia technology.

2.4 多媒体技术的应用领域

如今各行各业中随处可见多媒体技术的应用，如计算机、通信、出版、广告等行业，而且它也正渐渐进入人们的家庭生活中，如多媒体电视等。同时，随着国际互联网的兴起，多媒体技术会随着网络的发展和延伸不断地发展和进步，进而开创多媒体技术的新时代，它的主要应用领域包含如下几个方面。

1. 教育领域

多媒体技术在教育领域的应用克服了传统教育的弊端，使人们以一种更自然的方式来接受教育。利用多媒体技术编制的教学课件，可以将图像、文本、声音和视频综合起来，营造出图文并茂、生动逼真的教学环境，从而可大大激发学生学习的积极性和主动性，使学生从被动接受知识转变为主动选择教学信息，提高了学习效率，改善了学习效果和学习环境。教师可根据课堂的实际情况及时得到学生的反馈，根据反馈信息对教学内容进行修改和补充。

远程教育是多媒体技术应用的典范。现代远程教育是随着现代信息技术的发展而产生的一种新型教育方式，是以现代远程教育为目的，多种媒体优化、有机组合的教育方式，是构筑 21 世纪终身学习体系的主要手段。现代远程教育可以有效发挥各种教育资源的优势，为各类教育提高其教育质量提供有力支持，可以为各种社会成员提供方便广泛的教育服务。现代远程教育手段的特点是教师和学生能够跨越时空进行实时或非实时的交互，这是现代远程教育与传统教育方式最显著的区别，也是其优势所在。

现代远程教育手段有以下优点。

（1）教师的讲授和学生的学习可以在不同的地点同时进行，师生之间可以进行充分的交流，学生能够根据自己的需要自主安排学习时间和地点，自主选择学习内容，自安排学习计划，随时提出学习中的问题并能及时获得解答。

（2）现代远程教育有利于个体化学习，它以学生自学为主，充分发挥学生自主学习的主动性、积极性及创造性。

（3）现代远程教育手段可以为学生提供优质的教学服务，教师可以及时了解学生的学习进度和对课程的理解程度，解答学生提出的问题。

（4）现代远程教育给教与学的概念赋予了新的内涵，将给教育带来深刻的变革，推动教育观念、教育思想、教育模式和教学方法的更新。

（5）通过远程教育，身处异地的学生可以通过计算机网络来上课，改变了上课必须到课堂这一传统模式，解决了某些学生没有时间去外地学习的问题，同时也促进了学术的交流。

2．商业

多媒体在商业方面应用主要包括以下几个方面。

1）办公自动化

各种先进的数字输入、输出设备（如数码相机、扫描仪、图文传真机、激光打印机等）构成了办公室自动化设备。多媒体办公系统是视听一体化的办公信息处理和通信系统。

它主要有以下功能：

（1）办公信息管理，将各种信息，包括文件、档案、报表、数据、图形、音像资料等进行加工、整理、存储，形成可共享的信息资源；

（2）召开可视的电话电视会议；

（3）进行多媒体邮件的传递。

多种办公设备与多媒体系统的集成可以真正实现办公自动化。办公自动化不仅可以帮助工作人员浏览和处理大量的信息和数据，而且可以节约大量的人力和物力资源。

2）产品广告

现代的社会是一个广告的社会，商品的销售离不开广告，要想制作出让人耳目一新的广告，商家就必须运用各种多媒体素材对自己的商品进行生动逼真的展示。

3）查询服务

在信息社会，商场、银行、医院、机场等服务行业可以利用多媒体计算机系统更好地为顾客提供方便、自由的交互式查询服务。

4）企业生产

现代化企业的综合信息管理和生产过程的自动化控制，都离不开对多媒体信息的采集、监视、存储、传输，以及综合分析处理和管理。应用多媒体技术来综合处理多种信息，可以做到信息处理综合化、智能化，从而提高工业生产和管理的自动化水平。多媒体技术在工业生产实时监控系统中，尤其在生产现场设备故障诊断和生产过程参数监测等方面有着非常重大的实际应用价值。在一些危险环境中，多媒体实时监控系统也将起到越来越重要的作用。将多媒体技术用于模拟实验和仿真研究，会大大促进科研与设计工作的发展。将多媒体技术用于科学计算可视化，可将本来抽象、枯燥的数据用三维图像动态显示，使研究对象的内因与其外形变化同步显示。

3．医疗

现代先进的医疗诊断技术的共同特点是，以现代物理技术为基础，借助于计算机技术对医疗影像进行数字化和重建处理。计算机在医学成像过程中起着至关重要的作用。随着临床要求的不断提高以及多媒体技术的发展，出现了新一代具有多媒体处理功能的医疗诊断系统。多媒体医疗影像系统在媒体种类、媒体介质、媒体存储及管理方式、诊断辅助信息、直观性和实时性等方面都使传统诊断技术相形见绌，尤其是借助于网络可以实现远程医疗，这将引起医疗领域的一场新的革命。

4．电子出版物

电子出版物是指将图、文、声、像等信息以数字方式存储在磁、光、电介质上，通过计算机来阅读和使用，并可复制发行的大众传播媒体。电子出版物具有容量大、体积小、价格低、保存时间长等优点，它不仅可以记录文字数据信息，而且可以存储图像、声音、动画等视听信息，同时还可以进行交互式阅读和检索。电子出版物的出版形式可以以软磁盘、只读光盘、交互式光盘、计算机的硬盘等作为载体。这种出版形式是将纸面的文字变为电子数据存储，当用户需要信息时，可以购买存有这些信息的磁盘或光盘，通过计算机来进行阅读。因此，这种电子出版形式改变了信息的存储方式。

电子出版物的出版形式还可以在网络上进行出版发行，即网络电子出版，它是电子出版的另一种主要方式。这种出版形式将信息内容发布在互联网上，完全改变了信息的传输方式，改变了信息与读者之间的关系，用户可以主动获取网上信息，也可被动接受网上信息。目前，在互联网上出版的电子报刊已达数百种。这种网络出版通过互联网技术实现了信息在网络上的传输，使信息发布者和读者之间建立了一种实时的、交互的、基于服务器/浏览器结构的网络结构。利用互联网和多媒体计算机，足不出户就可以查阅世界各大图书馆的书库和所需的各种资料，其特点是信息的传播速度快、更新快。与传统媒体方式（比如报纸、书、杂志等）和磁、光盘类电子出版物相比，网络出版物不受时间和地域的限制，它面向所有允许的互联网用户；这些允许的用户可以在任何时间、任何地点阅读网上的信息，可将自己的信息反馈给信息发布者，有一定权限的用户还可以随时发布最新新闻，所有信息均存储在服务器上，信息既可以及时补充，也可以进行修改和删除。最重要的是，网络出版使信息的检索方便而且灵活，用户可按时间、关键词，甚至加上某些算法进行综合信息检索。

5．多媒体通信

多媒体技术的一个重要应用领域就是多媒体通信。多媒体通信是 20 世纪 90 年代迅速发展起来的一项技术。一方面，多媒体技术使计算机能同时处理视频、音频和文本等多种信息，提高了信息的多样性；另一方面，网络通信技术取消了地域限制，提高了信息的瞬时性。二者结合所产生的多媒体通信技术把计算机的交互性、通信的分布性及电视的实效性有效地融为一体，成为当前信息社会的一个重要标志。

多媒体通信涉及的技术面极为广泛，如人机界面、数字信号处理、大容量存储装置、数据库管理系统、数据压缩/解压缩、多媒体操作系统、高速网络、通信方式、计算机网络及相关的各种软件工程技术。随着互联网的普及和计算机网络技术的发展，网络如今已成为人类社会生活中不可缺少的一个组成部分。目前，人们利用网络主要进行网络聊天、网上购物、可视电话、视频会议、远程文件传输、浏览与检索多媒体信息资源、多媒体邮件以及远程教学等方面。人们在网络上传递和获取各种多媒体信息，而且已经不满足于由文字组成的交谈了，希望能够通过语音甚至动态图像进行实时通信。在购物网站上，越来越多的人足不出户就可购买到自己所喜欢的商品。目前进入千家万户的数字电视将可以让观众根据需要选取电视台节目库中的信息。

总之，多媒体网络为多媒体通信提供了一个传输环境，使计算机的交互性、网络的分布性和多媒体信息的综合性有机结合，突破了计算机与通信行业的界限，为人们提供了全新的信息服务。

6．家庭娱乐

从某种意义上来说，多媒体技术是一种界面技术，数字化音频和视频技术的日益成熟以及大容量光盘的出现，给计算机的多媒体化奠定了物质基础，发展到今天，不仅包括 3D 游戏，还加入了欣赏 CD 音乐、观看 VCD/DVD 电影、制作和聆听计算机数字音乐（MP3 和 MIDI）等内容。计算机的娱乐性也加速了计算机进入家庭的进程，计算机在娱乐方面可以成为家庭音响设备。多媒体计算机均配有音质和音色俱佳的立体声声卡，而这种立体声的声卡还配有音乐设备数字接口（即 MIDI 接口），用户可以通过这种接口将各种音乐设备和计算机连接起来，自己演奏喜欢的乐曲，也可以亲自编曲演奏，而且编出的乐曲可以存储起来供以后随时播放或编辑使用。计算机在娱乐方面还可成为一个全功能的高清晰彩电，用户能在计算机工作的同时在屏幕上开一个窗口来欣赏多姿多彩的电视节目，其高清晰度的画面使普通彩电望尘莫及。而且所显示的电视画面既可以以图形的形式储存在硬盘上，也可以用连接到电视卡上的录像机录制下来，再进行编辑和处理，当然也能看录像节目，可以说是全功能的"家庭影院"。

以上所述都是多媒体应用的不同形式，这些应用在很多方面具有相同的特点：①它们采用的信息不再局限于文字，而是扩展到了多种媒体信息；②它们与人的各种活动密切相关，甚至直接面向人进行工作；③它们采用的技术形式大致相同，但所实现的应用却很不一样。总之，多媒体网络正向着信息社会的各个领域迅速渗透，它将给人类带来的变化目前是不可估量的。在飞速发展的信息时代，在日新月异的互联时代，技术的发展不断改变着人们的交流与沟通方式。

2.4　Areas Applied Multimedia Technology

Nowadays, the multimedia technology is everywhere. In the areas of computer, communication, publishing, commercial, and etc, it's easy to find the multimedia technology. Further more, it's getting into people's daily life quickly. One of the examples is the multimedia TV. At the same time, along with the fast growing up of internet, multimedia technology will continuously quickly grow up and developed. We may see a new era of the multimedia technology. The major areas the multimedia applied include the followings.

1. Education area

The appearance of the multimedia technology in the education area changed the traditional education way. It is the multimedia technology that makes the education closer to the natural. The teaching slides made based on the multimedia technology could put images, texts, audio, and video together and create an almost real world teaching environment with images and texts mixed together to greatly encourage the students' leaning interest. Before this stage, the learners accept the teaching contents passively without any options. Now, the learners could choice whatever needed to learn based what has been provided. This is a big change. It increases the learning efficiency and improves the learning result and learning environment. Teachers could get the feedback from the students on time based on the real teaching situation happening in the class and then modify and improve the teaching contents.

Remote education is a typical multimedia technology application example. Remote education itself is a new education format developed along with the information technology development. It vividly combines and optimizes the multiple Medias together to serve the remote education. It is the major mean for people to learn through the life time in 21st century. Modern remote education could efficiently use the different education resources to provide strong support for varies kinds of education to improve their service quality. It could provide convenient broad services to different group of people in the community. The major character and advantage of the modern remote education are that the teacher and the students could interact real-time or not real-time over the distance. It is also the major difference between the traditional education the modern remote education.

The remote education has following advantages.

(1) The teacher's teaching activity and the students' learning activity could happen at the different locations at the same time. The teacher and the students could well communicate each other. The students could arrange the learning time, learning location, learning contents, and learning objects based on their own schedules. The students could ask questions anytime and get the answers right away.

(2) The remote education benefits the individual studying. It is based on the self learning and explores the students' learning initiative, anxiety, and creativity.

(3) The modern remote education method could provide the students with high quality education services. Teacher could know the students' learning process, how they understood the course on time, and answer the questions they have.

(4) The modern remote education created new ideas for the traditional teaching and learning. It's bringing the deep revolution in education and pushing the education idea, education guideline, education format, and education method to change. Via the remote education, the students located far away could attend the classes via computer network.

(5) This changed the traditional way attending the classes in the local classrooms and solved the problems some students can't leave the hometown and go to another city/place to study for some reason. It also pushes the academic communication.

2. Business

The multimedia technology application in business is shown as follows.

1) Office automation

Sorts of modern input and output equipment such as digital camera, scanner, photo and text fax machine, laser printer, and etc. forms the office automation equipment. Multimedia office working system is an integration of vision and hearing system processing office stuffs and communication.

It has the functions like managing the office information.

(1) It takes the diverse information including files, archives, report forms, data, graphics, audio, video, and etc. and creates the shared information source by processing, organizing, and storage the above information.

(2) By using multimedia technology, we can have visible telephone meetings.

(3) send or receive the multimedia emails.

The integration of kinds of office equipment and multimedia system could make the real office automation. This real office automation not only can help office staffs to view and process tons of information, but also save a lot of man power and physical resources.

2) Product commercials

Nowadays, product almost can't survive without commercials. To make vivid and attractive commercials, one needs the multimedia technology and materials to present the products.

3) Inquiring services

In today's information society, service industrials such as big department stores, banks, hospitals, airports, and etc. could take advantage of the multimedia computer system and better provide the customers convenient and freedom interacting inquiring services.

4) Manufacturing

In modern enterprises, both integrated information management and manufacturing automatic control are based on the collection, monitoring, storage, transmission, general analyzing, and management of multimedia information. Apply the multimedia technology to compositely process diverse information could make the information processing synthesized and intelligent and therefore improve manufacturing automatic management. Multimedia technology also applied in real time manufacture process monitoring. It's very useful especially in equipment fault reason analyzing and manufacture process parameters checking. In some hazard locations, the multimedia real time monitoring system is playing a more and more crucial role. In simulation experiments and emulation research, the multimedia technology is pushing the research and design catching a big step. In scientific visible calculations, the multimedia technology presents the abstract and boring data in 3-Dimension dynamic images and synchronizes the object's inside changing reason and outside affection in the research process.

3. Medical treatment

The common character of the modern medical treatment is based on the modern physics, applies the computer technology on the medical images to digitalize and rebuild them. Computer is playing a key role in medical imaging. Along with the higher requirement form the clinics and the development of multimedia technology, the new medical examination system with the ability to process multimedia information appeared. The traditional medical examination system can't compare with the new with multimedia technology built in one in many ways including media, media storage, media management, diagnostic aid information, direct observation, real-time, and etc. Especially, the new system could take advantage of the internet and do the remote medical treatment. For this reason, there may be a new revolution in the medical treatment area.

4. Electronic publishing

The electronic publishing means to save the graphics, images, texts, audios, and videos on the media of magnetic, optic, electronics. Then copy and distribute to the public. The reader read them through the computers or computer based equipment. The electronic reading materials have more capacity, smaller occupancy, cheaper, and storage longer characters. It not only can record text files, but also can record images, sounds, cartoons, and etc. audiovisual contents. At the same time, it can proceed interact reading and indexing. The way the electronic publishing could have the format of soft disk, CD, interact CD, hard disk, and etc. This way changed the paper record format to the electronic record format. When users need it, they could purchase the related magnetic disk or CD with the desired information saved and read the information on the computer. We could see the way changed the traditional information storage format.

The electronic distribution could happen in the internet as well. That is directly publishes the reading materials in the internet. It is the other important format of electronic publishing. This format totally changed the information transportation method, changed the relationship between the wanted information and the readers. The readers/users could actively search the information wanted on line or passively accept the information from the internet. Nowadays, there are hundreds of newspapers, academic papers, and magazines published in the internet. This internet publishing is based on the internet technology, realizes the information internet transportation, and created a real-time, interactive, and server based network structure between the information publisher and the readers. With the internet and the multimedia computer, we could search the books and other needed reading materials stored in the libraries all over the world. The search is fast and the contents searched updates quickly. Compares with the traditional media such as newspapers, books, magazines, and etc. and magnetic disk or CD electronic publishing, the internet publishing is not limited to time and areas. It opens to all desired internet users. Those permitted users could get into the information anywhere anytime. Furthermore, they could get back to the information publisher via the internet and upload the news they know with certain authority. All the information saved on the server and can be updated, modified, and deleted anytime wanted. More importantly, the internet publishing provided a convenient and flexible search way. The users could quickly get the wanted information indexed in time, keywords, and even composited calculations.

5. Multimedia communication

One of very important application areas of the multimedia technology is the multimedia communication. Multimedia communication has been a fast pace technology since 90's 20 century. Multimedia technology made the computer could deal with video, audio, text, and etc. diverse information at the same time. On the other hand, network communication technology overcomes the area limit and greatly improved information transmission speed. Multimedia communication technology combines the above two together and puts the computer's interactive, communication's distribution, and TV's direct vision into one union. This created a stone mark for the modern society.

The multimedia communication technology has a very broad technical range. It includes human-machine interface, digital signal processing, high capacity storage device, Database management system, data compression/decompression, multimedia operating system, high speed internet, communication styles, computer network, and related diverse software engineering. Along with the acceptance of the internet and the development of computer network, the network or internet has been part of the people's routine life. Nowadays, people majorly use internet to talk on line, shop on line, make visible telephone, have visible telephone conference, have remote file transmission, view and search multimedia information source, deal with multimedia email, have remote education, and etc. People could get much different multimedia information on line. People are not satisfied with the pure text message anymore. They want voice and even dynamic image communication. In shopping webpage, more and more people like to get the staffs they want while staying at home. The digital TV owned by many families already provide the options to people to chose their favourite programs from program database.

In conclusion, multimedia network provides the multimedia communication a transmission environment. It dynamically combines the computer's interactivity, network's distributives, and multimedia information's synthesis, breaks the boundaries between computer and communication industries, and provides people a brand new information service.

6. Home entertainment

Multimedia technology is an interface technology. The gradual mature of the digital audio and video technologies and the appearance of the high capacity CD sets up the base of computer multimedia interface. Up to now, it not only includes 3D games, but also includes CD music, VCD/DVD movie, make and listen to digital music (MP3 and MIDI), and etc. The entertainment of the computer pushed it get into people's home sooner. It could be part of the home theatre. Multimedia computer usually comes with better sound tone and quality stereo sound card. The card usually has music equipment digital interface, which is MIDI interface. By those interfaces, user could connect different music equipment with the computer, play the music he/she likes, compose new music and play it, and save the composed new music and play or edit later. Computer could be a fully functional high resolution TV in entertainment. User could open a window to enjoy the interested TV program while the computer is working on something else. Its high resolution screen is better than normal TV's. The showing TV screens could be either saved in the hard disk or recorded into the VCR connected to the computer. The recorded images in VCR could be edited or processed later. Sure, they could be replayed later as well. It is a fully functional home theatre.

Showing above are the different ways of multimedia application and they have some common characters. First of all, the information format they are using is not limited to the text only but extends to many different information formats. Secondly, they are closely related to people's different activities and even directly face people to work. Thirdly, they take similar technologies and realize quite different applications. In conclusion, the multimedia network is getting into all the areas of information society. The big changes it could bring to our life are not known yet. In current quickly developed information era and changes daily internet era, the new technology continuously changes people's opinion exchange format and communication way.

2.5　多媒体关键技术

多媒体的应用离不开多媒体技术，多媒体技术涉及很多领域，主要有计算机领域、通信领域和电视领域等。在计算机领域包括计算机硬件、软件、数据的压缩/解压缩算法、数值处理方法、计算机图形图像处理学、人工智能、计算机网络等；在通信领域包括通信技术、光电子技术、调制/解调技术、微波技术等；在电视领域包括声音和信号处理方法、集成电路技术等。下面简要介绍其中的几种。

2.5.1　多媒体压缩和存储

1．多媒体压缩技术

在多媒体应用系统中，若要表示、传输和处理大量数字化的声音、图片、影像视频信息等，其数据量非常大。例如，一幅具有中等分辨率（640×480 像素）的真彩色图像（24bit/像素），它的数据量约为每帧 7.37MB。若要达到 25 帧/s 的全动态显示要求，每秒所需的数据量约为 184MB，因此，要求系统的数据传输速率必须达到 184Mb/s，这在目前是无法达到的。对于声音也是如此，若用 16bit/样值的 PCM 编码，采样速率选为 44.1kHz，则对于双声道立体声，每秒将有 176kB 的数据量。由此可见音频、视频的数据量之大，如果不进行处理，那么计算机系统几乎无法对它进行存取和实时交换。因此，在多媒体计算机系统中，为了达到令人满意的图像、视频画面质量和听觉效果，必须解决视频、图像、音频信号数据的大容量存储和实时传输问题。解决的方法，除了提高计算机本身的性能及通信信道的带宽外，更重要的是对多媒体进行有效的数据压缩。

多媒体数据之所以能够压缩，是因为视频、图像、声音这些媒体数据具有很大的可压缩性。以目前常用的位图格式的图像存储方式为例，在这种形式的图像数据中，像素与像素之间无论在行方向，还是在列方向都具有很大的相关性，因而整体上数据的冗余度很大，在允许一定限度失真的前提下，能对图像数据进行很大程度的压缩。

根据解码后数据与原始数据是否完全一样进行分类，压缩方法可分为有损压缩和无损压缩两大类。有损压缩法会减少信息量，而损失的信息是不能再恢复的，因此这种压缩法是不可逆的。无损压缩法可去掉或减少数据中的冗余，但这些冗余值是可以重新插入到数据中的，因此，冗余压缩是可逆的过程。

2. 数据冗余

冗余的基本概念和冗余的分类。

简单地说冗余就是多余的东西，不是表达信息必不可少的部分。或者说冗余是指信息所具有的各种性质中多余的无用空间，其多余的程度称为"冗余度"，一般而言，图像和语音的数据冗余度很大，冗余可分为多种类型，这部分内容将在第三章中讲述。

3. 如何进行数据压缩

数据压缩的核心是计算方法。

数据冗余类型和数据压缩算法是对应的，一般根据不同的冗余类型采用不同的编码形式来进行数据压缩。

首先来看看什么是数据压缩。

简单而概括地说，数据压缩就是按照某种数学方法对计算机系统中的数据，去除冗余的算法。

1）几种常用数据的数据量分析

（1）图像的数据量。

图像数据量＝图像水平分辨率×图像垂直分辨率×像素深度/8

结果的单位为字节数。

假定一幅图像的分辨率为1024×768，图像深度为24位，则该图像所占用的空间为

$$1024×768×24/8=2304KB$$

（2）文本的数据量。

用于演示的文本，假设屏幕的分辨率为1024×768，屏幕上的字符为16×16点阵，每个字符用32B表示。则显示一屏字符所需的存储空间为

$$(1024/16)×(768/16)×32B=98304B(96KB)$$

（3）音频的数据量。

数字音频的数据量由采样频率、采样精度、是否是立体声三个因素决定，即

数字音频的数据量=采样频率×采样精度×声道数

假定采样频率为44100Hz，采样精度为16bit，双声道立体声模式。1s所产生的数据量为

$$44100×16×2=1411200bit$$

如果转成字节有1411200/8 =176400B。

传输时是按位传输的，所以说数据量是 1411200b/s；存储时是按字节存储的，所以说数据量是 176400B/s，约为 172KB/s。

每分钟数据量为

$$172KB×60s 约为 10MB/min$$

一首音乐曲或歌曲的长度大约为 4min，所以其对应的音频数据量约为 40MB。换句话说，一首音乐曲或歌曲的数据量约为 40MB。

（4）视频的数据量。

以电视信号为例，我国采用 PAL 制视频信号，扫描速度为每秒 25 帧。每帧信号数据量为 12MB。按照一秒钟显示 25 帧画面计算．每秒钟的数据量为

$$12MB×25＝300MB$$

2）为什么要进行语音编码

语音编码的目的同于其他数据编码一样，主要在于压缩数据。在多媒体语音数据的存储和传输中，数据压缩是必不可少的，这一点从上面的讲述中已看到了。

在数据通信及多媒体通信中，无论哪种数据编码均可分为信源编码；信道编码。

（1）信源编码：是指通过编码将信号源中信息的冗余度尽可能多的除去，形成一个适合用来传输的信号，主要解决有效性问题。

（2）信道编码：为了使处理过的信号在传输过程中不出错或少出错，以及即使出了错也能自动检错或尽量纠错而进行的编码，主要解决可靠性问题。

3）"编码"的分类

"编码"大体上可分为两类，字符编码和数据编码。

字符编码主要解决字符信息的交互问题，具体如下。

（1）信息加密编码——经过加密的内容，不知道编码标准的人很难识别，这种编码已经有数千年的历史，如众所周知的：电报码。

（2）信息交换编码——如邮政编码、身份证编码等。

（3）通过计算机处理和传输编码——如计算机输入编码等。

4）音频编码的主要技术指标

如前所述，对数字信息包括音频信息进行编码的目的是在不影响人们使用的情况下使数字音频信息的数据量达到最少，以便于存储和传输。编码的好坏程度通常用如下 5 个属性衡量：

（1）编码速率（比特率）——小；

（2）语音质量——高；

（3）计算复杂程度——低；

（4）延迟——少；

（5）适应能力（坚韧性，Robustness）——强。

上列 5 个指标中越接近右边所列的编码越好。换句话说，在这个领域中的研究人员和工程师们是朝着右边所列的方向努力的。

4．数字音频、视频处理技术

数字音频、视频技术的主要特征为在多媒体技术中的音频、视频技术采用的是全数字

技术。

数字音频、视频处理技术的主要组成有如下几个部分。

1）模拟音频、视频信号的数字化（模/数转换 A/D）

A/D 转换是将人所能接受的模拟音频、视频信号转换为计算机能够识别的数字信息，它是数字音频、视频处理技术中的第一步，也是其基础技术。

2）数字音频、视频信息的压缩编码（信源编码）

信源编码则是将数字化后的音频、视频信号的数据根据不同的应用，按照不同的标准及其算法进行压缩处理，从而达到降低传输码率的目的，它是数字音频、视频处理技术的关键。

3）数字音频、视频信息的存储与传输编码（信道编码）

信道编码是将压缩后的音频、视频数据根据存储与传输的介质不同进行相应的调制，使其符合该介质的要求或者达到提高频率资源利用率的目的；信道编码的另一个作用就是对存储与传输的数据进行容错技术处理，以确保重放数据的准确性，因此，信道编码是数字音频、视频处理技术的保证。

5. 多媒体存储技术

高效快速的存储设备是多媒体系统的基本部件之一。多媒体的音频、视频、图像等信息虽经过压缩处理，但仍需相当大的存储空间，在大容量只读光盘存储器 CD—ROM 问世后，才真正解决了多媒体信息存储空间问题。1996 年推出的 DVD（Digital Video Disc）新一代光盘标准，使得基于计算机的数字视盘驱动器从单个盘面上读取 4.7 GB 增加到 17 GB 的数据量。另外，作为数据备份的存储设备也有了发展。常用的备份设备有磁带、磁盘和活动式硬盘等。

由于存储在 PC 服务器上的数据量越来越大，使得对 PC 服务器的硬盘容量需求越来越高。为了避免磁盘损坏而造成的数据丢失，采用了相应的磁盘管理技术，磁盘阵列（Disk Array）就是在这种情况下诞生的一种数据存储技术。这些大容量存储设备为多媒体应用提供了便利条件。

2.5　Multimedia Key Technologies

The application of multimedia needs the support of multimedia technologies. Many areas are referred in the multimedia technologies. Those areas include computer, communication, and TV. Within the computer area, there are computer hardware and software, data compression and decompression algorithms, discrete-data processing methods, computer photograph and image processing principles, human organization, computer network and etc. Within the communication area, there are digital communication technology, photo-electronic technology, modulate and demodulate technology, microwave technology, and etc. Within TV area, there are sound and signal processing methods, integrate circuit technology, and etc. The following introduce several of those technologies.

2.5.1　Multimedia Compression and Storage

1. Multimedia Compression Technology

In multimedia application, to represent, transmit, and process a lot of digitalized sound, photos, audios, and videos, the quantity of the data is huge. For example, one frame's data quantity of one true colour medium resolution image with 640×480 pixels and 24 bits/pixel is roughly 7.37 MB. To meet the requirement of 25 frames/second fully dynamic displaying, the data quantity per second is about 184 MB. This requires the system's data transmission speed has to reach 180Mbps, which is not currently available yet. The similar situation happens to audio signal as well. For a 16bits/sample PCM encoding, if the sampling rate is 44.1kHz, and dual channel stereo, the data quantity per second will be 176kB. From this we know that audio and video have huge data quantities and need to be processed before the computer system can normally save, take, and real-time exchange them. In multimedia computer systems, to get the satisfied image, audio, and video, we must solve the huge image, audio, and video signal data storage and real-time transmission issues. To do so, besides increasing the computer's processing ability and communication channel's bandwidth, the more important things need to be done is to compress the multimedia data.

The reason the multimedia data could be compressed is that there are compressing spaces inside the image, audio, and video data. Take the currently common image bitmap storage format as an example; the pixels inside the data in this format are both horizontally and vertically closely correlated each other. This makes the whole data's redundancy is big which allows us to greatly compress the data within a limited distortion.

Based on whether the decoded data is the same as the original one, the compression method could be classified into two different kinds. One of them is called lossless and the other is called lossy. The lossy compression will reduce the information contained inside the original data and it can't be recovered. So this method is irreversible. The lossless compression could take off or reduce the redundancy in the original data which could be reinserted back to the original data. So this kind of compression is reversible.

2. Data Redundancy

The Basic Idea of Redundancy.

Simple speaking, the redundancy is something too much and is not really necessary to keep. In another word, the redundancy is a useless character inside many characters the information we need has, which occupies the unnecessary space. Redundancy is also to tell how much the useless content the information has. Generally speaking, the image and speech data has a lot redundancy. Entropy or information content is a measure of predictability or redundancy. In situations where there is redundancy in a body of information, it should be possible to adopt some form of coding which exploits the redundancy in order to reduce the space which the information occupies. This is the idea underlying approaches to data compression. The redundancy could be classified into several types.

The detailed description please refer to chapter3.

3. How to Compress the Data

Data compression is the process of converting an input data stream (the source stream or the original raw data) into another data stream (the output, or the compressed, stream) that has a smaller size. A stream is either a file or a buffer in memory. Data compression is popular for two reasons:

(1) People like to accumulate data and hate to throw anything away. No matter how big a storage device one has, sooner or later it is going to overflow. Data compression seems useful because it delays this inevitability.

(2) People hate to wait a long time for data transfers. When sitting at the computer, waiting for a Web page to come in or for a file to download, we naturally feel that anything longer than a few seconds is a long time to wait.

There are many known methods for data compression. They are based on different ideas, are suitable for different types of data, and produce different results, but they are all based on the same principle, namely they compress data by removing redundancy from the original data in the source file. Any nonrandom data has some structure, and this structure can be exploited to achieve a smaller representation of the data, a representation where no structure is discernible.

Simply speaking, data compression is to find a method or algorithm to reduce the redundancy existing inside the giving data.

The main aim of the field of data compression is, of course, to develop methods for better and better compression. However, one of the main dilemmas of the art of data compression is when to stop looking for better compression. Experience shows that fine-tuning an algorithm to squeeze out the last remaining bits of redundancy from the data gives diminishing returns. Modifying an algorithm to improve compression by 1% may increase the run time by 10% and the complexity of the program by more than that.

1) Data Quantity Analysing of Several Common Data Type

(1) Images' Data Quantity.

$$IDA = IHR \times IVR \times IQL/8$$

Where IDA is the Image Data Quantity, IHR is the Image Horizontal Resolution, IVR is the Image Vertical Resolution, and IQL is the Image Quantization Level. The unit of the result is byte.

Suppose we have an image with the horizontal and vertical resolution as 1024×768. Its quantization level is 24. Then this image will occupy the space of

$$1024 \times 768 \times 24 / 8 = 2304KB.$$

(2) Text's Data Quantity.

Let's have a look at the texts used in displaying. Suppose the screen's resolution is 1024×768. Each character showing is 16×16 pixels array. Each character occupies 32 bytes space. Then the storage spaces needed to display one whole screen characters will be as follows.

$$(1024/16) \times (768/16) \times 32B = 98304B(96KB)$$

(3) Audio Data Quantity.

The data quantity of the digitalized audio signal is depends on three factors and they are sampling frequency, quantization levels, and whether or not stereo channels. It could be calculated by the following formula.

Audio Data Quantity = Sampling frequency × Quantization levels×Number of Channels

If we use sampling frequency 44100 Hz, quantization levels 16 bits, and double channels stereo mode, the audio data quantity produced every second will be as showing below.

$$44100×16×2 = 1411200 \text{ bits}$$

Convert it to bytes:

$$1411200 / 8 = 176400 \text{ Bytes}$$

Since the data is transmitted in bit by bit, we say the data quantity is 1411200 bits/s when related to data transmission. On the other hand, the data is saved in bytes, so we say the data quantity is 176400 bytes/s, roughly 172 KB when related to data storage.

The data quantity in a minute is,

$$172 \text{ KB}×60 \text{ s roughly equal to } 10 \text{ MB} / \text{min}$$

The length of one music or song is about 4 minutes. Its correspondent audio data quantity is about 40 MB. In another word, the data quantity of one music or song is about 40 MB.

(4) Video Data Quantity.

We take TV signal as an example. In PAL mode which is used in China, the scan speed is 25 frames per second. In each frame, the data quantity is 12M. Based on the fact that 25 frames will be scanned, the data quantity for every second will be,

$$12 \text{ MB}×25 = 300 \text{ MB}$$

2) Why the Speech Coding is Necessary?

The major purpose of speech coding is the same as other data coding to compress the data. In the storage and transmission of multimedia speech data, data compression is a must. This could be seen from above.

In data communication and multimedia communication, all the data coding could be classified into two types.

① Source coding;

② Channel coding.

(1) Source coding is to reduce the redundancy from the signal source data as much as possible. It involves changing the message source to a suitable code to be transmitted through the channel. An example of source coding is the ASCII code, which converts each English character to a byte of 8 bits. It majorly solves the information transmission efficiency issues.

(2) On the other hand, channel coding is to encode the message again after the source coding by introducing some form of redundancy so that errors can be detected or even corrected. It's the opposite of source coding. In the reality, they need to be adjusted based on the application. Channel majorly solves the information transmission reliability issues.

3) Different Coding

There are character coding and data coding available.

Character coding solves the problems existed in the character information exchange. The typical ones are listed below.

(1) Information encryption coding - the information contained in the encrypted code is hard to be find by others who don't know how the information was encoded. This kind of coding has been

used for thousands of years. A very good example is the telegraph code.

(2) Information exchange coding - some codes like post code and personal identification code.

(3) Computer process and transport coding - the codes like computer input code and etc.

4) Speech Signal Coding Major Characters

As seen before, the purpose to code the digital information including digital speech signal is to get as less as possible data quantity without affecting the normal applications. This way it'll be easier for storage and transmission. There are usually five characters used to judge whether the coding is good or not.

(1) Coding rate (bit rate) — small;

(2) Decoded voice quality — high;

(3) Computing complexity — low;

(4) Delay — short;

(5) Robustness — strong.

If the coding is as listed on the right side, then it is a good one. In another word, the researchers and engineers in the field are seeking the coding algorithms to meet the requirement listed on the right side.

4. Digital Audio and Video Processing Technology

Major characters of digital audio and video technology include the following. In multimedia technology, the audio and video technology applied pure digital technology.

The digital audio and video processing technology usually consist the following several parts.

1) Digitalize the analogue audio and video signals (Analogue to Digital Converting A/D)

A/D is the technique to convert the analogue audio and video signals to the digital signals that the computer systems could recognize. It is the first step of digital audio and video signal

processing. It is also the basic of the digital audio and video signal processing.

2) Digital audio and video signals compress coding (source coding)

Source coding is to apply certain algorithms based on the different applications and corresponding standards to the digital audio and video signals to reduce the redundancy and compress the data. The purpose is to reduce the transmission ratio. It is the key technology in digital audio and video signal processing.

3) Digital audio and video signals storage and transmission coding (Channel coding)

The essence of channel coding is based on two rules: introduction of information redundancy and averaging the noise influence. Introduction of information redundancy is realized by attaching an additional symbol sequence to the information block representing a given message. This sequence is selected in such a way that the transmitted message could be easily distinguished from other messages that could potentially be transmitted. Messages are represented by the symbol sequences in such a way that it is very unlikely that channel perturbations distort so high a number of symbols in the sequence that these erroneous symbols would destroy the possibility of a unique association of the received symbol sequence with the transmitted message. The effect of noise averaging, in turn, is achieved by association of the redundant symbols with a few different information symbols representing a given message. The channel coding is a kind of guarantee of digital audio and video signal processing.

5. Multimedia Storage Technology

High speed and high efficiency storage equipment is one of the basic components of multimedia system. Even after the compression processing, the multimedia image, audio, video, and etc. data still needs quite size of storage space. This problem was not really solved until the appearance of CD-ROM. The new DVD (Digital Video Disk) standard proposed in 1966 made it possible for computer's digital video driver to increase the reading ability from 4.7GB to 17GB from one side of the disk. Beside this, the backup storage equipment has been getting better as well. The common equipment includes disk tape, disk, portable hard disk, and etc.

Along with the increase of the data quantity saved in the PC server, the server's storage capacity is getting bigger and bigger. To avoid the data loss duo to the disk damage, the correspondent disk management technology has been used. Disk Array is one of such data storage technologies appeared at this point. The high capacity storage equipment has helped the multimedia application a lot.

2.5.2 多媒体数据库与检索技术

1. 多媒体数据库

用于管理多媒体信息的数据库称为多媒体数据库（MDBMS），它是多媒体技术与数据库相结合的产物。由于多媒体数据具有复合性、分散性、时序性等特点，因此，多媒体数据库技术除包含一般数据库技术（如数据存储管理、数据共享、并发控制、事务处理等）之外，还应解决以下问题。

（1）支持图形、图像、声音、动态视频、文字等多种媒体的一般类型及用户定义的特殊类型。

（2）支持定长数据和非定长数据的集成管理。

（3）具有表示和处理对象间复杂关系的能力，以及保持让复杂对象完整性和一致性的机制。

（4）保证具有时序性的信息单元之间在时间、空间上的衔接与同步。

（5）多媒体数据的巨额数据量的存储。

（6）支持多媒体操作的用户界面等。

2. 多媒体检索技术

1）图像检索

20 世纪 90 年代，随着网络资源的迅速增长和大规模数字图像库的出现，基于内容的图像检索技术（CBIR）应运而生。在基于内容的图像检索系统中，图像利用其本身的视觉信息（如颜色、纹理和形状等高维特征向量）进行描述，利用图像之间的视觉相似性度量来实现查询。检索时，由于用户难以直接输入目标图像对应的特征向量，因此系统要求用户提供具有代表性的示例图像或者手绘的草图，然后系统利用该图像的特征向量在图像数据库中查找与其在视觉内容上比较相似的图像，并按照相似度的大小提取前面的多幅图像作为检索结果返回给用户，如图 2-1 所示。

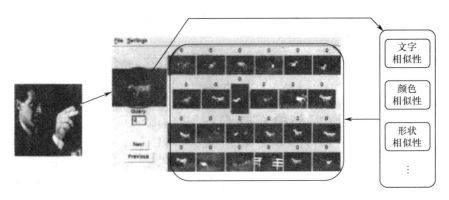

图 2-1　基于内容的图像检索示例

由于基于内容的图像检索系统可自动完成图像视觉内容的提取和匹配，因此，CBIR 技术克服了手工标注的低效性和主观性，并在过去的十几年中取得了大量的科研成果，建立起了一些研究性或者商用性的图像检索原型系统。

2）视频检索

视频检索是指使用视频内容分析技术，通过对视觉、听觉以及文本等内容的分析，将非结构化的视频数据结构化，如图 2-2 所示，并提取有效的特征来描述结构化的内容单元，在此基础上建立视频的索引、浏览和检索系统，旨在提供给用户方便的视频信息获取方式。

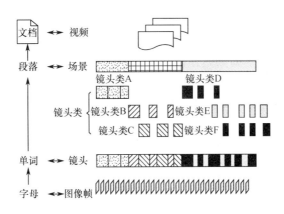

图 2-2　视频内容的层次组织结构

目前，视频检索系统大多是根据用户提交的例子（应用相关的内容模式）进行查询，利用例子的视觉特征进行相似性匹配，之后系统返回检索的结果。然而，由于视觉特征在内容表征方面的有限性，检索结果即使通过精确匹配算法也不能完全与人对视频的理解相吻合。为此，人们试图运用相关反馈技术，即让用户参与到检索过程中来，利用用户提供的交互信息，进行有指导的学习，以得到较为满意的检索效果。事实上，视觉特征对用户的意义不大，对用户来说，更自然的检索方式是基于自然语言的检索，用户也少有耐心通过与机器多次的交互才得到查询结果。由此可以看出，视频检索面临着一个困境，而走出这个困境的途径在于实现图像内容的理解。因此，视频内容表示的研究将提高计算机的信息理解能力，为用户提供一个友好的信息交互的基础。

2.5.2　Multimedia DataBase and Index Technology

1. Multimedia DataBases

The DataBase used to manage multimedia data is called Multimedia DataBase. More precisely, we want to store pictures, motion pictures, videos, sounds, music and parts of these, and perhaps we want to do it in a way that allows us to formulate questions about impressions, feelings or sense of something. Questions that, for example, sort out pictures that contain pieces of interest (a small part in it that interests us) or let us listen to parts of music that is similar to some tune you have heard somewhere. Databases that can manage tasks of this kind are usually called multimedia databases. Besides the normal DataBase's techniques, they shall have the following functions.

(1) Support graphics, images, speeches, dynamic videos, texts, and etc normal types and other customized types.

(2) Support the fixed length data and non fixed length data management.

(3) Could present and process the complicated relationship between the objects. Meanwhile could keep the complicated objects' integration and consistency.

(4) Guarantee the connection and synchronization among the sequenced information units.

(5) Able to store the huge quantity of multimedia data.

(6) Support the multimedia operating user interface.

2. Multimedia Retrieval Technology

1) Image Retrieval

During the 90's of 20 century, along with the fast development of network sources and the appearance of large scale digital image database, content based image retrieval (CBIR) technique produced. "Content-based" means that the search analyzes the contents of the image rather than the metadata such as keywords, tags, or descriptions associated with the image. The term "content" in this context might refer to colors, shapes, textures, or any other information that can be derived from the image itself. CBIR is desirable because most web-based image search engines rely purely on metadata and this produces a lot of garbage in the results. Also having humans manually enter keywords for images in a large database can be inefficient, expensive and may not capture every keyword that describes the image. Thus a system that can filter images based on their content would provide better indexing and return more accurate results.

Different implementations of CBIR make use of different types of user queries.

Query by example is a query technique that involves providing the CBIR system with an example image that it will then base its search upon. The underlying search algorithms may vary depending on the application, but result images should all share common elements with the provided example.

Options for providing example images to the system include:

- A preexisting image may be supplied by the user or chosen from a random set.
- The user draws a rough approximation of the image they are looking for, for example with

blobs of color or general shapes.

There is a growing interest in CBIR because of the limitations inherent in metadata-based systems, as well as the large range of possible uses for efficient image retrieval. Textual information about images can be easily searched using existing technology, but this requires humans to manually describe each image in the database. This is impractical for very large databases or for images that are generated automatically. It is also possible to miss images that use different synonyms in their descriptions. Systems based on categorizing images in semantic classes like "cat" as a subclass of "animal" avoid this problem but still face the same scaling issues. The process is shown in Fig.2-1.

2) Video Retrieval

The retrieval of a video clip based on its contents is a much more challenging problem than in image retrieval, because more features, often with different importance, are involved as shown in Fig.2-2. This is an open and very active research topic. Existing work typically fall within one of these two categories:

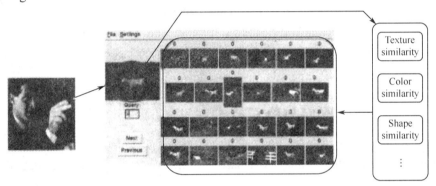

Fig. 2-1　Content Based Image Retrieval, CBIR Schema

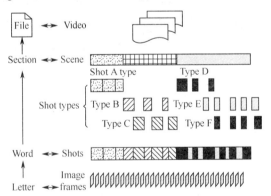

Fig. 2-2　The Video Contents Organization Structure

- Use content-based image retrieval techniques applied to the key-frames of the video sequences. Although easy to implement, it is limited for lack of temporal information.
- Incorporate motion information (sometimes object tracking) into the retrieval process. Richer queries can be formulated, because now temporal information is available, at the

expense of higher computational cost.

From the VideoQ, we can get a better idea.

Querying. VideoQ is a content-based video search system. It allows the user to search for a particular object, scene, or subject in a large video database. VideoQ provides the user with a wide range of search methods including standard text search based on keywords and content based search based on color, shape, texture, and motion. The visual query is effective when it is combined with text search. The text search is used to pre-filter videos and to reduce the number of videos to be searched using content-based search.

Features. VideoQ automatically extracts coherent visual features such as color, texture, shape, and motion. These features are then grouped into higher semantic knowledge forming video objects and associated spatio-temporal relationships. VideoQ performs automatic video object segmentation and tracking and also includes query with multiple objects.

Matching. The user typically formulates visual query by sketch. The user can draw objects with a particular shape, paint color and texture, and specify motion. The video objects of the sketch are then matched against those in the database and a ranked list of video shots is returned. Result Presentation. The retrieved video shots are presented on a separate window, where each shot is identified by a thumbnail version of its key-frame.

Applications. VideoQ is designed to serve as Web-based interactive video engine, where the user queries the system using animated sketches. It includes search and manipulation of MPEG-compressed videos.

2.5.3 多媒体网络与通信

1. 多媒体通信

多媒体网络应用的基本要求是能够在计算机网络上传送多媒体数据，所以多媒体通信技术也是多媒体的关键技术之一。多媒体通信技术是指通信技术、计算机技术和电视技术的相互渗透。从通信的角度来看，多媒体通信是继电报、电话、传真以后的新一代（第四代）通信手段，是多媒体技术与通信技术的结合，是两者应用的拓展和延伸。多媒体通信的特点如下。

1）数据量大

多媒体的数据量大（尤其是图像、视频），存储容量大，传输带宽要求高。虽然可以压缩，但高倍压缩往往以牺牲图像质量为代价。

2）实时性强

多媒体中的声音、动画、视频等实时媒体对多媒体传输设备的要求很高。即使带宽充足，如果通信协议不合适，也会影响多媒体数据的实时性。电路交换方式时延短，但占用专门信道，不易共享；而分组交换方式时延偏长，且不适于数据量变化较大的业务使用。

3）时空约束

多媒体中各媒体彼此相互关联、相互约束，这种约束既存在于空间上，也存在于时间上。而通信系统的传输又具有串行性，因此，必须采取延迟同步的方法进行再合成，这种合成包括时间合成、空间合成及时空合成。

4）多媒体交互性

多媒体系统的关键特点是交互性。这就要求多媒体通信网络提供双向的数据传输能力，这种双向传输通道从功能和带宽来讲是不对称的。

支持多媒体通信的网络有电信网络（电话网、分组交换网、数字数据网和综合业务数据网等）、计算机网络和电视传播网（有线、无线）。宽带综合业务数字网（B-ISDN），通常是基于异步传输模式（ATM）的光纤网，具有电路交换延迟小、分组交换速率高及速率可变等特点，可传输高保真的立体声、普通和高清晰度的电视节目，是多媒体通信的理想传输信道。

多媒体通信技术中有许多特殊问题需要解决，如带宽、相关数据类型的同步、多媒体设备的控制、不同终端和网络服务器的动态适应、超媒体信息的实时性、可变视频数据流处理、网络频谱及信道分配等问题。

2．多媒体网络

多媒体网络系统本质上是一种计算机网络，它与普通网络的主要区别在于能够为多媒体服务，如对多媒体数据进行获取、存储、处理、传输等操作。

互联网上的媒体应用包括以下几个方面。

1）音频点播

在这类应用中，客户请求传送经过压缩并存放在服务器的声音文件，这些文件可包含任何声音内容。客户可在任何时间、任何地点从声音点播服务器中读取声音文件。使用这类软件时，用户启动播放器几秒钟后，就可以一边播放一边接收数据文件，而不是整个文件下载之后才开始播放。许多产品也为用户提供了交互功能。

2）视频点播

这是一类典型的交互式多媒体服务系统。因为存放和播放视频文件比声音文件需要大得多的存储空间和传输带宽，所以视频点播系统一般运行在宽带网中。

3）IP电话

IP电话是指在IP网络上进行呼叫和通话，这种应用支持人们在Internet上进行通话。目前IP电话用于长途通信时，价格比PSTN电话的价格便宜很多，质量略差。然而，随着质量的改善,IP电话已经逐渐开始占用较大的话音通信份额。

4）分组实时视频会议

这类应用系统与IP电话类似，但可传输视频图像，并允许多人参加。

5）流媒体技术

流媒体又称为流式媒体，就是应用流技术在网络上传输多媒体文件，而流技术就是把连续的影像和声音信息经过压缩处理后放上网站服务器，让用户一边下载一边观看、收听，而不需要等整个压缩文件下载到自己机器后才可以观看的网络传输技术。这种技术使时间延迟大大减少。

流媒体实际指的是一种新的媒体传送方式和技术，而不是一种新的媒体。流媒体系统包括编码工具（用于创建、捕捉和编辑多媒体数据，形成流媒体格式）、流媒体数据、服务器（存放和控制流媒体的数据）、网络、播放器（供用户端浏览流媒体文件），这5个部分有些是服务器需要的，有些是用户端需要的。

（1）流媒体技术的特点。

① 时间大大缩短；

② 对系统缓存容量的需求大大降低。

（2）流媒体技术的分类。

流式传输技术可分为两种：

① 顺序流式传输：顺序流式传输就是顺序下载，在下载文件的同时用户可以观看，但是，用户的观看与服务器上的传输并不是同步进行的，用户是在一段延时后才能看到服务器上传出来的信息，或者说用户看到的总是服务器在若干时间以前传出来的信息。在这过程中，用户只能观看已下载的那部分，而不能要求跳到还未下载的部分。顺序流式传输比较适合高质量的短片段，因为它可以较好地保证节目播放的最终质量。它适合于在网站上发布的供用户点播的音视频节目。

② 实时流式传输。顾名思义，在这种传输方式中，用户看到的是服务器上传来的实时信息，没有时延。

2.5.3　Multimedia Network and Communications

1. Multimedia Communications

The basic requirement of the application of multimedia network is to transmit the multimedia data over the computer networks. So the multimedia communication technology is also one of the key technologies of multimedia. The multimedia communication technology is an integrated technology of communication technology, computer technology, and TV technology. From the view of communication, multimedia communication is the fourth communication generation after telegraph, telephone, and fax. It's also a combination and extension of multimedia technology and communication technology. It could be described as follows.

1) Huge quantity of data

Multimedia has big quantity of data, especially for images and videos, occupies big storage spaces, and requires a wide bandage channel to transmit. Although the data could be compressed somehow, the more compression will get the poorer image quality.

2) Real-time

Inside the multimedia data, the voice, cartoon, video, and etc. real-time media have very high demand to the multimedia transmission equipment. Without the proper communication protocol, even with the adequate bandwidth, the real-time character of multimedia data will still be affected. The circuit exchange format has shorter latency. But it occupies a particular transmission channel and not easy to share. On the other hand, package exchange format has longer latency and not suitable for the applications with wide data quantity variety.

3) Time and Space Restriction

The different media inside the multimedia are related and restricted each other. This restriction is existed both in time and space zone. Since the communication system is usually series, the multimedia data must be re-synthesised with delay synchronous method. This kind of synthesis includes time synthesis, space synthesis, and time-space synthesis.

4) Multimedia Interactivity

Interactivity is the multimedia systems' key character. This means that the multimedia communication network shall provide two-way data transmission channels which are not equally functioned and have same bandwidth.

The networks supporting multimedia communication include telecom network which has telephony network, package exchange network, digital data network, integrated services digital network, and etc., computer network, and TV broadcasting network which has wired and wireless. The broadband integrated services digital network (B-ISDN) is usually based on asynchronous transmission mode (ATM) fibre network which has the characters of shorter circuit exchange latency, higher package exchange rate, variable transmission speed, and etc. This network could transport both high quality stereo and normal and high definition TV programs. It is the multimedia communication's ideal transmission channel.

There are many special issues staying with the multimedia communications need to be solved. Those issues include bandwidth, related data type synchronization, multimedia equipment control, different terminals and network servers dynamically match each other, the super media information's real-time, variable video data streaming process, network frequency distribution, communication channel allocation, and etc.

2. Multimedia Network

Multimedia network is actually a kind of computer network. The main difference from other networks is that it can serve the multimedia, such as gain, store, process, and transmit multimedia data.

The application of media in internet includes the followings.

1) Audio demand

In such application, user at the user terminal requests the compressed audio files saved in the server, which could contain any type of audio contents. User can read the audio files from the audio ordering server at any time and any place. When using this kind of software, several seconds after the audio player started, user can receive the audio file data while hearing the music. One doesn't have to wait for the whole file downloaded before playing it. Many products provide the users interactive function.

2) Video demand

This is a typical interactive multimedia service system. Since saving and playing the video files need much more saving spaces and transmission bandwidth than doing the audio files, the video ordering system usually runs in the wide bandwidth networks.

3) IP Telephony

IP telephony means to make phone calls over the IP network. This application supports people making phone calls by internet. Currently, when making long distance calls, the IP phone's cost is much lower than the PSTN phone with a little poor quality. However, along with the IP phone's quality improvement, it has been occupying bigger telephony market.

4) Group Real-time Video Conferencing

This kind of application is similar to IP telephony. But it allows more people to join and can transmit video images.

5) Streaming Media Technology

The streaming media technology is also called streamed media technology. It applies the streaming technology to transmit multimedia files over the internet. The streaming technology is basically to convert the continuous video and audio files into the digital ones via the compressing and encoding technology and put those digital ones on the internet. The users could watch and listen to them while downloading. They don't have to wait for the whole piece completely downloaded into their computers like they did. This is a new internet transmitting technology and greatly reduced the normal delay time.

The streaming media is actually means a new kind of media transmission format and technology and not a new media. The streaming media system includes encoding tools (used for creating, catching and editing multimedia data and forming the streaming media format), streaming media data, server (to store and control the streaming media data), network, and player (provided for user to view the steaming media files). Some of those five parts are for server and others for user.

(1) The Characteristics of the Streaming Media.

① The duration is much shorter;

② The system storage capacity requirement is much lower.

(2) The Types of the Streaming Media.

The streaming transmission technologies could be classified into two types.

① Sequential streaming transmission technology: The sequential streaming transmission is sequential downloading. User could view the files while downloading. But the watching and the transmission on the server are not synchronized. User is actually watching the contents delayed a short time by the server. In another word, the contents the user could view are the ones the server sent out at an earlier time. During this short period of time, the user only can watch the already downloaded part and can't jump to the undownloaded part. The sequential streaming transmission is suitable to the high quality short videos since it could guarantee the final broadcasted programs` quality. It is good for the network stations on demanding audio video programs.

② Real-time streaming transmission. From the name we could tell that in this kind of streaming transmission, users could view the contents transmitted from server in real-time without delay.

2.5.4 虚拟现实与交互技术

1. 虚拟现实

虚拟现实（Virtual Reality，VR）通常是指用立体眼镜、传感手套、三维鼠标等一系列传感设备来实现一种三维现实。人们通过这些设备向计算机传送各种动作信息，并通过视觉、听觉、触觉及嗅觉等设施来体验身临其境的自然感觉。虚拟现实技术是利用计算机生成一种模拟环境，通过各种传感设备，令使用者获得置身于现实情境或场景的幻觉，实现用户与该环境直接进行自然交互的技术。

目前，这种技术已得到广泛应用，如各种模拟训练系统、3D 电影等。

2. 交互技术

在传统计算机与人的交互中，大多数都采用文本信息实现，交互手段比较单一。在多媒体环境下，各种媒体并存，用户与系统的交互包括视觉、听觉、触觉、味觉和嗅觉等各种手段。各种媒体的时空安排和效应，相互之间的同步和合成效果，相互作用的解释和描述等都是表达信息时所考虑的问题。因此，多媒体系统中各种媒体信息的时空合成以及人机之间灵活的交互方法等仍是多媒体领域需要研究和解决的棘手问题。

2.5.4　Virtual Reality and Interactive Technology

1. Vir tual Reality

Virtual reality (VR) is a computer-simulated environment that can simulate physical presence in places in the real world or imagined worlds. Most current virtual reality environments are primarily visual experiences, displayed either on a computer screen or through special stereoscopic displays, but some simulations include additional sensory information, such as sound through speakers or headphones. Some advanced, haptic systems now include tactile information, generally known as force feedback, in medical and gaming applications. Furthermore, virtual reality covers remote communication environments which provide virtual presence of users with the concepts of telepresence and telexistence or a virtual artifact (VA) either through the use of standard input devices such as a keyboard and mouse, or through multimodal devices such as a wired glove, the Polhemus, and omnidirectional treadmills. The simulated environment can be similar to the real world in order to create a lifelike experience—for example, in simulations for pilot or combat training—or it can differ significantly from reality, such as in VR games.

Virtual reality is often used to describe a wide variety of applications commonly associated with immersive, highly visual, 3D environments.

Currently, it has been had a wide applications such as simulated training systems, 3D movies, and etc.

2. Interactive technology

In traditional computer interactive format, usually takes text single way as the information conveyer. Under the circumstances of multimedia, multiple media exist at the same time. The interactivities between the user and the system contain vision, hearing, touching, tasting, smelling, and etc. different ways. The issues like different media's timing and effectiveness, synthesis and integration effectiveness, the explanation and description of interactivities, and etc. need to be considered when present the information. Therefore, multiple media's time and space zone integration and the flexible interactive ways between the user and the system are still the complicated research topics in multimedia areas and need to be solved.

练　习　题

1. 简述多媒体发展的历程。
2. 什么是媒体?什么是多媒体?
3. 多媒体可分为哪几类?
4. 多媒体技术有什么特点?
5. 什么是多媒体技术?其处理的对象有哪些?

6. 简述图形与图像的区别。

7. 多媒体关键技术有哪些?

8. 简述虚拟现实技术的概念及其特征。

9. 在设计人机界面时要注意什么?

Exercises

1. Briefly describe the multimedia's developing history.

2. What is the media? What is the multimedia?

3. How the multimedia could be classified?

4. What kind of characters does the multimedia have?

5. What is the multimedia technology? What is it majorly deal with?

6. Briefly tell the difference between graphic and image.

7. What are the multimedia's key technologies?

8. Simply describe the virtual reality technology and its characters.

9. What do you need to pay attention to when design the interactive interface?

第三章 多媒体数据编码

多媒体数据编码的主要目的就是要对多媒体数据进行压缩，以便于存储和传输。

3.1 数据压缩的基本概念

数据压缩要解决三个基本问题。

（1）什么是数据压缩。

（2）为什么要进行数据压缩。

（3）如何进行数据压缩。

3.1.1 什么是数据压缩

简单地说，数据压缩是一种算法或者技术。这种算法或者技术能够以一种紧凑的数据方式来表达同样的信息。在数字多媒体几十年来革新式的发展过程中，它已成为至关重要的推动性的技术之一。

如今，人们在存储和通过媒体传输之前广泛地使用数据压缩软件来减小文件的尺寸。这些数据压缩软件包括有：zip、gzip、WinZip 等。数据压缩技术已经嵌入到了越来越多的应用软件当中。很多情况下，在人们还没有意识到的时候，数据已经被应用软件压缩了。

到目前为止，数据压缩已经发展成为很多应用软件中的一个基本特性。同时它也成为计算机科学及工程中的一个既重要又很活跃的研究领域。可以说，如果没有数据压缩，所有和其相关的领域，例如，互联网、数字电视、移动通信、视屏通信等就不可能有近年来如此迅速地发展。

下面列出了一些和数据压缩有关的一些典型的领域。

（1）个人通信系统。如传真、声音短信、电话等。

（2）计算机系统。如存储结构、磁盘、磁带等。

（3）移动计算。

（4）分布式计算机系统。

（5）计算机网络，特别是互联网。

（6）多媒体的逐步更新换代，图像及信号处理。

（7）图像存档和视频会议。

（8）数字和卫星电视。

3.1.2 为什么要进行数据压缩

数据压缩的对象是数据，目的就是要用一种相对紧凑的数据方式来表达包含在源数据中的信息。

多媒体技术是所有计算机应用领域中信息量最大的领域，音频、文本和图像都是动辄几十兆甚至上百兆字节的大文件，多媒体技术所面临的最大难题就是海量数据问题，如果不采用数据压缩技术，将在很大程度上限制多媒体技术的发展，也不可能具备良好的应用前景。

进一步说，数据和信息有着不同的概念。数据是用来记录和传送信息的，是信息的载体。对我们真正有用的不是数据本身，而是数据所携带的信息。所以，在所要信息不变的前提下，数据量越小则越易于存储和传输。

简而言之，数据压缩的目的是：在传送和处理信息时，尽量减少数据量。

3.1.3　如何进行数据压缩

如何进行数据压缩是本章要回答的问题。也是多年来，学者们研究的一个热门话题。许多编码算法应运而生。

Chapter 3　Multimedia Data Coding

The major purpose of the multimedia data coding is to compress the source multimedia data to reduce the data size and make the storage and transmission easier.

3.1　The basic idea of data compression

The data compression should answer the following three questions.

(1) What is the data compression.

(2) Why should process the data compression.

(3) How to process the data compression.

3.1.1　What the data compression is

Simply speaking, the data compression is the science or art of representing the same information in a compact way. It has been one of the critical enabling technologies for the ongoing digital multimedia revolution for decades.

Most people frequently used data compression software such as zip, gzip, WinZip, and etc. to reduce the original data file size before storing and transferring it in the media. Compression techniques have been embedded in more and more software. Data are often compressed without people even knowing it.

Furthermore, data compression has become a common requirement for most application software as well as an important and active research area in computer and electrical engineering area. Without compression techniques, none of the ever growing internet, digital TV, high definition TV, mobile communication or increasing video communication techniques would have been practically developed.

Typical examples of application areas that are relevant to and motivated by the data compression include the followings.

(1) Personal communication systems such as facsimile, voice mail, and telephony.

(2) Computer systems and their accessories such as memory structures, disks, tapes, and compact disks.

(3) Mobile computing.

(4) Distributed computer systems.

(5) Computer networks, especially the Internet.

(6) Multimedia evolution, imaging, signal processing.

(7) Image archival and video conferencing.

(8) Digital TV, high definition TV, and satellite TV.

3.1.2　Why process the data compression

Let's have another look at the data compression. The object of the data compression is the original data. The purpose is to find a compact way to represent the information contained in the original data.

Multimedia technology area contains the biggest amount of information among all the computer application areas. The size of all the multimedia files including audio, text, video, and etc. is very easy to get up to tens of megabytes or even hundreds of megabytes. One of the toughest problems that the multimedia technology application is facing is the tones of data processing. Obviously, without data compression, it'll be almost impossible for the multimedia technology develops well either currently or in the future.

Furthermore, the data we mentioned is different from the information we need. The logical relationship is that the information is contained in the data. In another words, the data is the vehicle to record and transport the information. Like we said, what we really need is the information contained in the data and not the data itself. So, as soon as the information we need is not distorted, the smaller the better for the data used to record and transport the information since the smaller the easier to store and transmit.

Conclusively, the purpose of the data compression is to reduce the data quantity without distorting the information contained while processing and transmitting it.

3.1.3　The Way to Process the Data Compression

We'll take care of this issue all through this chapter. This has been a hot topic of the scholars and researchers for years in this field. Many algorithms have been produced around this topic.

3.2　数据冗余

3.2.1　冗余的基本概念

如在第二章已经讲到的，冗余是指信息所具有的各种性质中多余的无用空间，其多余的程度称为"冗余度"，一般而言，图像和语音的数据冗余度很大。

信息量、数据量和冗余量之间的关系为

$$I=D-du$$

式中：I 为信息量；D 为数据量；du 为冗余量，冗余量 du 包含在 D 中，冗余量 du 应在数据进行存储和传输之前去掉。这也就是数据压缩要做的事情。

重新看一下上一章的例子。

播音员的播音语速一般为每分钟 180 字，由于计算机中用两个字节表示一个汉字，因此，播音员一分钟阅读的汉字共占用 360B。为了把播音员的声音数字化，需要以高出播音员声音频率一倍的频率进行采样。这就是说，一般播音员的播音语音信号频率为 4kHz，所需的采样频率即为 8kHz。当采用 8bit 的采样精度进行采样时，得到的一秒钟数字音频信号的数据量为

$$8kHz \times 8bit = 64kb/s$$

则一分钟的数据量为

$$64kb/s \times 60s/min = 3840kb/min$$

$$3840(kb/min)/8 = 480kB$$

比较二者，播音员一分钟阅读的汉字共占用 360B，折合 0.36kB，而数字化后每分钟的语音数据占用了 480kB，两者的数据量相差一千多倍，而要表达的信息量是相同的，可见数据冗余现象很严重。同时也说明数据压缩是有余地的。

3.2.2　数据冗余分类

从上面的例子可见一斑。其实，对于多媒体数据来说，有着很大的压缩潜力。因为多媒体数据中存在着多种数据冗余。

信息中数据冗余的现象比较普遍，数据冗余的种类也不尽相同。可以主要将它们归纳为以下几种。①空间冗余；②时间冗余；③信息熵冗余；④结构冗余；⑤知识冗余；⑥感知冗余；⑦其他冗余。

1．空间冗余

这种冗余指的是规则物体的表面具有物理相关性，将其表面数字化后表现为数据冗余。例如，一面白墙，上面挂着一幅画，拍成数字照片后，墙面上除了挂画的地方，其余的地方全部是相同的颜色（白色）。这就是说，墙面所有像素与相邻颜色信息完全相同，在统计上是冗余的。冗余的像素数据可以压缩，甚至相邻颜色极为接近的像素数据也可以压缩，只要掌握适度就可以保证图像良好的视觉效果。

2．时间冗余

指在时间序列上的数据冗余。视频信号和动画等有序排列的图像很容易产生数据冗余现象。在播放有序排列图像时，相邻画面中同一位置的内容有变化，则这一位置的内容是"活动"的。而相邻画面中的其余内容没有变化，画面视觉效果相对静止，这时，相邻画面无变化的内容构成了时间上的冗余。

3．信息熵冗余

信息熵的定义为

$$E = -\sum_{i=0}^{k-1} p_i \cdot \log_2 p_i$$

式中：E 为信息熵；k 为某集合中数据类数或码元的个数；p_i 为对应码元的发生概率。

信息熵的概念最早由信息论的创始人香农提出，它定量的表述了数据中所含信息量的大小。一般情况下，数据量 D 的值必然大于信息熵 E，产生信息熵冗余。

4．结构冗余

在数字化图像中，具有规则纹理的表面、大面积相互重叠的相同图案，规则有序排列的图形等结构，都存在数据冗余，这种结构上的冗余称为"结构冗余"。

5．知识冗余

知识是人类独有的，是认知自然、总结规律而得到的。人类一旦掌握了知识，凭借经验就可辨别事物，无须进行全面的比较和辨别。由于这种特性，如果继续全面的表达一些能够通过经验而得到的一些信息，就会产生知识冗余。比如说，图像的理解就与某些基础知识有关。

例如，人脸的图像有同样的结构：嘴的上方有鼻子，鼻子上方有眼睛，鼻子在中线上……

6．感知冗余

例如：人类的视觉敏感度一般小于图像的表现力，图像的微小色彩变化、亮度层次的细腻变化以及轮廓的细微差别不易察觉，如果把上述提到的部分仍然表示出来，这就产生了感知冗余。

3.2　Data Redundancy

3.2.1　The Basic Idea of Redundancy

Simple speaking, the redundancy is something too much and is not really necessary to keep. In another word, the redundancy is a useless character inside many characters the information we need has, which occupies the unnecessary space. Redundancy is also to tell how much the useless content the information has. Generally speaking, the image and speech data has a lot redundancy. Entropy or information content is a measure of predictability or redundancy. In situations where there is redundancy in a body of information, it should be possible to adopt some form of coding which exploits the redundancy in order to reduce the space which the information occupies. This is the idea underlying approaches to data compression.

The following formula expresses the relationship among information content, data quantity, and redundancy.

$$I=D–du$$

Where I is stand for information content, D is stand for data quantity, and the du is stand for redundancy. The redundancy is inside the data quantity which should be removed before the transmission or storage.

Let's have a look at the following example. The speaking speed of a broadcasting speaker is about 180 words per minutes. Based on the fact that in computer, each Chinese character needs two bytes to store and present, the words spoke by the broadcasting speaker occupy 360 bytes space. To digitalize the speaking signal, we need equal to or more than two times the speaker's sound signal frequency as the sampling frequency. The broadcasting speaker's normal speaking signal's frequency is roughly 4KHz. So the sampling frequency shall not less than 8KHz. Suppose we use 8 bits quantization level, then we get each second the digitalized sound signal data quantity is:

$$8kHz \times 8bit = 64kb/s$$

Then every minute the data quantity is:

$$64kb/s \times 60s/min = 3840kb/min$$

Covert to bytes, we get:

$$3840(kb/min)/8 = 480kB$$

Compare the above two of them, the Chinese words the broadcasting speaker spoke in one minute occupy 360 bytes space or 0.36kbytes. After digitalization, each minute the digitalized speech signal data occupies 480kbytes. The difference between the two data quantity is more than 1000 times. The information represented here is the same. We could know that there is a lot redundancy exists in the digitalized data.

3.2.2 Redundancy classifications

From the example above we know that there are plenty of potential rooms inside the multimedia data to compress. This is because of the kinds of redundancies existed in the multimedia data.

The redundancies existed in the multimedia data could be classified as follows. ①Space redundancy; ②Time redundancy; ③Entropy redundancy; ④Structure redundancy; ⑤Knowledge redundancy; ⑥Sensitive redundancy; ⑦Other redundancy.

Let's have a closer look at each one of them.

1. Space Redundancy

This kind of redundancy refers to the physical corresponding relationship existed on the regular physical surfaces. Digitalize those kinds of surface data will be represented as data redundancy. For example, take a digital picture of the paint hung on a clear white wall. Except where the paint is, all other places are the same colour, white. This means that the places around the paint have same colour pixels or same colour information. This statically is space redundancy. Those redundant pixels could be compress somehow. Actuarially, it doesn't have to be the same colour, as soon as the similar colour, the pixels could be compressed as well. The image's vision quality will be affected only if the compress overdone.

2. Time Redundancy

This kind of data redundancy is on time series. Video signals, animations, and etc. series images are easy to produce data redundancy. This is because that when display series images, some contents inside the consecutive image are not changing at all. From the visible images, this part is still. Only the parts changed from the consecutive images are dynamic or active. In this situation, the non-change parts form the time redundancy.

3. Entropy redundancy

The definition of Entropy is as follows.

$$E = -\sum_{i=0}^{k-1} p_i \cdot \log_2 \frac{p_i}{}$$

Where E is the Entropy, k is the total element number inside the collection, and p_i is the correspondent element occur probability.

The idea of Entropy is proposed by C.E.Shannon, the creator of the foundation of information theory in his published paper, A Mathematical Theory of Communication, in 1948. In his paper, Shannon first time applied the idea of Entropy to the communication. Entropy quantitatively measures the information contained in the certain data. Normally, the quantity of the data, D is always greater than the Entropy, E. There is redundancy existing. This kind of redundancy is called entropy redundancy.

4. Structure Redundancy

In digitalized images, the structures like regular line surfaces, big areas of cross folded same patens, regular ranged series graphics, and etc. have data redundancy. This kind of redundancy is called structure redundancy.

5. Knowledge Redundancy

Knowledge is owned by human being only. It is achieved by knowing the nature and the experiences human being had in the real life. As soon as one has it, one could tell something based on the knowledge one had without physically touching or doing it. In this situation, if still describes the unnecessary details which could be known by knowledge, will produce redundancy known as structure redundancy.

A good example is the understanding of an image. It is related to some knowledge. Such as some parts of the human being face are similar, there is a nose above the mouth; there are eyes above the nose; the nose is in the vertical middle of the face...

6. Sensitive Redundancy

We can know this from an example. Human's sensitivity to the images is smaller than the image's representation. This means that the image's minor changes with colour, very detailed changes with brightness level, and tiny differences with the outline are not that sensitive to human being. If still represent the parts mentioned will produce sensitive redundancy.

3.3 无损数据压缩

如前所述，数据压缩就是要去除冗余度，以尽少的数据表达所要的信息。具体来说，数据压缩可又分成两种类型：一种称为无损压缩，另一种称为有损压缩。

无损压缩是指使用压缩后的数据进行重构（或者称为还原，解压缩）时，重构后的数据与原始的数据完全相同；因而所包含的信息丝毫无损。无损压缩用于要求重构的信号与原始信号完全一致的场合。一个很常见的例子是磁盘文件的压缩。根据目前的技术水平，无损压缩算法一般可以把普通文件的数据压缩到原来的 1/2～1/4。一些常用的无损压缩算法有哈夫曼算法和 LZW 算法。

有损压缩是指使用压缩后的数据进行重构时，重构后的数据与原来的数据有所不同，因而所包含的信息有所丢失，但不会引起人们对原始资料所表达的信息造成误解。有损压缩适用于重构信号不一定非要和原始信号完全相同的场合。例如，图像和声音的压缩就可以采用有损压缩，因为其中包含的数据往往多于人们的视觉系统和听觉系统所能接收的信息，丢掉一些数据而不至于对声音或者图像所表达的意思产生误解，但可大大提高压缩比，减少数据量。

这里主要介绍目前用得最多和技术最成熟的无损压缩编码技术。包含哈夫曼编码、算术编码、RLE 编码和词典编码。不打算开发压缩技术和编写压缩程序的读者可不必深究编译码的详细过程。

3.3　Lossless Data Compression

As shown above, the data compression is to get rid of the redundancy and use as less as possible data the represent the information we wanted. Furthermore, the data compression could be classified into two catalogues. One is the lossless compression and the other is the lossy compression.

The lossless compression consists of those techniques guaranteed to generate an exact duplicate of the input data stream after a compression/decompression cycle. This is the type of compression used when storing disk files, database records, spreadsheets, or word processing files. In these applications, the loss of even a single bit could be catastrophic. Based on current techniques, the lossless compression could reduce the file size to the 1/2 to 1/4 of the original one's. Some of frequently applied lossless compression algorithms include Huffman and LZW (Lempel-Ziv & Welch).

Lossy data compression concedes a certain loss of accuracy in exchange for greatly increased compression. Lossy compression proves effective when applied to graphics images and digitized voice. By their very nature, these digitized representations of analog phenomena are not perfect to begin with, so the idea of output and input not matching exactly is a little more acceptable. Most lossy compression techniques can be adjusted to different quality levels, gaining higher accuracy in exchange for less effective compression. Until recently, lossy compression has been primarily implemented using dedicated hardware. In the past few years, powerful lossy-compression programs have been moved to desktop CPUs, but even so the field is still dominated by hardware implementations.

The techniques discussed in this book will be lossless ones including Huffman coding, Arithmetic coding, RLE coding, and Dictionary coding. For those who are not going to write programming codes don't have to go to the details of the decoding process.

3.3.1 香农—范诺与哈夫曼编码

1. 香农—范诺编码

香农—范诺编码算法需要用到下面两个基本概念。

1）熵的概念

熵的概念是由信息论的创始人香农 1948 年在他的著名论文 "A Mathematical Theory of Communication"中首次提出的，它奠定了信息论的基础。

（1）熵是信息量的度量方法，它表示某一事件出现的消息越多，事件发生的可能性就越小，数学上就是概率越小。

（2）某个事件的信息量用 $I_i = -\log_2 p_i$ 表示，其中 p_i 为第 i 个事件发生的概率。

$$0 < p_i \leqslant 1$$

2）信源 S 的熵的定义

按照香农的理论，信源 S 的熵定义为

$$H(S) = \eta = \sum p_i \log_2 (1/p_i)$$

式中：p_i 为符号 S_i 在 S 中出现的概率；$\log_2(1/p_i)$ 为包含在 S_i 中的信息量，也就是编码 S_i 所需要的位数。例如，一幅用 256 级灰度表示的图像，如果每一个像素点灰度的概率均为 $p_i = 1/256$，编码每一个像素点就需要 8 位。

3）"变长编码"

"变长编码"：频繁出现的符号用较短的代码代替，较少使用的符号用较长的代码代替，每个符号的代码各不相同，但总的编码长度会减少。

4）"前缀编码"

"前缀编码"的主导思想是，任何一个字符的编码，都不是另一个字符编码的前缀。反过来说就是，任何一个字符的编码，都不是由另一个字符的编码加上若干位 0 或 1 组成。以下是一个最简单的前缀编码的例子。

5）"符号编码"

A 0   B 10   C 110   D 1110   E 11110

有了上面的码表，你一定可以轻松地从下面这串二进制流中分辨出真正的信息内容了，如

11100101011101101111100010

DABBDCEAAB

为了使用不固定的码长表示单个字符，编码必须符合"前缀编码"的要求，即较短的编码决不能是较长编码的前缀。要构造符合这一要求的二进制编码体系，二叉树是最理想的选择。

那么什么是二叉树呢？

树是一种重要的非线性数据结构，直观地看，它是数据元素（在树中称为结点）按分支关系组织起来的结构，很像自然界中的树那样。

树结构在客观世界中广泛存在，如人类社会的族谱和各种社会组织机构都可用树结构来形象的表示。

树在计算机领域中也得到广泛应用，如在编译源程序时，可用树表示源程序的语法结构。又如在数据库系统中，树型结构也是信息的重要组织形式之一。一切具有层次关系的问题都可用树来描述。

图 3-1 所示是一个典型的二叉树结构。

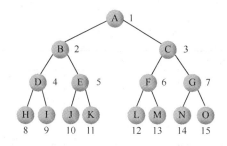

图 3-1 典型树结构

图 3-2 所示为用于实际编码的二叉树。

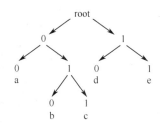

图 3-2 用于实际编码的二叉树

从这棵二叉树，可得到其相应的符号编码为：a - 00，b - 010，c - 011，d - 10，e - 11。

观察这棵二叉树，可以发现要编码的字符总是出现在树叶上，假定从根向树叶行走的过程中，左转为 0，右转为 1，则一个字符的编码就是从根走到该字符所在树叶的路径。正因为字符只能出现在树叶上，任何一个字符的路径都不会是另一字符路径的前缀路径，符合要求的前缀编码也就构造成功了。

[例 3.1] 有一幅 40 个像素组成的灰度图像，灰度共有 5 级，分别用符号 A，B，C，D 和 E 表示，40 个像素中出现灰度 A 的像素数有 15 个，出现灰度 B 的像素数有 7 个，出现灰度 C 的像素数有 7 个，出现灰度 D 的像素数有 6 个，出现灰度 E 的像素数有 5 个，如表 3-1 所列。如果用 3 个位表示 5 个等级的灰度值，也就是每个像素用 3 位表示，编码这幅图像总共需要 120 位。

按照香农理论，这幅图像的熵为

$$H(S) = (15/40) \times \log_2(40/15) + (7/40) \times \log_2(40/7) + \cdots + (5/40) \times \log_2(40/5)$$
$$= 2.196$$

这就是说每个符号用 2.196 位表示，40 个像素需用 87.84 位。

表 3-1 符号在图像中出现的数目

符号	出现的次数	符号	出现的次数
A	15	D	6
B	7	E	5
C	7		

最早阐述和实现这种编码的是 Claude Shannon（1948 年）和 R.M.Fano（1949 年），时间是 1952 年。因此被称为香农—范诺算法。这种方法采用从上到下的方法进行编码。首先按照符号出现的频度或概率排序，例如，A，B，C，D 和 E，如表 3-2 所列。然后使用递归方法

分成两个部分，每一部分具有近似相同的次数，如图3-3所示。按照这种方法进行编码得到的总位数为91，实际的压缩比约为1.3∶1。

表3-2　香农—范诺算法举例表

符号	出现的次数(p_i)	$\log_2(1/p_i)$	分配的代码	需要的位数
A	15(0.375)	1.4150	00	30
B	7(0.175)	2.5145	01	14
C	7(0.175)	2.5145	10	14
D	6(0.150)	2.7369	110	18
E	5(0.125)	3.0000	111	15

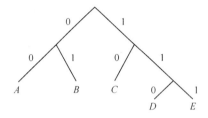

图3-3　香农—范诺算法编码举例

再列举一个例子：下面这串出现了5种字符的信息（40个字符长）：

cabcedeacacdeddaaabaaabababaaabbacdebaceada5种字符的出现次数分别：a - 16，b - 7，c - 6，d - 6，e - 5。

a）将给定符号按照其频率从大到小排序，对上面的例子，应该得到：

a - 16　　b - 7　　c - 6　　d - 6　　e - 5

b）将序列分成上下两部分，使得上部概率总和尽可能接近下部概率总和，则有：

a - 16　　　b - 7

c - 6　　d - 6　　e - 5

c）把第二步中划分出的上部作为二叉树的左子树，记0，下部作为二叉树的右子树，记1。

d）分别对左右子树重复2和3两步，直到所有的符号都成为二叉树的树叶为止，则有如图3-3所示的二叉树：

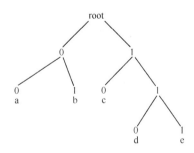

图3-4　编码后的二叉树

于是我们得到了此信息的编码表：

a - 00　　b - 01　　c - 10　　d - 110　　e – 111

可以将例子中的信息编码为：

cabcedeacacdeddaaabaaabababaaabbacdebaceada　　10 00 01 10 111 110 111 00 10 00 10……

86

码长共 91 位。考虑用 ASCII 码表示上述信息需要 8×40=320 位，我们确实实现了数据压缩。

3.3.1　Shannon–Fano Coding and Huffman Coding

1. Shannon-Fano Coding

Before we go further, we need to introduce two information theory ideas.

1) Entropy

Entropy is introduced by Shannon who is the creator of information theory. He first proposed the entropy in his professional paper, a mathematical theory of communication in 1948. The paper had become the foundation of the information theory.

(1) The entropy is the quantification of information. When the news about one event occurring is smaller, the probability about the event happening is smaller. Mathematically, the probability is smaller.

(2) The entropy of one event is expressed as $I_i=-\log_2 p_i$, Where pi is the happening probability of the event and

$$0<p_i\leqslant 1$$

2) The definition of the entropy of a source collection S

Based on the Shannon's theory, the definition of the entropy of a source collection S is as follows:

$$H(S)=\eta=\sum p_i\log_2(1/p_i)$$

Where the p_i is the happening probability of S_i within the source collection S; log $(1/p_i)$ represents the information quantity included inside S_i. The calculation result does also infer to the minimum bits needed to encode S_i.

For example, having an image represented in 256 grey levels. Suppose the probability of the grey level of each pixel is $p_i=1/256$. Then, to encode each pixel will need at least 8 bits.

3) Variable Length Coding

Variable Length Coding means uses the shorter codes to represent the symbols more frequently applied and uses the longer code to represent the symbols less frequently applied. Each symbol's code is different from the others and the total length of the whole coding will be shorter because of the efficiency of the encoding.

4) Prefix Encoding

The idea of prefix encoding is that no encoded symbol of this code is a prefix of any part or whole code of anther symbol. In another word, you can't get any symbol's code by adding any or some 1 or 0 on another symbol's code.

5) Symbol Encoding

If we give the code as

> A 0,　　B 10,　　C 110,　　D 1110,　E 11110

Then we could get the symbols from the 0 and 1 sequence

> 1110010101110110111100010

as　　　　　　　　　　DABBDCEAAB

which are the original symbols we encoded.

In order to represent the individual symbol with variable length code, the code shall meet the requirement of prefix encoding. This way, the shorter code will not be the prefix of the longer code. To form such a kind of binary codes, two separate tree structure is an ideal option.

What actually the two separate tree structure is?

The tree structure is an important nonlinear data structure. Apparently, it is the data elements which are called notes in the tree structure connected each other with branch structure. The structure looks like the natural tree.

The tree structure exists everywhere in our real world. Such as Chinese family generations record and diverse Chinese social organization structure could be represented as a tree structure.

The tree structure has broad application in the computer field as well. When programming a source code, one could use the tree structure to describe the programming grammar structure. Another useful example is the database structure. Inside the database, the tree structure is one of the most important information organizing structure. Actually, all levels related objects could be described in tree structure.

The picture in Fig.3-1 shows a typical tree structure.

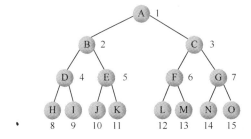

Fig. 3-1   Typical Tree Structure

Next, let's have a look at the two branch tree structure used in the real encoding as shown.in Fig. 3-2.

Front the showing two branch tree, we could get the encoded codes as: a - 00, b - 010, c - 011, d - 10, e - 11.

If we have a closer look at this tree, we could find that each symbol's encoded code is at the end of the branch where the leaf is. On this tree, if we're moving from the root to the leaf, when we turn left, we give 0, when we turn right, we give 1, then we will get each symbol's encoded code from the root to this symbol's leaf. Just for the reason that each symbol only can be found at the end of the branch where the leaf is, it is guaranteed that any symbol's path won't be the prefix of another symbol's path. This way, we got the prefix coding.

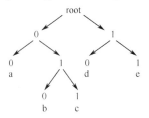

Fig. 3-2   Two Branch Tree Seructure

[Example 3.1] There is a grey level image consisted by 40 pixels. It has 5 grey levels and represented as *A*, *B*, *C*, *D*, *E*. Within the 40 pixels, grey level *A* appeared 15 times, *B* appeared 7 times, *C* appeared 7 times as well, *D* appeared 6 times, and *E* appeared 5 times as shown in Table 3-1. If we use 3 digits to represent the values of 5 grey levels or in another word, we use 3 digits group to represent any one of the grey level values, we'll need total 120 digits to represent the whole image's 40 pixels.

Table3-1　Symbol appears inside image

Symbol	Appeared times	Symbol	Appeared times
A	15	*D*	6
B	7	*E*	5
C	7		

Based on the Shannon's theory, the entropy of the image is

$$H(S) = (15/40) \times \log_2(40/15) + (7/40) \times \log_2(40/7) + \cdots + (5/40) \times \log_2(40/5)$$
$$= 2.196$$

This means that each symbol representing the grey level value and appearing at the position of each pixel will need minimum 2.196 digits to be encoded. The whole image with total 40 pixels will need 87.84 digits to be encoded.

This encoding method was first proposed in 1952 by Claude Shannon (1948) and R.M. Fano (1949) and so it's called Shannon-Fano coding. The coding uses top to bottom method encoding. It firstly sequences the symbols based on their appearing times or in another word probability. In the example, the symbols sequenced as *A*, *B*, *C*, *D*, and *E* as shown in Table 3-2. Then we put the symbols into two groups based on closer appearing probability. The process is showing in Figure 3-3. Finally, we got totally 91 digits encoded codes. The actual compression ratio is 1.3 : 1.

Table3-2　Shannon-Fano Coding Example

Symbol	Appearing times(p_i)	$\log_2(1/p_i)$	Codes	Total Digits
A	15(0.375)	1.4150	00	30
B	7(0.175)	2.5145	01	14
C	7(0.175)	2.5145	10	14
D	6(0.150)	2.7369	110	18
E	5(0.125)	3.0000	111	15

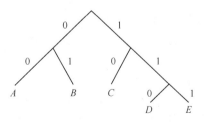

Fig. 3-3　Shannon-Fano Coding

Let's have a look at another example. In the following string, there are total 40 symbols which include a, b, c, d, e five symbols.

cabcedeacacdeddaaabaababababaaabbacdebaceada

Each symbol appeared times/probability is: a - 16，b - 7，c - 6，d - 6，e - 5.

To get the Shannon-Fano code, we follow the steps stated above.

a) Sequence the symbols based on their appearing probability from bigger to smaller and we get:

a - 16   b - 7   c - 6   d - 6   e - 5

b) Divide the above sequence into two groups and make the two groups' total probability as close as possible.

a - 16   b - 7
c - 6   d - 6   e - 5

c) We take the first group as the left branch noted as 0 and the second group as the right branch noted as 1 of the two-branch tree.

d) Repeat the step b and c to form the left and right branches of the two-branch tree until every symbol becomes one of the leaves of the tree. We could get the following two-branch tree shown in Fig 3-4.

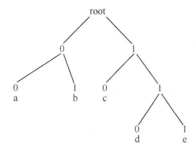

Fig. 3-4   Encoded Two Branch Tree

From the tree, we could get the encoded Shannon-Fano codes as follows.

a - 00   b - 01   c - 10   d - 110   e - 111

At this point, the information string "cabcedeacacdeddaaabaababababaaabbacdebaceada" will be coded as 10 00 01 10 111 110 111 00 10 00 10 ⋯ Totally there are 91 digits. Imaging if we use ASCII code, the given information string will need 8×40 = 320 digits. Comparing the two results, we know that we did realize the data compression.

2．哈夫曼编码

变长编码可以实现数据压缩。变长编码的关键问题是每个字符的编码要符合"前缀编码"——使得接收方在收到报文后能正确地译码，也就是能正确判断现在收到的一个二进制位是前一字符的末位还是一个新的字符的首位。

哈夫曼成功地解决了这个难题，使变长编码得以被实际采用。

哈夫曼计算法步骤进行：

（1）将信号源的符号按照出现概率递减的顺序排列。

（2）将两个最小出现概率进行合并相加，得到的结果作为新符号的出现概率。

（3）重复进行步骤1和2直到概率相加的结果等于1为止。

（4）在合并运算时，概率大的符号用编码0表示，概率小的符号用编码1表示。

（5）记录下概率为1处到当前信号源符号之间的0，1序列，从而得到每个符号的编码。

现在，仍然使用上面的例子来说明哈夫曼编码方法。

a - 16   b - 7   c - 6   d - 6   e - 5

其相应的二叉树如图 3-5 所示。

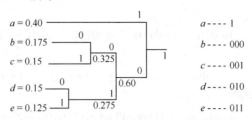

图 3-5　哈夫曼编码二叉树

回顾信息熵的知识，使用学到的计算方法，可得上面的例子中，每个字符的信息熵为

$E_a = -\log_2(16/40) = 1.322$

$E_b = -\log_2(7/40) = 2.515$

$E_c = -\log_2(6/40) = 2.737$

$E_d = -\log_2(6/40) = 2.737$

$E_e = -\log_2(5/40) = 3.000$　信息熵为

$E = E_a \times 16 + E_b \times 7 + E_c \times 6 + E_d \times 6 + E_e \times 5 = 86.601$

也就是说，表示该条信息最少需要 86.601 位。可以看出，香农·范诺编码和哈夫曼编码都已经比较接近该信息熵值了。

同时，也能看出，无论是香农还是哈夫曼，都只能用近似的整数位来表示单个符号，而不是理想的小数位。可以将它们做一个对比，如表 3-3 所列。

表 3-3　理想与实际编码位数

符号	理想位数	S-F 编码	哈夫曼编码
$a=16$	1.322	2	1
$b=7$	2.515	2	3
$c=6$	2.737	2	3
$d=6$	2.737	3	3
$e=5$	3.000	3	3
总计	86.601	91	88

这就是像哈夫曼这样的整数位编码方式无法达到最理想的压缩效果的原因之一。

再列举一个例子：求下列字符串的哈夫曼编码

$\dfrac{aaaa}{4}$	$\dfrac{bbb}{3}$	$\dfrac{cc}{2}$	$\dfrac{d}{1}$	$\dfrac{eeee}{5}$	$\dfrac{fffffff}{7}$

（用原始的 ASCII 码表示时，共 22*8=176 bits）

与前面类似，可得到此序列的哈夫曼树，如图 3-6 所示。

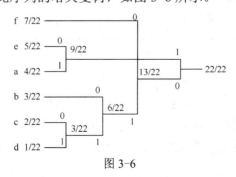

图 3-6

由此树得出的 Huffman 编码为

$$f=00 \quad e=10 \quad a=11 \quad b=010 \quad c=0110 \quad d=0111$$

按顺序排列为

$$a=11 \quad b=010 \quad c=0110 \quad d=0111 \quad e=10 \quad f=00$$

原数据序列为

<u>aaaa</u>   <u>bbb</u>   <u>cc</u>   <u>d</u>   <u>eeeee</u>   <u>ffffff</u>

经过哈夫曼编码之后的整个数据为

$$11111111010010010011001100111101010101000000000000000000$$

（共 7×2+5×2+4×2+3×3+2×4+1×4=53 bits 注：按字母出现次数的多少从大到小排列。）

其压缩比为

$$176：53=3.32：1$$

列举另外一个例子。

信源符号及其概率如下：

符号	a1	a2	a3	a4	a5	a6	a7
概率	0.20	0.19	0.18	0.17	0.15	0.10	0.01

求其哈夫曼编码。

解题过程及相应的二叉树如下面的图 3-7 所示。

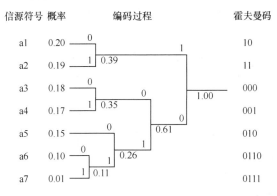

图 3-7

再列举一个例子，进一步熟悉哈夫曼编码的方法。

信源符号及其概率如下表所示：

符号	x1	x2	x3	x4	x5	x6	x7
概率	0.35	0.20	0.15	0.10	0.10	0.06	0.04

求其哈夫曼编码。

解题过程及相应的二叉树如下面的图 3-8 所示。

哈夫曼编码是哈夫曼在 1952 年提出来的一种编码方法，即从下到上的编码方法。现再以一个具体的例子说明它的编码步骤。

（1）初始化，根据符号概率的大小按由大到小顺序对符号进行排序，如表 3-4 和图 3-9 所示。

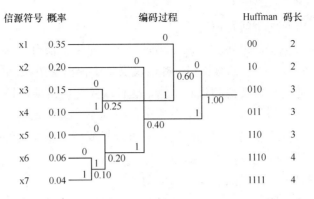

图 3-8

表 3-4　哈夫曼编码举例

符号	出现的次数	$\log_2(1/p_i)$	分配的代码	需要的位数
A	15(0.3846)	1.38	0	15
B	7(0.1795)	2.48	100	21
C	6(0.1538)	2.70	101	18
D	6(0.1538)	2.70	110	18
E	5(0.1282)	2.96	111	15

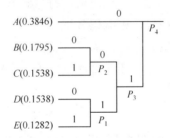

图 3-9　哈夫曼编码方法

（2）把概率最小的两个符号组成一个节点，如图 3-9 中的 D 和 E 组成节点 P_1。

（3）重复步骤 2，得到节点 P_2、P_3 和 P_4，形成一棵"树"，其中的 P_4 称为根节点。

（4）从根节点 P_4 开始到相应于每个符号的"树叶"。从上到下标上"0"（上枝）或者"1"（下枝），至于哪个为"1"哪个为"0"则无关紧要，最后的结果仅仅是分配的代码不同，而代码的平均长度是相同的。

（5）从根节点 P_4 开始顺着树枝到每个叶子分别写出每个符号的代码，如表 3-3 所列。

（6）按照香农理论，这幅图像的熵为

$$H(S) = (15/39) \times \log_2(39/15) + (7/39) \times \log_2(39/7) + \cdots + (5/39) \times \log_2(39/5)$$
$$= 2.1859$$

压缩比为 1.37：1。

哈夫曼码的码长虽然是可变的，但却不需要另外附加同步代码。例如，码串中的第 1 位为 0，那么肯定是符号 A，因为表示其他符号的代码没有一个是以 0 开始的，因此下一位就表示下一个符号代码的第 1 位。同样．如果出现"110"。那么它就代表符号 D。如果事先编写出一本解释各种代码意义的"词典"，即码簿．则可以根据码簿一个码一个码地依次进行译码。

采用哈夫曼编码时有两个问题值得注意。

（1）哈夫曼码没有错误保护功能，在译码时，如果码串中没有错误，那么就能一个接一个地正确译出代码。但如果码串中有错误，哪怕仅仅是 1 位出现错误，不但这个码本身译错，更糟糕的是一个错误还会导致其他的代码出错，这种现象称为错误传播（Error Propagation）。计算机对这种错误也无能为力，说不出错在哪里，更谈不上去纠正它。

（2）哈夫曼码是可变长度码，因此，很难随意查找或调用压缩文件中间的内容，然后再译码，这就需要在存储代码之前加以考虑。

尽管如此，哈夫曼码还是得到了广泛的应用。

与香农—范诺编码相比，这两种方法都自含同步码，在编码之后的码串中都不需要另外添加标记符号，即在译码时分割符号的特殊代码。此外，哈夫曼编码方法的编码效率比香农—范诺编码效率高一些，请读者自行验证。

哈夫曼编码的特点。

（1）编码必须是整数位长度。因此当某个字符出现的概率非常大之时，哈夫曼编码可能给出比最优编码大得多的编码。举例来说，如果某个统计方法能将 90%的概率分配给一个字符，那么其最优编码应该为$-\log_2（0.9）=0.15$bits，但哈夫曼编码至少会分配 1bit，可以看到这种情况下是实际需要的 6 倍长度。

（2）对不同信号源的编码效率不同，当信号源的符号概率为 2 的负幂次方时，达到 100％的编码效率；若信号源符号的概率相等，则编码效率最低。

（3）哈夫曼编码表是编码的重要依据，为了节省编码时间，通常把哈夫曼编码表存储在发送端和接收端。否则，在进行编码时还要传送编码表，在很大程度上延长了编码时间。

2. Huffman Coding

As we know already that the variable-length Encoding could compress data. The key of the variable-length Encoding is that every symbol's code shall meet the requirement of prefix coding to guarantee the receiver could properly decode the received data. In another word, the receiver could properly tell whether the received data is the beginning of new symbol's code or the end of the current symbol's code.

Huffman successfully solved this tuff topic and proposed a practical coding algorithm.

The steps of Huffman algorithm is as follows.

(1) Sequence the source symbols based on their appearance probabilities in a descend order.

(2) Add the two smallest probabilities together and put the result as the appearance probability of an imagined new symbol.

(3) Repeat the steps (1) and (2) until the sum of the two closest probabilities equals to 1.

(4) When processing the summary of the two probabilities, set 0 to the bigger probability branch and 1 to the smaller probability branch.

(5) Start from where the probability is 1, count the 0 or 1 along the branches, and get the correspondent symbol's code.

We could use the same example showing above to further explain the Huffman algorithm.

$$a - 16 \quad b - 7 \quad c - 6 \quad d - 6 \quad e - 5$$

Its relative two-branch tree is shown in Fig 3-5.

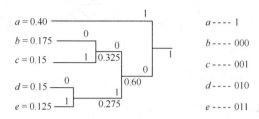

Fig 3-5   Huffman Coding Two Branch Tree

As indicated in the information theory, the entropy of each symbol could be calculated as:

$E_a = -\log_2(16/40) = 1.322$

$E_b = -\log_2(7/40) = 2.515$

$E_c = -\log_2(6/40) = 2.737$

$E_d = -\log_2(6/40) = 2.737$

$E_e = -\log_2(5/40) = 3.000$

The total average entropy could be calculated as:

$E = E_a \times 16 + E_b \times 7 + E_c \times 6 + E_d \times 6 + E_e \times 5 = 86.601$

This means that the least digits needed to represent the given information is 86.601. As we have seen that both Shannon-Fano and Huffman coding have been very close to the entropy of the given information.

From the above samples, we could also see that both Shannon-Fano and Huffman coding can only use the approached integrals instead of the ideal decimal digits to represent the symbols. The following Table 3-3 gives a comparison.

Table 3-3   The Ideal&Real Loding Digits

Symbols	Ideal digits	S-F coding	Huffman coding
a=16	1.322	2	1
b=7	2.515	2	3
c=6	2.737	2	3
d=6	2.737	3	3
e=5	3.000	3	3
Total	86.601	91	88

This is why the coding algorithms like Huffman coding can't get the best compression result.

Let's have a look at another example:

What is the Huffman coding of following character sequence?

$$\frac{aaaa}{4} \quad \frac{bbb}{3} \quad \frac{cc}{2} \quad \frac{d}{1} \quad \frac{eeee}{5} \quad \frac{fffffff}{7}$$

(When represent the sequence in original ASCII codes, we will need 22×8 = 176 bits.)

Similar as before, we could get its Huffman tree as shown in.Fig. 3-6.

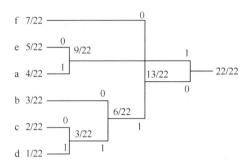

Fig. 3-6

From the tree, we could get its Huffman coding as:

f=00   e=10   a=11   b=010   c=0110   d=0111

If we put it in descending order, we get:

a=11   b=010   c=0110   d=0111   e=10   f=00

The original character sequence is:

aaaa   bbb   cc   d   eeeee   fffffff

After the Huffman algorithm encoding we get:

1111111101001001001100110011110101010101000000000000000

Total there are 7×2+5×2+4×2+3×3+2×4+1×4=53 bits.

The compression ratio is

$$176 : 53 = 3.32 : 1.$$

Furthermore, let's have a look at another example:

The source symbols and their corresponding probabilities are listed as follows:

Symbols	a1	a2	a3	a4	a5	a6	a7
Probabilities	0.20	0.19	0.18	0.17	0.15	0.10	0.01

Question: What's their Huffman coding?

The solution process and the corresponding two-branch tree are as shown in Fig. 3-7.

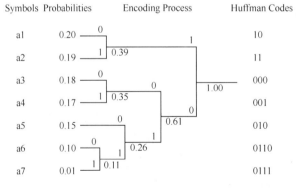

Fig. 3-7

We have another Huffman encoding example to help you further familiar with the Huffman algorithm.

The source symbols and their corresponding probabilities are listed as follows:

Symbols	x1	x2	x3	x4	x5	x6	x7
Probabilities	0.35	0.20	0.15	0.10	0.10	0.06	0.04

The question is asking their Huffman codes.

The corresponding two-branch tree and the process of finding the answers are as shown in Fig. 3 - 8.

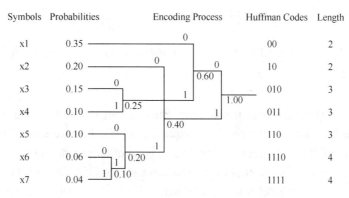

Fig. 3-8

Huffman coding is an algorithm proposed by Huffman in 1952. It is from top to bottom encoding method. Here we give another example step by step to understand its procedure.

(1) Initialization. Sequence the source symbols based on the appearance probabilities in descending order as shown in Table 3-4 and Fig. 3-9.

Table3-4   Huffman Coding Example

Symbols	Appearing Times	$\log_2(1/p_i)$	Assigned Codes	Needed digits
A	15(0.3846)	1.38	0	15
B	7(0.1795)	2.48	100	21
C	6(0.1538)	2.70	101	18
D	6(0.1538)	2.70	110	18
E	5(0.1282)	2.96	111	15

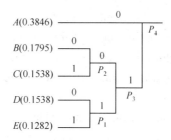

Fig.3-9   Huffman coding

(2) Put the two symbols with the smallest probabilities into one note such as the symbols D and E into note P_1 shown in Fig. 3-9.

(3) Repeat the step (2), get the notes P_2, P_3, and P_4, and form the two-branch tree. The note P_4 is called the root note.

(4) Start from the root note P_4 to each leaf corresponding to each symbol, assign "0" or "1" to

the above branch and "1" or "0" to the lower branch. Actually, it doesn't matter which branch is "0" or "1". But it's better to keep it consecutively either way from the beginning to the end. The only difference is that the final encoded codes will be different. But the average codes length will be the same.

(5) Start from the root note P_4, along the branches reach the leaves, and write each symbol's code as shown in Table 3-3.

(6) According to the Shannon information theory, the entropy of this given image (refer to Table 3-3) is:

$$H(S) = (15/39) \times \log_2(39/15) + (7/39) \times \log_2(39/7) + \cdots + (5/39) \times \log_2(39/5)$$
$$= 2.1859$$

So the compression ratio is 1.37 : 1.

The length of the Huffman code is changing. But it doesn't need the extra synchronizing code. For example, if the first digit of the received sequence is "0", then we know for sure that it is the code of the first symbol "A". The reason is that there is no other symbol's code started with "0". So the next received digit must be the first digit of another symbol's code. For the same reason, if the received code is "110", it definitely the code of symbol "D". This way, if we have a code book of the source symbols witch is called dictionary at the receiver side, we could decode the receive codes one by one and get the original symbols.

There are two things should catch our attention when we use Huffman coding.

(1) Huffman coding doesn't have error bit protection function. When we decoding, if the codes don't have any error bit, we could decode the codes and get the original symbols one by one without any issue. But if there is any error bit in the received bits stream, even only one bit is wrong, we may decode this code which includes the wrong bit into a wrong source symbol. Even worse, we may decode the others into the wrong source symbols as well since this one error bit will change the whole bit stream's structure. This is called error propagation. Computer can do nothing to this kind of error. It can't tell where the error is and certainly can't correct it.

(2) Since the Huffman coding is a variable length coding, it's almost impossible to find or catch any randomly wanted content inside the compressed file and decode from this particular point. This should be aware in advance.

Huffman coding is not perfect as indicated above, but it's still applied broadly.

Compared with Shannon-Fano coding, both of them have their own synchronization bits and don't need the extra mark bits which is used to separate the symbols in the decoding process. Huffman coding has relatively higher coding efficiency ratio. Readers could find this out by themselves.

The major characters of Huffman coding could be abstracted as follows.

(1) The length of the encoded codes must be an integer. For this reason, when one symbol's appearing probability is very big, Huffman encoding may give much longer length code than the optimum one. For example, if in one situation, one source symbol's appearing probability is 90%, based on the information theory, it's optimum code length shall be $-\log_2(0.9)=0.15$ bits. But Huffman encoding will assign at least one bit which is the six times of the optimum one's to the symbol.

(2) It has different coding ratio efficiency to the different source symbol structures. It could reach 100% coding ratio efficiency when the source symbols have the appearing probabilities of

power of minus two. On the other hand, it has the lowest coding ratio efficiency when the source symbols have equal appearing probabilities.

(3) The Huffman coding table is a very important decoding base. Usually save the same coding table at both the sending and receiving sides to save coding time. Otherwise, we have to transfer the coding table while we encoding. It will increase the encoding time in a big scale.

3.3.2 算术编码

算术编码在图像数据压缩标准（如 JPEG，JBIG）中扮演了重要的角色。在算术编码中，消息用 0 到 1 之间的实数进行编码。算术编码用到两个基本的参数：符号的概率和它的编码间隔。信源符号概率决定压缩编码的效率，也决定编码过程中信源符号的间隔，而这些间隔包含在 0 到 1 之间。编码过程中的间隔决定了符号压缩后的输出。

算术编码对整条信息（无论信息有多么长），其输出仅仅是一个数，而且是一个介于 0 和 1 之间的二进制小数。例如算术编码对某条信息的输出为 1010001111，那么它表示小数 0.1010001111，也即十进制数 0.64。从后面的叙述中可以更清楚地看到这一点。

算术编码器的编码过程可用下面的例子加以解释。

[例 3.2] 假设信源符号为 {00, 01, 10, 11}，这些符号的概率分别为 {0.1, 0.4, 0.2, 0.3}，根据这些概率可把间隔[0, 1）分成 4 个子间隔：[0, 0.1），[0.1, 0.5），[0.5, 0.7），[0.7, 1），其中[x, y）表示半开放间隔，即包含 x 不包含 y。上面的信息可综合在表 3-5 中。

表 3-5　信源符号、概率和初始编码间隔

符号	概率	初始编码间隔
00	0.1	[0, 0.1)
01	0.4	[0.1, 0.5)
10	0.2	[0.5, 0.7)
11	0.3	[0.7, 1)

如果二进制消息序列的输入为：10 00 11 00 10 11 01，编码时首先输入的符号是 10. 找到它的编码范围是[0.5, 0.7)。由于消息中第二个符号 00 的编码范围是[0, 0.1)，因此，它的间隔就取[0.5, 0.7)的第一个 1/10 作为新间隔[0.5, 0.52)。依此类推，编码第 3 个符号 11 时取新间隔为[0.514, 0.52)，编码第 4 个符号 00 时，取新间隔为[0.514, 0.5146)，…消息的编码输出可以是最后一个间隔中的任意数。整个编码过程如图 3-10 所示。

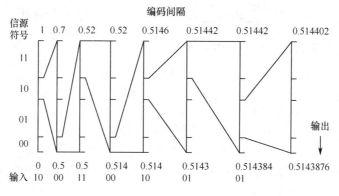

图 3-10　算术编码过程举例

99

这个例子的编码和译码的全过程分别表示在表 3-6 和表 3-7 中。根据上面所举的例子，可把计算过程总结如下。

表 3-6　编码过程

步骤	输入符号	编码间隔	编码判决
1	10	[0.5, 0.7)	符号的间隔范围[0.5, 0.7)
2	00	[0.5, 0.52)	[0.5, 0.7)间隔的第一个 1/10
3	11	[0.514, 0.52)	[0.5, 0.52)间隔的最后三个 1/10
4	00	[0.514, 0.5146)	[0.514, 0.52)间隔的第一个 1/10
5	10	[0.5143, 0.51442)	[0.514, 0.5146)间隔的第五个 1/10 开始，二个 1/10
6	11	[0.514384, 0.51442)	[0.5143, 0.51442)间隔的最后 3 个 1/10
7	01	[0.5143876, 0.514402)	[0.514384, 0.51442)间隔的 4 个 1/10，从第 1 个 1/10 开始
8	从[0.5143876, 0.514402]中选择一个数作为输出：0.5143876		

表 3-7　译码过程

步骤	间隔	译码符号	译码判决
1	[0.5, 0.7)	10	0.51439 在间隔[0.5, 0.7)
2	[0.5, 0.52)	00	0.51439 在间隔[0.5, 0.7)的第 1 个 1/10
3	[0.514, 0.52)	11	0.51439 在间隔[0.5, 0.52)的第 7 个 1/10
4	[0.514, 0.5146)	00	0.51439 在间隔[0.514, 0.52)的第 1 个 1/10
5	[0.5143, 0.51442)	10	0.51439 在间隔[0.514, 0.5146)的第 5 个 1/10
6	[0.514384, 0.51442)	11	0.51439 在间隔[0.5143, 0.51442)的第 7 个 1/10
7	[0.51439, 0.5143948)	01	0.51439 在间隔[0.51439, 0.5143948)的第 1 个 1/10
8	译码的消息：10 00 11 00 10 11 01		

考虑一个有 M 个符号 $a_i = (1, 2, \cdots, M)$ 的字符表集，假设概率 $p(a_i) = p_i$，而 $\sum p_i(a_i) = p_1 + p_2 + \cdots p_m = 1$。输入符号用 x_n 表示，第 n 个子间隔的范围用 $I_n = [l_n, r_n] = \left[l_{n-1} + d_{n-1} \sum_{k=1}^{i} p_{k-1}, l_{n-1} + d_{n-1} \sum_{k=1}^{i} p_k \right)$ 表示。$l_0 = 0$，$d_0 = 1$ 和 $p_0 = 0$，l_n 表示间隔左边界的值，r_n 表示间隔右边界的值，$d_n = r_n - I_n$ 表示间隔长度，编码步骤如下。

（1）首先在 1 和 0 之间给每个符号分配一个初始子间隔，子间隔的长度等于它的概率，初始子间隔的范围用 $I_1 = [l_1, r_1] = \left[\sum_{k=1}^{i} p_{k-1}, \sum_{k=1}^{i} p_k \right)$ 表示。令 $d_1 = r_1 - l_1$，$L = l_1$ 和 $R = r_1$。

（2）L 和 R 的二进制表达式分别表示为

$$L = \sum_{k=1}^{\infty} u_k 2^{-k} \text{ 和 } R = \sum_{k=1}^{\infty} V_k 2^{-k}$$

式中：u_k 和 v_k 等于 "1" 或者 "0"。

比较 u_1 和 v_1：①如果 $u_1 \neq v_1$，不发送任何数据，转到步骤（3）；②如果 $u_1 = v_1$，就发送二进制符号 u_1。

比较 u_2 和 v_2：①如果 $u_2 \neq v_2$，不发送任何数据，转到步骤（3）；②如果 $u_2 = v_2$，就发送二进制符号 u_2。

……

这种比较一直进行到两个符号不相同为止，然后进入步骤（3）。

（3）n 加 1，读下一个符号。假设第 n 个输入符号为 $x_n = a_i$，按照以前的步骤把这个间隔分成如下所示的子间隔：

$$I_n = [l_n, r_n) = \left[l_{n-1} + d_{n-1} \sum_{k=1}^{i} p_{k-1}, I_{n-1} + d_{n-1} \sum_{k=1}^{i} p_k \right)$$

令 $L = l_n$，$R = r_n$，和 $d_n = r_n - l_n$，然后转到步骤（2）。

[例 3.3] 假设有 4 个符号的信源，它们的概率如表 3-8 所列。

输入序列为：x_n: $a_2, a_1, a_3, \cdots$。它的编码过程如图 3-11 所示，现说明如下。

表 3-8 符号概率

信源符号 a_i	概率 p_i	初始编码间隔
a_1	$p_1 = 0.5$	[0, 0.5)
a_2	$p_2 = 0.25$	[0.5, 0.75)
a_3	$p_3 = 0.125$	[0.75, 0.875)
a_4	$p_4 = 0.125$	[0.875, 1)

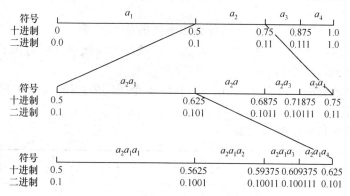

图 3-11 算术编码概念

输入第 1 个符号是 $x_i = a_2$：，可知 $i = 2$，定义起始间隔 $I_1 = [l_1, r_1) = \left(\sum_{k=1}^{i} p_{i-k}, \sum_{k=1}^{i} p_k \right) = \left(\sum_{k=1}^{2} p_{k-1}, \sum_{k=1}^{2} p_k \right)$

$= [0.5, 0.75)$，由此可知 $d_1 = 0.25$，左右边界的二进制数分别表示为：$L = 0.5 = 0.1(B)$，$R = 0.75 = 0.11(B)$。按照步骤（2），$u_1 = v_1$，发送 1。因 $u_2 \neq v_2$，因此转到步骤（3）。

输入第 2 个字符 $x_2 = a_1$，$i = 1$。它的子间隔 $I_2 = [l_2, r_2) = \left[l_1 + d_1 \sum_{k=1}^{i} p_{k-1}, l_1 + d_1 \sum_{k=1}^{i} p_k \right) = [0.5, 0.625)$，由此可得 $d_2 = 0.125$。左右边界的二进制数分别表示为：$L = 0.5 = 0.100 \cdots (B)$，$R = 0.101 \cdots (B)$。按照步骤（2），$u_2 = v_2 = 0$，发送 0，而 u_3 和 v_3 不相同，因此在发送 0 之后就转到步骤（3）。

输入第 3 个字符 $x_i = a_3$，$i = 3$，它的子间隔 $I_3 = [l_3, r_3) = \left[l_2 + d_2 \sum_{k=1}^{i} p_{k-1}, l_2 + d_2 \sum_{k=1}^{i} p_k \right) = [0.59375, 0.609375)$，由此可得 $d_3 = 0.015625$。左右边界的二进制数分别表示为：$L = 0.59375 = 0.10011(B)$，$R = 0.609375 = 0.100111(B)$。按照步骤（2），$u_3 = v_3 = 0$，$u_4 = v_4 = 1$，$u_5 = v_5 = 1$，但 u_6 和 v_6 不相同，因此，在发送 011 之后转到步骤（3）。

……

发送的符号是：$10011\cdots$。被编码的最后的符号是结束符号。

101

就这个例子而言，算术编码器接受的第 1 位是"1"。它的间隔范围就限制在[0.5，1)，但在这个范围里有 3 种可能的码符 a_2、a_3 和 a_4，因此，第 1 位没有包含足够的译码信息。在接受第 2 位之后就变成"10"，它落在[0.5, 0.75)的间隔里，由于这两位表示的符号都指向 a_2 开始的间隔，因此，就可断定第一个符号是 a_2。在接受每位信息之后的译码情况如表 3-9 所列。

表 3-9　译码过程表

接受的数字	间隔	译码输出
1	[0.5, 1)	⋯
0	[0.5, 0.75)	a_2
0	[0.5, 0.609375)	a_1
1	[0.5625, 0.609375)	⋯
1	[0.59375, 0.609375)	a_3
⋮	⋮	⋮

在上面的例子中，假定编码器和译码器都知道消息的长度，因此译码器的译码过程不会无限制地运行下去。实际上在译码器中需要添加一个专门的终止符，当译码器看到终止符时就停止译码。

借助下面的另一个例子来进一步阐释算术编码的基本原理。

[例 3.4]：假设信源符号为{A，B，C，D}，这些符号的概率分别为{0.1，0.4，0.2，0.3}求信源 CADACDB 的算术编码。

解：根据题目所给概率可把间隔[0，1）分成 4 个子间隔：[0，0.1），[0.1，0.5），[0.5，0.7），[0.7，1），其中[x，y)表示半开放间隔，即包含 x 不包含 y。

信源符号、概率和初始编码间如下：

符号	A	B	C	D
概率	0.1	0.4	0.2	0.3
初始编码间隔	[0,0.1)	[0.1,0.5)	[0.5,0.7)	[0.7,1.0)

输入序列：CADACDB

其编码计算过程如图 3-12 所示。

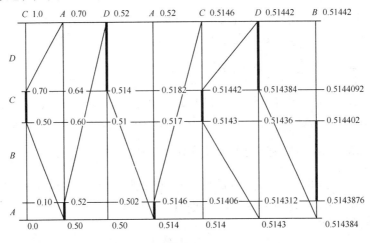

图 3-12

表 3-10 中更详尽地显示了其编码计算过程。

紧跟着的表 3-11 中列出了其译码过程。

表 3-10 编码过程

步骤	输入符号	编码间隔	编码判决
1	C	[0.5, 0.7)	符号的间隔范围[0.5, 0.7)
2	A	[0.5, 0.52)	[0.5, 0.7)间隔的第一个 1/10
3	D	[0.514, 0.52)	[0.5, 0.52)间隔的最后三个 1/10
4	A	[0.514, 0.5146)	[0.514, 0.52)间隔的第一个 1/10
5	C	[0.5143, 0.51442)	[0.514, 0.5146)间隔的第五个 1/10 开始，二个 1/10
6	D	[0.514384, 0.51442)	[0.5143, 0.51442)间隔的最后 3 个 1/10
7	B	[0.5143876, 0.514402)	[0.514384, 0.51442)间隔的 4 个 1/10，从第 1 个 1/10 开始
8	从[0.5143876, 0.514402)中选择一个数作为输出：0.5143876		

表 3-11 译码过程

步骤	间隔	译码符号	译码判决
1	[0.5, 0.7)	C	0.51439 在间隔[0.5, 0.7)
2	[0.5, 0.52)	A	0.51439 在间隔[0.5, 0.7)的第 1 个 1/10
3	[0.514, 0.52)	D	0.51439 在间隔[0.5, 0.52)的第 7 个 1/10
4	[0.514, 0.5146)	A	0.51439 在间隔[0.514, 0.52)的第 1 个 1/10
5	[0.5143, 0.51442)	C	0.51439 在间隔[0.514, 0.5146)的第 5 个 1/10
6	[0.514384, 0.51442)	D	0.51439 在间隔[0.5143, 0.51442)的第 7 个 1/10
7	[0.51439, 0.5143948)	B	0.51439 在间隔[0.51439, 0.5143948)的第 1 个 1/10
8	译码的消息：C A D A C D B		

对比编码和译码的两张表，它们没什么区别，只是运算顺序不同而已。

相应的解码过程示于图 3-13 中。

算术编码步骤可简洁归纳如下：

1. 编码器在开始时将"当前间隔"[L，H)设置为[0，1)。

2. 对每一事件，编码器按步骤（a）和（b）进行处理。

（a）编码器将"当前间隔"分为子间隔，每一个事件一个。

（b）编码器选择子间隔应与下一个确切发生的事件相对应，并使它成为新的"当前间隔"。

3. 最后输出的"当前间隔"的下边界就是该给定事件序列的算术编码。

解码过程　0.5143876

（具体过程示于图 3-13 中）

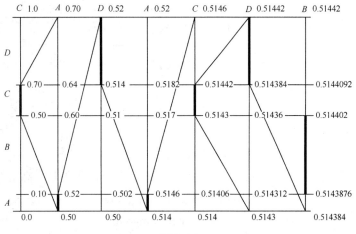

图 3-13

算术编码的代码描述为

```
set Low to 0
set High to 1
while there are input symbols do
    take a symbol
    Range = High - Low
    High = Low + Range * Range High
    Low  = Low  + Range * Range Low
end of while
output Low
```

再看一个例子。

[例 3.5]假设一则消息"state_tree",用算术编码方法给该消息编码。

概率分布:

字符	概率
t	0.3
s	0.1
r	0.1
e	0.3
a	0.1
_	0.1

沿着"概率线"为每一个单独的符号设定一个范围,哪一个被设定到哪一段范围并不重要,只要编码和解码都以同样方式进行就可以,这里所用的 6 个字符被分配的范围如图 3-14 所示。

符号	_	a	e	r	s	t
概率	0.1	0.1	0.3	0.1	0.1	0.3
初始化区间	[0,0.1)	[0.1,0.2)	[0.2,0.5)	[0.5,0.6)	[0.6,0.7)	[0.7,1.0)
Range Low	0	0.1	0.2	0.5	0.6	0.7
Range High	0.1	0.2	0.5	0.6	0.7	1

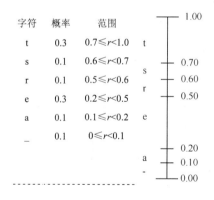

图 3-14

(1)初始化时,被分割的范围 range=high-low=1,下一个范围的低、高端分别由下式计算:

$$Low=low+range\times range\ low$$

$$High=low+range\times range\ high$$

其中等号右边的 low 为上一个被编码字符的范围低；range low 和 range high 分别为被编码符号已给定的字符出现概率范围的 low 和 high。

（2）对消息第一字符 s 编码：s 的 range low=0.6，s 的 range high=0.7 因此，下一个区间的 low 和 high 为：

Low=low+range×range low=0+1×0.6=0.6

High=low+range×range high=0+1×0.7=0.7

Range=high-low=0.7-0.6=0.1

S 将区间[0,1）=>[0.6,0.7），如图 3-15 所示。

（3）对第二个字符 t 编码，使用的新生范围为[0.6, 0.7），因为 t 的 range low=0.7，range high=1.0，因此下一个 low，high 分别为

Low=0.6+0.1×0.7＝0.67

High=0.6+0.1×1.0＝0.70

Range=0.7-0.67=0.03

t 将[0.6,0.7）=>[0.67,0.70），如图 3-16 所示。

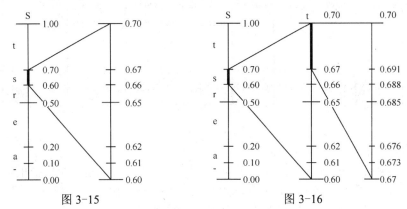

图 3-15   图 3-16

（4）对第三个字符 a 编码，在新生成的[0.67,0.70）中进行分割，因为 a 的 range low=0.10，range high =0.2,因此下一个 low，high 分别为

Low=0.67+0.03×0.1＝0.673

High=0.67+0.03×0.2＝0.676

Range=0.676-0.673=0.003

a 将[0.67,0.70）=>[0.673,0.676），如图 3-17 所示。

（5）对第四个字符 t 编码，在新生成的[0.673,0.676）上进行分割。因为 t 的 range low=0.70，range high=1.0，则下一个 low，high 分别为

Low=0.673+0.003×0.7＝0.6751

High=0.673+0.003×1.0＝0.676

Range=0.0009

t 将[0.673,0.676）=>[0.6751,0.676），如图 3-18 所示。

同理得到下面各字符 e, ＿,t,r,e,e 编码所得到的范围分别为 [0.67528,0.67555），[0.67528,0.675307），[0.675 298 9,0.675 307），[0.675 302 95,0.675 303 76），[0.675 303 112,0.675 303 355），[0.675 303 160 6,0.675 303 233 5），如图 3-19 所示。

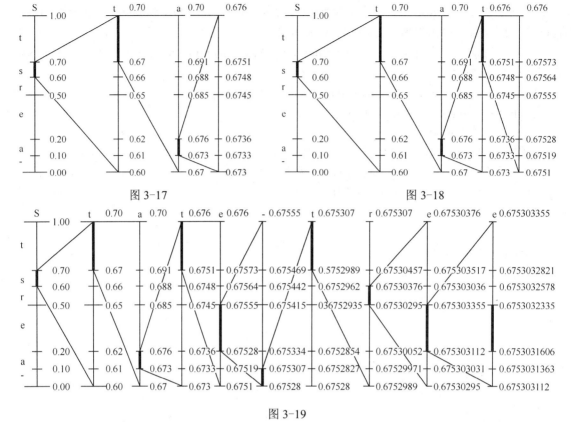

图 3-17 图 3-18

图 3-19

在算术编码中需要注意的几个问题：

（1）由于实际计算机的精度不可能无限长．运算中出现溢出是一个明显的问题，但多数机器都有 16 位、32 位或者 64 位的精度，因此这个问题可使用比例缩放方法解决。

（2）算术编码器对整个消息只产生一个码字，这个码字是在间隔[0，1)中的一个实数，因此译码器在接收到表示这个实数的所有位之前不能进行译码。

（3）算术编码也是一种对错误很敏感的编码方法，如果有一位发生错误就会导致整个消息译错。

算术编码可以是静态的或者自适应的。在静态算术编码中，信源符号的概率是固定的，在自适应算术编码中，信源符号的概率根据编码时符号出现的频繁程度动态地进行修改，在编码期间估算信源符号概率的过程称为建模。需要开发动态算术编码的原因是因为事先知道精确的信源概率是很难的，而且是不切实际的。当压缩消息时，不能期待一个算术编码器获得最大的效率，所能做的最有效的方法是在编码过程中估算概率。因此动态建模就成为确定编码器压缩效率的关键。

在算术编码的使用中还存在版权问题。

3.3.2 Arithmetic Coding

The arithmetic coding plays a very important role in image data compression algorithms such as JPEG, JBIG, and etc. In arithmetic coding, the messages are encoded by the real decimal numbers between 0 and 1. There are two basic coefficients used in the arithmetic coding. One is the symbol's probability and the other is the encoding interval. The symbol probability decides both the encoding efficiency and the processing intervals of the source symbols. Those intervals are always falling in between 0 and 1. The intervals created in the encoding process give out the encoding output result.

The arithmetic encoding only output one number to the whole message no matter how long the message is. The number is always one of the binary numbers between 0 and 1. For example, if after the encoding, the arithmetic output 1010001111 for one message, it represents the decimal 0.1010001111 which is also decimal 0.64. We'll have a closer look at this later on.

Let's have a look at the following example to understand the arithmetic encoding.

[Example 3.2] Suppose the source symbols are {00, 01, 10, 11}. Their corresponding probabilities are {0.1, 0.4, 0.2, 0.3}. Based on those probabilities, could divide the interval [0, 1) into 4 sub-intervals and they are [0, 0.1), [0.1, 0.5), [0.5, 0.7), [0.7, 1) where the [x, y) representing the half open interval which means that the x is included in the interval and y not. The info described could be expressed in the following Table 3-5.

Table3-5　Source Symbols, Probabilities, and Initiate Intervals

Symbols	Probabilities	Initiate Intervals
00	0.1	[0, 0.1)
01	0.4	[0.1, 0.5)
10	0.2	[0.5, 0.7)
11	0.3	[0.7, 1)

If the binary input message series are 10 00 11 00 10 11 01, the first encoded symbol will be 10. From the above description and the Table 3-5, we know that its corresponding interval is [0.5, 0.7). Since the second symbol, 00's initial encoding interval is [0, 0.1) as shown in Table 3-5, its new encoding interval will take [0, 0.1) based on the interval [0.5, 0.7) that gives out [0.5, 0.52). Using the similar method, we could get the third symbol, 11's new interval as [0.514, 0.52), the fourth symbol, 00's new interval as [0.514, 0.5146), and etc... The final encoding output could be one of any numbers within the final interval. The whole process is as shown in Fig. 3-10.

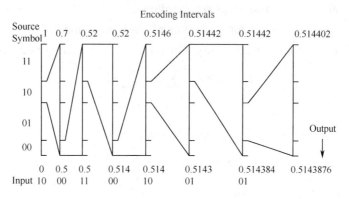

Fig.3-10　The Process of Arithmetic Encoding

The whole process of the encoding the decoding are shown in Table 3-6 and Table 3-7. The calculation process could be concluded as following.

Table 3-6   Encoding Process

Steps	Input Symbols	Encoding intervals	Encoding Judging
1	10	[0.5, 0.7)	Range of the symbol[0.5, 0.7)
2	00	[0.5, 0.52)	[0.5, 0.7)The first 1/10 of the interval
3	11	[0.514, 0.52)	[0.5, 0.52)The last three 1/10 of the interval
4	00	[0.514, 0.5146)	[0.514, 0.52)The first 1/10 of the interval
5	10	[0.5143, 0.51442)	[0.514, 0.5146)Two 1/10, start from the fifth 1/10
6	11	[0.514384, 0.51442)	[0.5143, 0.51442)The last three 1/10 of the interval
7	01	[0.5143836, 0.514402)	[0.514384, 0.51442)Four 1/10, start from the first 1/10
8	Pick one number from the interval[0.5143836, 0.514402)as output: 0.5143876		

Table 3-7   Decoding Process

Steps	Intervals	Decoded Symbols	Decode Judging
1	[0.5, 0.7)	10	0.51439 Falls in the interval [0.5, 0.7)
2	[0.5, 0.52)	00	0.51439 Falls in the first 1/10 of interval [0.5, 0.7)
3	[0.514, 0.52)	11	0.51439 Falls in the 7th 1/10 of interval [0.5, 0.52)
4	[0.514, 0.5146)	00	0.51439 Falls in the first 1/10 of interval [0.514, 0.52)
5	[0.5143, 0.51442)	10	0.51439 Falls in the 5th 1/10 of interval [0.514, 0.5146)
6	[0.514384, 0.51442)	11	0.51439 Falls in the 7th 1/10 of interval [0.5143, 0.51442)
7	[0.51439, 0.5143948)	01	0.51439 in the first 1/10 of interval [0.51439, 0.5143948)
8	After decoded message: 10 00 11 00 10 11 01		

Suppose there are M symbols in a source and could be expressed as $a_i=(1, 2, \cdots, M)$. Suppose the corresponding probability is $p(a_i) = p_i$ and $\sum p_i(a_i) = p_1 + p_2 + \cdots p_m = 1$. We represent the input sequence as x_n and the n's interval as

$$I_n = [l_n, r_n) = \left[l_{n-1} + d_{n-1} \sum_{k=1}^{i} p_{k-1}, l_{n-1} + d_{n-1} \sum_{k=1}^{i} p_k \right)$$

$l_0 = 0$, $d_0 = 1$ 和 $p_0 = 0$. l_n is the left limit of the interval and r_n right limit and the interval. $d_n = r_n - l_n$ represents the interval. The encoding steps are as follows.

(1) Initially, assign an interval for each symbol between 0 and 1. The range of the interval equals the symbol's correspondent probability. The range of the initiating interval is

$$I_1 = [l_1, r_1) = \left[\sum_{k=1}^{i} p_{k-1}, \sum_{k=1}^{i} p_k \right)$$

Let $d_1 = r_1 - l_1$, $L = l_1$ and $R = r_1$.

(2) The binary expression of L and R are as follows.

$$L = \sum_{k=1}^{\infty} u_k 2^{-k} \text{ and } R = \sum_{k=1}^{\infty} V_k 2^{-k}$$

Where u_k and v_k is either "1" or "0".

Compare u_1 and v_1. ① If $u_1 \neq v_1$, won't send any data and go to step (3). ② If $u_1 = v_1$, send the binary symbol u_1.

Compare u_2 and v_2. ① If $u_2 \neq v_2$, won't send any data and go to step (3). ② If $u_2 = v_2$, send the binary symbol u_2.

....

Process the compression until the two symbols are not equal to each other. Go to step (3).

(3) Add 1 to n and read the next symbol. Suppose the nth input symbol is $x_n = a_i$, the sub-interval will be as follows.

$$I_n = [l_n, r_n) = \left[l_{n-1} + d_{n-1} \sum_{k=1}^{i} p_{k-1}, I_{n-1} + d_{n-1} \sum_{k=1}^{i} p_k \right)$$

Let $L = l_n$, $R = r_n$, and $d_n = r_n - l_n$. Go to step (2).

[Example 3.3] There are four symbols inside a source and their correspondent probabilities are showing in Table 3-8.

Table 3-8 Symbol Probabilities

Source Symbol a_i	Probability p_i	Initial Intervals
a_1	$p_1=0.5$	[0, 0.5)
a_2	$p_2=0.25$	[0.5, 0.75)
a_3	$p_3=0.125$	[0.75, 0.875)
a_4	$p_4=0.125$	[0.875, 1)

The input sequence is X_n: $a_2, a_1, a_3, \cdots$. The encoding process is as shown in Fig. 3-11.

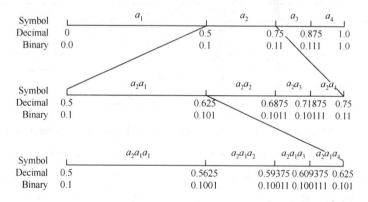

Fig.3-11   The Idea of Arithmetic Encoding

The first input symbol is $x_i = a_2$, where $i=2$. From the above, the start interval will be as follows.

$$I_1 = [l_1, r_1) = \left[\sum_{k=1}^{i} p_{k-1}, \sum_{k=1}^{i} p_k \right) = \left[\sum_{k=1}^{2} p_{k-1}, \sum_{k=1}^{2} p_k \right) = [0.5, 0.75)$$

From this, we get $d_1=0.25$. The binary left and right limits of the range will be: $L=0.5=0.1$(B) and $R = 0.75 = 0.11$(B). Following the step (2), $u_1=v_1=1$, send 1. Since $u_2 \neq v_2$, go to step (3).

Input the second symbol $x_2 = a_1, i = 1$. Its sub-interval is

$$I_2 = [l_2, r_2) = \left[l_1 + d_1 \sum_{k=1}^{i} p_{k-1}, l_1 + d_1 \sum_{k=1}^{i} p_k \right) = [0.5, 0.625).$$

From this we could get $d_2 = 0.125$. The binary numbers of left and right limits of the range are expressed as $L=0.5=0.100\cdots$(B) and $R = 0.101\cdots$(B). Abide by step (2), since $u_2=v_2=0$, send 0 and

u_3 is not equal to v_3, go to step (3).

Input the third symbol $x_i = a_3, i=3$. Its sub-interval is

$$I_3 = [l_3, r_3] = \left[l_2 + d_2 \sum_{k=1}^{i} p_{k-1}, l_2 + d_2 \sum_{k-1}^{i} p_k \right] = [0.59375, \ 0.609375).$$

Same as above, we know that $d_3 = 0.015625$. The left and right binary limits are $L= 0.59375 = 0.10011(B)$ and $R= 0.609375 = 0.100111(B)$. Abide by step (2), $u_3 = v_3 = 0$, $u_4 = v_4 = 1$, $u_5 = v_5 = 1$, and $u_6 \neq v_6$, then send out 011 and go to step (3).

......

The sent symbols are $10011\cdots$ The last encoded symbol is the end symbol.

For this same sample, the first digit the decode side receiver gets is 1. It's falling in the range of $[0.5, 1)$. But there are three possibilities in this range which are symbol a_2, a_3 and a_4. This means that the first received digit 1 doesn't have enough information to be decoded into one of the original symbols. After the receiver gets the second digit 0, the received number becomes 10 which is falling in the range of $[0.5, 0.75)$. Since both of the two digits fall in the interval of a_2, we could for certain know that the first original symbol is a_2. The whole decoding process is shown in Table 3-9.

Table 3-9　Decoding Process

Received digits	Intervals	Decoded Outputs
1	[0.5, 1)	...
0	[0.5, 0.75)	a_2
0	[0.5, 0.609375)	a_1
1	[0.5625, 0.609375)	...
1	[0.59375, 0.609375)	a_3
⋮	⋮	⋮

In the above example, we suppose both the encoder side and the decoder side know the length of the original message. So the decoder could stop the decoding process at certain point. In reality, there will be a terminate digit added at the decoder side, as soon as the decoder sees this digit, it will stop the decoding process.

We have another arithmetic coding example as follows.

[Example3.4] Suppose we have a source as $\{A, B, C, D\}$. The symbols' correspondent probabilities are $\{0.1, 0.4, 0.2, 0.3\}$. Asking what is the arithmetic codes of the input sequence CADACDB.

Solution: Based on the probabilities we have, divide the interval $[0, 1)$ into four sub-intervals, $[0, 0.1)$, $[0.1, 0.5)$, $[0.5, 0.7)$, $[0.7, 1)$. $[x, y)$ is half open interval which means that x is included inside the interval and y is not.

The symbols, the correspondent probabilities, and the initial encoding intervals are shown as follows:

Symbols	A	B	C	D
Probabilities	0.1	0.4	0.2	0.3
Initial intervals	[0,0.1)	[0.1,0.5)	[0.5,0.7)	[0.7,1.0)

The input sequence is CADACDB.

The encoding process is showing in Fig. 3-12.

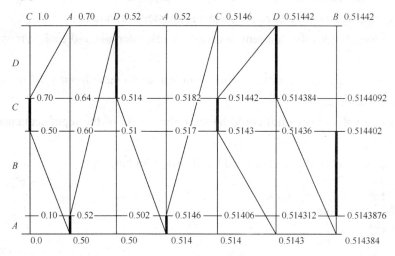

Fig. 3-12

The following two tables Table 3-10 and Table 3-11 give more detailed descriptions for the encoding and decoding process.

Compare the above encoding and decoding tables, there is not that much difference. The only thing they are not the same is that they have different computation sequences.

Table 3-10   Enioding precess

Steps	Input Symbols	Coding Intervals	Encoding Judgement
1	C	[0.5, 0.7)	The symbol's interval range [0.5, 0.7)
2	A	[0.5, 0.52)	The first 1/10 of interval [0.5, 0.7)
3	D	[0.514, 0.52)	The last three 1/10 of interval [0.5, 0.52)
4	A	[0.514, 0.5146)	The first 1/10 of interval [0.514, 0.52)
5	C	[0.5143, 0.51442)	Start from the fifth 1/10 of interval [0.514, 0.5146) and count two consecutive 1/10
6	D	[0.514384, 0.51442)	The last three 1/10 of interval [0.5143, 0.51442)
7	B	[0.5143836, 0.514402)	Start from the first 1/10 of interval [0.514384, 0.51442) and count four consecutive 1/10
8	Pick up one number from the interval [0.5143876, 0.514402) as the output: 0.5143876		

Table 3-11   Decoding precess

Steps	Intervals	Decoded symbols	Decoding Judgement
1	[0.5, 0.7)	C	0.51439 is within the interval [0.5, 0.7)
2	[0.5, 0.52)	A	0.51439 is within the first 1/10 of interval [0.5, 0.7)
3	[0.514, 0.52)	D	0.51439 is within the seventh 1/10 of interval [0.5, 0.52)
4	[0.514, 0.5146)	A	0.51439 is within the first 1/10 of interval [0.514, 052)
5	[0.5143, 0.51442)	C	0.51439 is within the fifth 1/10 of interval [0.514, 0.5146)
6	[0.514384, 0.51442)	D	0.51439 is within the seventh 1/10 of interval [0.5143, 0.51442)
7	[0.51439, 0.5143948)	B	0.51439 is within the first 1/10 of interval [0.51439, 0.5143948)
8	The final decoded message is: C A D A C D B		

The arithmetic coding could be concluded as follows.

1. The encoder divide the initial interval [L, H) as [0, 1).

2. To every input symbol, the encoder performs the steps (a) and (b).

(a). The encoder divides the "current interval" as sub-intervals and each symbol corresponds to one of them.

(b). The encoder chooses the interval corresponding to the next input symbol and makes it to be the new "current interval".

3. The last interval's lower limit could be the output code of the input sequence's arithmetic code, 0.5143876.

The showing process could be expressed by the program language.

```
set Low to 0
set High to 1
while there are input symbols do
    take a symbol
    Range = High - Low
    High = Low + Range * Range High
    Low  = Low  +  Range * Range Low
end of while
output Low
```

This example's decoding process is showing in the following Fig. 3-13.

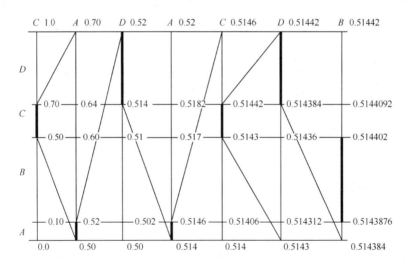

Fig. 3-13

Let's have a look at another example.

[Example 3.5] Suppose the original source message is "state_tree". What is the arithmetic code of this message?

The symbol(character) 's probabilities are showing as follows.

Symbols(Charaters)	Probabilities
t	0.3
s	0.1
r	0.1

e	0.3
a	0.1
_	0.1

Solution: Along "the probability line", set an initial interval with the interval [0, 1) for each symbol(character). It doesn't matter that the symbol's interval falls inside which part of the interval [0, 1) as soon as it's the same for both encoding and decoding sides. The following Fig. 3-14 is the allocation in this example. The table followed shows the allocation and the initial low and high limits of the ranges.

Symbols	_	a	e	r	s	t
Probabilities	0.1	0.1	0.3	0.1	0.1	0.3
Initial range	[0,0.1)	[0.1,0.2)	[0.2,0.5)	[0.5,0.6)	[0.6,0.7)	[0.7,1.0)
Range Low	0	0.1	0.2	0.5	0.6	0.7
Range High	0.1	0.2	0.5	0.6	0.7	1

Symbols(Characters)	Probabilities	Ranges
t	0.3	$0.7 \leqslant r < 1.0$
s	0.1	$0.6 \leqslant r < 0.7$
r	0.1	$0.5 \leqslant r < 0.6$
e	0.3	$0.2 \leqslant r < 0.5$
a	0.1	$0.1 \leqslant r < 0.2$
_	0.1	$0 \leqslant r < 0.1$

Fig. 3-14

The steps are as follows.

(1) At the beginning, the range of the interval is range = high-low = 1. The limits of next range are calculated as:

Low=low+range×range low

High=low+range×range high

The low at the right is the lower limit of last input symbol's encoding interval. The range low and range high are the low and high limits of current input symbol's interval.

(2) Encodes the first symbol "s" of the input message. Its range low = 0.6 and range high = 0.7. So the next low and high will be:

Low = low + range×range low = 0 + 1×0.6 = 0.6

High = low + range×range high = 0 + 1×0.7 = 0.7

Range = high-low = 0.7-0.6 = 0.1

s transfers the range from last [0, 1) to the new one [0.6, 0.7)as shown in Fig. 3-15.

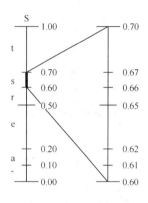

Fig. 3-15

The following figure shows the process of this transfer encoding process from the initial range to the new range at this step.

(3) Encodes the second symbol "t" of the input message. Last range at this stage is [0.6, 0.7). Since the t's range low = 0.7 and range high = 1, the next low and high could be calculated as:

Low = 0.6 + 0.1×0.7 = 0.67

High = 0.6 + 0.1×1.0 = 0.70

Range = 0.7 - 0.67 = 0.03

At this step, t transfers the range from [0.6, 0.7) to [0.67, 0.70)as shown in Fig. 3-16.

The process is as shown in the following continued figure.

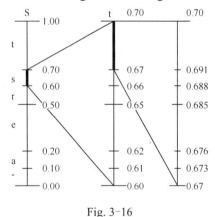

Fig. 3-16

(4) Encodes the third symbol "a" of the input message. Divides the transferred range [0.67, 0.70) based on the range low 0.10 and range high 0.20 of the input symbol "a". The next low and high will be as follows.

Low = 0.67 + 0.03×0.1 = 0.673

High = 0.67 + 0.03×0.2 = 0.676

Range = 0.676 - 0.673 = 0.003

"a" changed the range from [0.67, 0.70) to [0.673, 0.676), as show in Fig. 3-17.

(5) Encodes the fourth symbol "t" of the input message. Divides the range created in step (4) based on the "t" s initial interval limits, range low = 0.70 and range high = 1.00. The current low

and high will be calculated as follows.

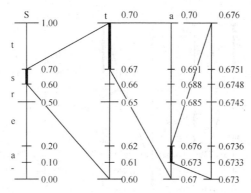

Fig. 3-17

Low = 0.673 + 0.003×0.7 = 0.6751

High = 0.673 + 0.003×1.0 = 0.676

Range = 0.676–0.6751 = 0.0009

This continued step is showing in Fig. 3-18.

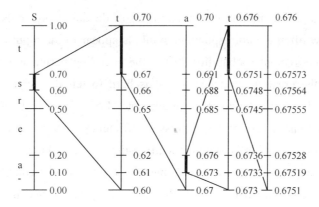

Fig. 3-18

Similarly, we could encode the rest symbols, "e", "_", "t", "r", "e", "e" in the input message. Their correspondent encoding ranges are: [0.67528, 0.67555), [0.67528, 0.675307), [0.675 298 9, 0.675 307), [0.675 302 95, 0.675 303 76), [0.675 303 112, 0.675 303 355), [0.675 303 160 6, 0.675 303 233 5).

Lastly, pick up one of any of the numbers within the final range [0.675 303 160 6, 0.675 303 233 5) as the encoding output. Here we take 0.675 303 160 6 as the encoding output.

The whole encoding process is showing in Fig. 3-19.

There are several issues list below should be mentioned in the arithmetic coding.

(1) Since the calculation precision within the computer is not unlimited, the encoding calculation overflow is obviously expected to see from time to time. Luckily, most computers have the precisions of 16, 32, or 64. The issue could be solved by using the ratio compress-expend method.

(2) The arithmetic encoder produce only one encode code to the whole input message which is one of the real numbers between [0, 1). So the decoder can't start the decoding until it receives

every digit of the whole encoded real number.

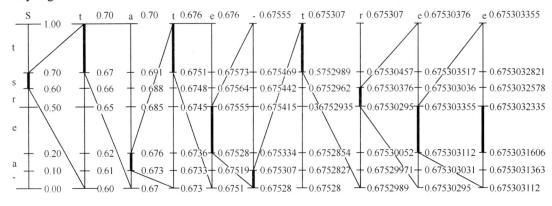

Fig. 3-19

(3) The arithmetic coding is very sensitive to the errors happened during encoding process. Any one digit error will results in the whole message being decode wrong.

The arithmetic coding could be either static or auto-adjustable. Within the static arithmetic calculation, the probabilities of the source symbols are fixed. On the other hand, within the auto-adjustable arithmetic calculation, the probabilities of the source symbols will be dynamically adjusted based on how often the correspondent symbols appear in the input message. The process estimating the dynamic symbol probability during the encoding period is called creating model. The reason to build the creating model is that it is hard to tell the correct probabilities of the symbols in advance. It's actually not that practicable. For this reason, you can't expect one arithmetic encoder could have very high compression efficiency when compress the given message. The most efficient thing we can do is to dynamically estimate the probabilities during the encoding process. So the dynamically creating model becomes the key of encoder compression efficiency.

The copyright issue is still exists in the arithmetic coding algorithm application.

3.3.3 RLE 编码

现实中有许多这样的图像，在一幅图像中具有许多颜色相同的图块。在这些图块中，许多行都具有相同的颜色，或者在一行中有许多连续的像素都具有相同的颜色值。在这种情况下就不需要存储每一个像素的颜色值，而仅仅存储一个像素的颜色值，以及具有相同颜色的像素数目就可以，或者存储一个像素的颜色值，以及具有相同颜色值的行数。这种压缩编码称为行程编码（run length encoding，RLE）。

为了叙述方便，假定有一幅灰度图像，第 n 行的像素值如图 3-20 所示。

图 3-20　RLE 编码的概念

用 RLE 编码方法得到的代码为：**8**0**3**1**50**8**4**180。代码中用黑体表示的数字是行程长度。

黑体字后面的数字代表像素的颜色值。例如，黑体字 50 代表有连续 50 个像素具有相同的颜色值，它的颜色值是 8。

对比 RLE 编码前后的代码数可以发现，在编码前要用 73 个代码表示这一行的数据，而编码后只要用 11 个代码表示代表原来的 73 个代码，压缩前后的数据量之比约为 7∶1，即压缩比为 7∶1。这说明 RLE 确实是一种压缩技术，而且这种编码技术相当直观，也非常经济。RLE 所能获得的压缩比有多大，主要取决于图像本身的特点。如果图像中具有相同颜色的图像块越大，图像块数目越少，获得的压缩比就越高。反之，压缩比就越小。

译码时按照与编码时采用的相同规则进行，还原后得到的数据与压缩前的数据完全相同。因此，RLE 是无损压缩技术。

RLE 压缩编码尤其适用于计算机生成的图像，对减少图像文件的存储空间非常有效。然而，RLE 对颜色丰富的自然图像就显得力不从心，在同一行上具有相同颜色的连续像素往往很少，而连续几行都具有相同颜色值的连续行数就更少。如果仍然使用 RLE 编码方法，不仅不能压缩图像数据，反而可能使原来的图像数据变得更大。请注意，这并不是说 RLE 编码方法不适用于自然图像的压缩，相反，在自然图像的压缩中还真少不了 RLE，只不过是不能单纯使用 RLE 的一种编码方法，需要和其他的压缩编码技术联合应用。

3.3.3　RLE Coding

There are many a kind of images in our real life. Many same color blocks exist in those kinds of images. Obviously, within those same color blocks, many lines have same color. In another words, many pixels within those lines have same color grey level values. In this situation, we don't need to save every pixel's color grey level value. Instead, we could only same one color grey level value for all of them and the quantity of total such kinds of pixels. Furthermore, we could only save one color grey value and the quantity of total lines with the same color grey value. Such kind of encoding compression algorithm is call Run Length Encoding, RLE.

To better understand the idea, let's have a look at an example. Suppose there is a grey levels image. The grey level values of the nth line pixels are shown in Fig. 3-20.

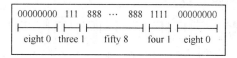

Fig. 3-20　RLE Coding Idea

The encoded code by RLE coding algorithm we could get is **80**3**15**0**84**1**80**. Within the code, the bold character represents the run length of the same grey level value. The digits follow the bold ones are the grey level values. For example, the bold digits 50 represent there are 50 same grey level value pixels together. The digit 8 followed represents that their grey level value is 8.

Compare the codes of before and after the RLE encoding, we can find that there are 73 digits to represent this line of data before the RLE encoding and there are only 11 digits to represent the same line of data after the RLE encoding. The compression ratio is roughly 7∶1. This means that the RLE algorithm is really a compression technique. This technique is easy to see and economically to apply. How big the compression ratio the RLE algorithm can get mainly depends on the structure and the characters the image has. Say the more the same color image block included in the image and the less the image block included in the image, the high the compression ratio gained and vice versa.

When decoding, use the same calculation method just in an opposite steps. The decoded message will be exact the same as the one before the encoding. So this RLE coding algorithm is lossless compression technique.

The RLE coding algorithm is especially good at computer created images. It's very good at reducing the image storage spaces. But it's not good anymore to deal with the colourful images. In those kinds of images, there is not much same color pixels in one line and even less same color pixels in several consecutive lines. If we still insist to use RLE algorithm in this situation, instead of getting compressed less quantity of data, we could get even bigger amount of data. One thing needs to mention though, this doesn't mean that the RLE algorithm can't be applied to the natural color images. Actually, it is widely used in the natural color image processing as long as it is corporately applied with other compression coding techniques.

3.3.4 词典编码

有许多场合，开始时不知道要编码数据的统计特性，也不一定允许事先知道它们的统计特性。因此，人们提出了许许多多的数据压缩方法，企图用来对这些数据进行压缩编码，在实际编码过程中以尽可能获得最大的压缩比。这些技术统称为通用编码技术。词典编码（Dictionary Encoding）技术就是属于这一类，这种技术属于无损压缩技术。

1．词典编码的思想

词典编码的根据是数据本身包含有重复代码这个特性。例如文本文件和光栅图像就具有这种特性。词典编码法的种类很多，归纳起来大致有两类。

第一类词典法编码的想法是企图查找正在压缩的字符序列是否在以前输入的数据中出现过，然后用已经出现过的字符串替代重复的部分，它的输出仅仅是指向早期出现过的字符串的"指针"。这种编码概念如图 3-21 所示。

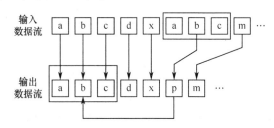

图 3-21　第一类词典法编码概念

这里所指的"词典"是指用以前处理过的数据来表示编码过程中遇到的重复部分。这类编码中的所有算法都是以 Abraham Lempel 和 Jakob Ziv 在 1977 年开发和发表的称为 LZ77 算法为基础的，例如 1982 年由 Storer 和 Szymanski 改进的称为 LZSS 算法就是属于这种情况。

第二类算法的想法是企图从输入的数据中创建一个"短语词典（Dictionary of the

Phrases)",这种短语不一定是像"严谨勤奋求实创新"和"国泰民安是坐稳总统宝座的根本"这类具有具体含义的短语,它可以是任意字符的组合。编码数据过程中当遇到已经在词典中出现的"短语"时,编码器就输出这个词典中的短语的"索引号",而不是短语本身。这个概念如图 3-22 所示。

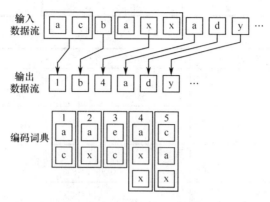

图 3-22 第二类词典法编码概念

J．Ziv 和 A．Lempel 在 1978 年首次发表了介绍这种编码方法的文章。在他们的研究基础上,TerryA．Wehch 在 1984 年发表了改进这种编码算法的文章,因此,把这种编码方法称为 LZW(Lempel—ZivWalch)压缩编码,首先在高速硬盘控制器上应用了这种算法。

2. LZ77 算法

为了更好地说明 LZ77 算法的原理,首先介绍算法中用到的几个术语。

(1)输入数据流(Input Stream):要被压缩的字符序列。

(2)字符(Character):输入数据流中的基本单元。

(3)编码位置(Coding Position):输入数据流中当前要编码的字符位置,指前向缓冲存储器中的开始字符。

(4)前向缓冲存储器(Look-ahead Buffer):存放从编码位置到输入数据流结束的字符序列的存储器。

(5)窗口(Window):指包含 W 个字符的窗口,字符是从编码位置开始向后数也就是最后处理的字符数。

(6)指针(Pointer):指向窗口中的匹配串且含长度的指针。

LZ77 编码算法的核心是查找从前向缓冲存储器开始的最长的匹配串。编码算法的具体执行步骤如下。

(1)把编码位置设置到输入数据流的开始位置。

(2)查找窗口中最长的匹配串。

(3)以"(Pointer,Length)Characters"的格式输出,其中 Pointer 是指向窗口中匹配串的指针,Length 表示匹配字符的长度,Characters 是前向缓冲存储器中的不匹配的第 1 个字符。

(4)如果前向缓冲存储器不是空的,则把编码位置和窗口向前移(Length + 1)个字符．然后返回到步骤(2)。

[例 3.6] 待编码的数据流如表 3-12 所列,编码过程如表 3-13 所列。

表 3-12　待编码的数据流

位置	字符	位置	字符	位置	字符
1	A	4	C	7	A
2	A	5	B	8	B
3	B	6	B	9	C

表 3-13　编码过程

步骤	位置	匹配串	字符	输出
1	1	—	A	(0, 0) A
2	2	A	B	(1, 1) B
3	4	—	C	(0, 0) C
4	5	B	B	(2, 1) B
5	7	A B	C	(5, 2) C

现作如下说明。

（1）"步骤"栏表示编码步骤。

（2）"位置"栏表示编码位置，输入数据流中的第 1 个字符为编码位置 1。

（3）"匹配串"栏表示窗口中找到的最长的匹配串。

（4）"字符"栏表示匹配之后在前向缓冲存储器中的第 1 个字符。

（5）"输出"栏以"（Back_chars，Chars_length）Explicit_character"格式输出。其中，（Back_chars，Chars_length）是指向匹配串的指针，告诉译码器"在这个窗口中向后退 Back_chars 个字符然后复制 Chars_length 个字符到输出"，Explicit_character 是真实字符。例如，表 3-10 中的输出"（5，2）C"告诉译码器回退 5 个字符，然后复制 2 个字符"AB"。

3.3.4　Dictionary Coding

In many cases, people don't know the static characters of the given data at the beginning and sometimes it's not allowed to know those characters of the given data. Under such situations, people proposed many compression calculation algorithms to compress the given data and gain as big as possible the data compression ratio. Those proposed techniques are called general encoding technique. The dictionary coding is one of them and it is a lossless data compression technique.

1. The basic idea of the dictionary coding

The basic idea of the dictionary coding is to take the advantage of data repeating characteristics in the given messages. The very common examples include the text files and fence images. There are many kinds of dictionary coding algorithms. But they could be classified into two types.

One type of the dictionary coding is trying to find out whether the current encoding symbols are the same as the ones have been inputted. If they have, the encoder will use the ones already received to replace the current symbols. Then its output will be the "pointer" pointed to the already received symbols instead of the current ones. This encoding idea is shown in Fig. 3-21.

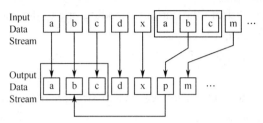

Fig. 3-21　The First Type Dictionary Coding Idea

The "dictionary" here means the symbols/characters have been received at an earlier time. All those kind of algorithms are based on the LZ77 algorithm proposed and developed by Abraham Lempel and Jakob Ziv in 1977. Such as LZSS algorithm improved by Storer and Szymanski in 1982 is one of them.

The other dictionary algorithm is trying to create a dictionary of the phrases from the input messages. Those kind of phrases don't like the normal phrases with full understandable meanings. They could be any combination of any characters. During the encoding process, as soon as the encoder sees the phrases contained in the dictionary, it will output the index pointed to this phrase instead of the phrase itself. This idea is drawing out in Fig. 3-22.

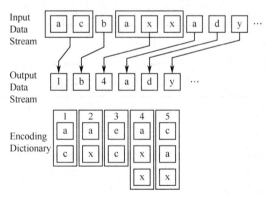

Fig. 3-22　The Second Type Dictionary Coding Idea

It were J. Ziv and A. Lempel who first proposed this coding algorithm in their published paper in 1978. Based on their research, Terry A. Wehch published his paper in 1984 and improved this algorithm. So the coding algorithm is called LZW(Lempel-ZivWalch) data compression algorithm.

2. LZ77 Algorithm

First of all, Let's introduce several particular phrases to better understand the principle of the LZ77 algorithm.

(1) Input Data Stream: The input symbols/characters sequence to be compressed.

(2) Character: The basic unit in the input data stream.

(3) Coding position: The position of current encoding symbol/character in the input data stream. It means the starting symbol inside the look-ahead buffer.

(4) Look-ahead buffer: The buffer used to store the data from the current encoding symbol/character to the end of the input data stream.

(5) Window: It means the window with the quantity of W symbols/characters inside. The symbols/characters are counted from the current encoding one to the end of the input data stream, which are the last processing symbols/characters.

(6) Pointer: Point to the match string within the window and include the length of the string as well.

The key of the LZ77 is to find the longest match string starting from the look-ahead buffer. The detailed steps are as follows.

(1) Set the encoding position to the start symbol/character of the input data stream.

(2) Search the longest match string within the window.

(3) In format "(Pointer, Length) character" output. Where Pointer is the pointer pointed to the match string within the window.

(4) If the look-ahead buffer is not empty, move the encoding position and widow forward Length+1 symbols/characters. Go back to step (2).

[Example 3. 6] The ready for encoding input data stream is shown in Table 3-12. The encoding process is shown in Table 3-13. The detailed explanation is as follows.

Table 3-12   Ready for Encoding Input Stream

Position	Synbols	Position	Synbols	Position	Synbols
1	A	4	C	7	A
2	A	5	B	8	B
3	B	6	B	9	C

Table 3-13   Encoding Process

Steps	Position	Match String	Symbols	Output
1	1	—	A	(0, 0) A
2	2	A	B	(1, 1) B
3	4	—	C	(0, 0) C
4	5	B	B	(2, 1) B
5	7	A B	C	(5, 2) C

(1) The "Step" column represents the encoding steps.

(2) The "Position" column represents the encoding position. The first symbol of the input data stream is position 1.

(3) The "Match String" column represents the longest match string found inside the window.

(4) The "Symbol" column represents the first symbol inside the look-ahead buffer after the

matching.

(5) The "Output" column has the output format of "(Back_chars ，Chars_length) Explicit_character". Where the (Back_chars，Chars_length) is the pointer pointed to the match string. It tells the decoder that backwards the quantity of Back_chars symbols and copy the quantity of Chars_length symbols to the output. The Explicit_character is a real symbol. For example, the output (5, 2) C tells the decoder that backwards 5 symbols and copy 2 symbols "AB".

3.3.5　LZW 算法

在介绍 LZW 算法之前，首先说明在 LZ78 算法中用到的几个术语和符号。

（1）字符流（Charstream）：要被编码的数据序列。

（2）字符（Character）：字符流中的基本数据单元。

（3）前缀（Prefix）：在一个字符之前的字符序列。

（4）缀—符串（String）：前缀+字符。

（5）码字（Code Word）：码字流中的基本数据单元，代表词典中的一串字符。

（6）码字流（Codestream）：码字和字符组成的序列，是编码器的输出。

（7）词典（Dictionary）：缀—符串表。按照词典中的索引号对每条缀—符串（String）指定一个码字（Code Word）。

（8）当前前缀（Current Prefix）：在编码算法中使用，指当前正在处理的前缀，用符号 P 表示。

（9）当前字符（Current Character）：在编码算法中使用，指当前前缀之后的字符，用符号 C 表示。

（10）当前码字（Current Code Word）：在译码算法中使用，指当前处理的码字，用 W 表示当前码字，String.W 表示当前码字的缀—符串。

在 LZW 算法中使用的术语与 LZ78 使用的相同，仅增加了一个术语——前缀根（root），它是由单个字符串组成的缀—符串。在编码原理上，LZW 与 LZ78 相比有如下差别。

① LZW 只输出代表词典中的缀—符串的码字。这就意味在开始时词典不能是空的，它必须包含可能在字符流出现中的所有单个字符，即前缀根。

② 由于所有可能出现的单个字符都事先包含在词典中。每个编码步骤开始时都使用一字符前缀（One-character Prefix），因此在词典中搜索的第 1 个缀—符串有两个字符。

现将 LZW 编码算法和译码算法介绍如下。

1. 编码算法

LZW 编码是围绕称为词典的转换表来完成的。这张转换表用来存放称为前缀的字符序列，并且为每个表项分配一个码字，或者称为序号，如表 3-14 所列。这张转换表实际上是把 8 位 ASCII 字符集进行扩充，增加的符号用来表示在文本或图像中出现的可变长度 ASCII 字符串。扩充后的代码可用 9 位、10 位，11 位、12 位甚至更多的位来表示。Welch 的论文中用了 12 位，12 位可以有 4096 个不同的 12 位代码. 这就是说，转换表有 4096 个表项，其中 256 个表项用来存放已定义的字符，剩下 3840 个表项用来存放前缀。

表 3-14　词典

码字	前缀	码字	前缀	码字	前缀
1	—	194	B	⋮	⋮
⋮	⋮	⋮	⋮	1305	a bcdefxy F01234
193	A	255	—	⋮	⋮

　　LZW 编码器（软件编码器或硬件编码器）就是通过管理这个词典完成输入与输出之间的转换。LZW 编码器的输入是字符流．字符流可以是用 8 位 ASCII 字符组成的字符串。而输出是用 n 位（例如 12 位）表示的码字流，码字代表单个字符或多个字符组成的字符串。

　　LZW 编码器使用了一种很实用的分析（Parsing）算法，称为贪婪分析算法（Greedy Parsing Algorithm）。在贪婪分析算法中．每一次分析都要串行地检查来自字符流的字符串，从中分解出已经识别的最长的字符串，也就是已经在词典中出现的最长的前缀。用已知的前缀加上下一个输入字符 C 也就是当前字符作为该前缀的扩展字符，形成新的扩展字符串——缀—符串：Prefix.C。这个新的缀—符串是否要加到词典中，还要看词典中是否存有和它相同的缀—符串 String。如果有，那么这个缀—符串就变成前缀．继续输入新的字符，否则就把这个缀—符串写到词典中生成一个新的前缀，并给一个代码。LZW 编码算法的具体执行步骤如下：

　　步骤 1：开始时的词典包含所有可能的根，而当前前缀 P 是空的；

　　步骤 2：当前字符（C）：= 字符流中的下一个字符；

　　步骤 3：判断缀—符串 P + C 是否在词典中

　　（A）如果"是"：P = P + C //（用 C 扩展 P）；

　　（B）如果"否"

　　　　a. 把代表当前前缀 P 的码字输出到码字流；

　　　　b. 把缀—符串 P + C 添加到词典；

　　　　c. 令 P：＝C //（现在的 P 仅包含一个字符 C）；

　　步骤 4：判断码字流中是否还有码字要译

　　（A）如果"是"，就返回到步骤 2；

　　（B）如果"否"

　　　　a. 把代表当前前缀 P 的码字输出到码字流；

　　　　b. 结束。

　　LZW 编码算法可用伪码表示。开始时假设编码词典包含若干个已经定义的单个码字，例如，256 个字符的码字，用伪码可以表示成：

```
Dictionary[j]←all n single-character, j=1, 2, …, n
j←n+1
Prefix←read first Character in Charstream
while((C←next Character)!=NUI.I.)
Begin
    If Prefix. C is in Dictionary
        Prefix←Prefix. C
    else
        Codestream←cW for Prefix
        Dictionary[j]←Prefix.C
        j←n+1
        Prefix←C
    end
Codestream←cW for Prefix
```

展开来说，如有数据"ACDBBCDBACCCBADBACCBA"

数据中只包含四种字符，即 A、B、C、D，这样可以分别用四个数字来代表 0、1、2、3，则原始数据就可表示为 023112310 2222103102210，共 22 位。

如果再定义 4 代表 BA、5 代表 CC、6 代表 CD，则压缩后的数据为：0611645543454，共 13 位，这样就比原来的少了 9 位！

[例 3.7]　有一个由字母 a、b、c 组成的输入字符流 abacaba，求其 LZW 编码。

解：1．初始化：#0=a,#1=b,#2=c。现在开始读取第一个字符 a，[.P.]a=a，可以在编译表中找到。不做任何事继续读取第二个字符 b，[.P.]b=ab。

0	1	2	3	4	5	6	7
a	b	c					

2．在编译表中不能找到，添加 ab 到编译表，同时输出第一个字符 a(0)，修改[.P.]=b。

0	1	2	3	4	5	6	7
a	b	c	ab				

输入码流：abacaba
输出码流：0

3．再读取下一个字符 a，[.P.]a=ba。

在编译表中不能找到，添加 ba 到编译表，同时输出第二个字符 b(1)，修改[.P.]=a。

0	1	2	3	4	5	6	7
a	b	c	ab	ba			

输入码流：abacaba
输出码流：01

4．再读取下一个字符 c，[.P.]c=ac。

在编译表中不能找到，添加 ac 到编译表，同时输出第三个字符 a(0)，修改[.P.]=c。

0	1	2	3	4	5	6	7
a	b	c	ab	ba	ac		

输入码流：abacaba
输出码流：010

5．再读取下一个字符 a，[.P.]a=ca。

在编译表中不能找到，添加 ca 到编译表，同时输出第四个字符 c(2)，修改[.P.]=a。

0	1	2	3	4	5	6	7
a	b	c	ab	ba	ac	ca	

输入码流：abacaba
输出码流：0102

6．再读取下一个字符 b，[.P.]b=ab。

在编译表找到 ab=3，修改[.P.]=ab，读取最后一个字符 a，[.P.]a=aba，在编译表中不能找到，添加 aba 到编译表，输出第五个字符 ab(3) ，同时修改[.P.]=a。

0	1	2	3	4	5	6	7
a	b	c	ab	ba	ac	ca	aba

输入码流：abacaba
输出码流：01023

7. 现在没有数据了，输出[.P.]的值 a（0）到编码流。

输入码流：abacaba
输出码流：01023 0

[例 3.8]　求字符串 ababcbaefbabaaa 的 LZW 编码。

解：1. 初始化：#0=a,#1=b,#2=c,#3=e,#4=f。

0	1	2	3	4	5	6	7	8	9	A	B	C	D	E
a	b	c	e	f										

现在开始读取第一个字符 a，[.P.]a=a，可以在编译表中找到。不做任何事继续读取第二个字符 b，[.P.]b=ab。

2. 在编译表中不能找到，那么添加 ab 到编译表，同时输出第一个字符 a(0)，修改[.P.]=b。

0	1	2	3	4	5	6	7	8	9	A	B	C	D	E
a	b	c	e	f	ab									

输入码流：ababcbaefbabaaa
输出码流：0

3. 再读取下一个字符 a，[.P.]a=ba。

在编译表中不能找到，那么添加 ba 到编译表，同时输出第二个字符 b(1)，修改[.P.]=a。

0	1	2	3	4	5	6	7	8	9	A	B	C	D	E
a	b	c	e	f	ab	ba								

输入码流：ababcbaefbabaaa
输出码流：01

4. 再读取下一个字符 b，[.P.]b=ab。

在编译表能找到 ab=5，修改[.P.]=ab，再读取下一个字符 c，[.P.]c=abc，在编译表中不能找到。添加 abc 到编译表，同时输出第三个字符 ab(5)，修改[.P.]=c。

0	1	2	3	4	5	6	7	8	9	A	B	C	D	E
a	b	c	e	f	ab	ba	abc							

输入码流：ababcbaefbabaaa
输出码流：015

5. 再读取下一个字符 b，[.P.]b=cb。

在编译表中不能找到，那么添加 cb 到编译表，同时输出第四个字符 c(2)，修改[.P.]=b。

0	1	2	3	4	5	6	7	8	9	A	B	C	D	E
a	b	c	e	f	ab	ba	abc	cb						

输入码流：ababcbaefbabaaa
输出码流：0152

6. 再读取下一个字符 a，[.P.]a=ba。在编译表找到 ba=6，修改[.P.]=ba，读取下一个字符 e，[.P.]b=bae，在编译表中不能找到，添加 bae 到编译表，修改[.P.]=e，同时输出第五个字符 ba(6)。

0	1	2	3	4	5	6	7	8	9	A	B	C	D	E
a	b	c	e	f	ab	ba	abc	cb	bae					

输入码流：ababcbaefbabaaa
输出码流：01526

7. 再读取下一个字符 f，[.P.]f=ef。在编译表中不能找到，添加 ef 到编译表，修改[.P.]=f，同时输出第六个字符 e(3)。

0	1	2	3	4	5	6	7	8	9	A	B	C	D	E
a	b	c	e	f	ab	ba	abc	cb	bae	ef				

输入码流：ababcbaefbabaaa
输出码流：015263

8. 再读取下一个字符 b，[.P.]b=fb。在编译表不能找到，添加 fb 到编译表，修改[.P.]=b，同时输出第七个字符 f(4)。

0	1	2	3	4	5	6	7	8	9	A	B	C	D	E
A	b	c	e	f	ab	ba	abc	cb	bae	ef	fb			

输入码流：ababcbaefbabaaa
输出码流：0152634

9. 再读取下一个字符 a，[.P.]a=ba。在编译表找到 ba=6，修改[.P.]=ba，读取下一个字符 b，[.P.]a=bab，在编译表中不能找到，添加 bab 到编译表，修改[.P.]=b，同时输出第八个字符 ba(6)。

0	1	2	3	4	5	6	7	8	9	A	B	C	D	E
a	b	c	e	f	ab	ba	abc	cb	bae	ef	fb	bab		

输入码流：ababcbaefbabaaa
输出码流：01526346

10．再读取下一个字符 a，[.P.]a=ba。在编译表找到 ba=6，修改[.P.]=ba，读取下一个字符 a，[.P.]a=baa，在编译表中不能找到，添加 baa 到编译表，修改[.P.]=a，同时输出第九个字符 ba(6)。

0	1	2	3	4	5	6	7	8	9	A	B	C	D	E
a	b	c	e	f	ab	ba	abc	cb	bae	ef	fb	bab	baa	

输入码流：ababcbaefbabaaa
输出码流：015263466

11．再读取下一个字符 a，[.P.]a=aa。在编译表中不能找到，添加 aa 到编译表，修改[.P.]=a，同时输出第十个字符 a(0)。

0	1	2	3	4	5	6	7	8	9	A	B	C	D	E
a	b	c	e	f	ab	ba	abc	cb	bae	ef	fb	bab	baa	aa

输入码流：ababcbaefbabaaa
输出码流：0152634660

12．输入数据结束，输出[.P.]的值 a（0）到编码流 。

输入码流：ababcbaefbabaaa
输出码流：01526346600

上面的例子中，编码都是从 0 开始的，其实从 1 开始原理是一样的。下面的例子说明了这一点。

[例 3.9] 输入字符流 ABABBABCABABBA，求其 LZW 编码。

解：1．初始化：#1=A,#2=B,#3=C。

现在开始读取第一个字符 A，[.P.] +A=A，可以在编译表中找到。不做任何事继续读取第二个字符 B，[.P.]+B=AB。

1	2	3	4	5	6	7	8	9	A	B
A	B	C								

2．在编译表中不能找到，添加 AB 到编译表，同时输出第一个字符 A(1)，修改[.P.]=B。

1	2	3	4	5	6	7	8	9	A	B
A	B	C	AB							

输入码流：ABABBABCABABBA
输出码流：1

3．再读取下一个字符 A，[.P.]+A=BA。

在编译表中不能找到，添加 BA 到编译表，同时输出第二个字符 B(2)，修改[.P.]=A。

1	2	3	4	5	6	7	8	9	A	B
A	B	C	AB	BA						

128

输入码流：ABABBABCABABBA
输出码流：12

4. 再读取下一个字符 B，[.P.]+B=AB。在编译表中能找到 AB=4，修改[.P.]=AB，再读取下一个字符 B，[.P.]+B=ABB，在编译表中不能找到，添加[.P.] =ABB 到编译表，同时输出第三个字符 AB(4)，修改[.P.]=B。

1	2	3	4	5	6	7	8	9	A	B
A	B	C	AB	BA	ABB					

输入码流：ABABBABCABABBA
输出码流：124

5. 再读取下一个字符 A，[.P.]+A=BA。在编译表能找到 BA=5，修改[.P.]=BA，再读取下一个字符 B，[.P.]+B=BAB，在编译表中不能找到，添加[.P.] =BAB 到编译表，同时输出第四个字符 BA(5)，修改[.P.]=B。

1	2	3	4	5	6	7	8	9	A	B
A	B	C	AB	BA	ABB	BAB				

输入码流：ABABBABCABABBA
输出码流：1245

6. 再读取下一个字符 C，[.P.]+C=BC。在编译表中找不到，添加[.P.] =BC 到编译表，同时输出第五个字符 B (2)，修改[.P.]=C。

1	2	3	4	5	6	7	8	9	A	B
A	B	C	AB	BA	ABB	BAB	BC			

输入码流：ABABBABCABABBA
输出码流：12452

7. 再读取下一个字符 A，[.P.]+A=CA。在编译表中找不到，添加[.P.] =CA 到编译表，同时输出第六个字符 C(3)，修改[.P.]=A。

1	2	3	4	5	6	7	8	9	A	B
A	B	C	AB	BA	ABB	BAB	BC	CA		

输入码流：ABABBABCABABBA
输出码流：124523

8. 再读取下一个字符 B，[.P.]+B=AB。在编译表能找到 AB=4，修改[.P.]=AB，再读取下一个字符 A，[.P.]+A=ABA，在编译表中不能找到，添加[.P.] =ABA 到编译表，同时输出第七个字符 AB(4)，修改[.P.]=A。

1	2	3	4	5	6	7	8	9	A	B
A	B	C	AB	BA	ABB	BAB	BC	CA	ABA	

输入码流：	ABABBABCABABBA
输出码流：	1245234

9. 再读取下一个字符 B，[.P.]+B=AB。在编译表能找到 AB=4，修改[.P.]=AB，再读取下一个字符 B，[.P.]+B=ABB，在编译表中能找到 ABB=6，修改[.P.]=ABB，再读取下一个字符 A，[.P.]+A=ABBA，在编译表中不能找到、添加[.P.] =ABBA 到编译表，同时输出第八个字符 ABB(6)，修改[.P.]=A。

1	2	3	4	5	6	7	8	9	A	B
A	B	C	AB	BA	ABB	BAB	BC	CA	ABA	ABBA

输入码流：	ABABBABCABABBA
输出码流：	12452346

10. 现在没有数据了，输出[.P.]的值 A（1）到编码流。

输入码流：	ABABBABCABABBA
最终结果输出码流：	124523461

再给出以下 LZW 编码算法步骤：

```
BEGIN
    s = 下一个要输入字符;
    while not EOF
    {
    c = 下一个要输入字符;
    If  s + c 存在于字典中;
        s = s + c;
    Else
        {
        输出对于 s 的编码;
        添加字符串 s + c 到字典中，并用新的编码符号标记;
        s = c;
        }
    }
    输出对于 s 的编码;
    END
```

2．译码算法

LZW 译码算法中还用到另外两个术语。

（1）当前码字（Current Code Word）：指当前正在处理的码字，用 cW 表示，用 strmg.cW 表示当前缀一符串；

（2）先前码字（Previous Code Word）：指先于当前码字的码字，用 pW 表示，用 string.pW 表示先前缀一符串。

LZW 译码算法开始时，译码词典与编码词典相同，它包含所有可能的前缀根。LZW 算法在译码过程中会记住先前码字（pW），从码字流中读当前码字（cW）之后输出当前缀一符串 string.cW。然后把用 string.cW 的第一个字符扩展的先前缀一符串 string.pW 添加到词典中。

LZW 译码算法的具体执行步骤如下：

步骤 1：在开始译码时词典包含所有可能的前缀根。

步骤 2：cW：＝码字流中的第一个码字。

步骤 3：输出当前前缀—符串 string.cW 到码字流。

步骤 4：先前码字 pW:＝当前码字 cW。

步骤 5：当前码字 cW:＝码字流中的下一个码字。

步骤 6：判断先前缀—符串 string.pW 是否在词典中

  （1）如果"是"，则

    a. 把先前缀—符串 string.pW 输出到字符流。

    b. 当前前缀 P:＝先前缀—符串 string.pW。

    c. 当前字符 C:＝当前前缀—符串 string.cW 的第一个字符。

    d. 把缀—符串 P＋C 添加到词典。

  （2）如果"否"，则

    a. 当前前缀 P: ＝先前缀—符串 string.pW。

    b. 当前字符 C: ＝当前缀—符串 string.cW 的第一个字符。

    c. 输出缀—符串 P＋C 到字符流，然后把它添加到词典中。

步骤 7： 判断码字流中是否还有码字要译

（1）如果"是"，就返回到步骤 4。

（2）如果"否"，结束。

LZW 译码算法可用伪码表示如下：

```
Dictionary[j]←all n single-character, j=1, 2, …, n
j←n+1
cW←first code from Codestream
Charstream←Dictionary[cW]
pW←cW
While((cW←next Code word)!=NULL)
    Begin
        If cW is in Dictionary
        Charstream←Dictionary[cW]
        Prefix←Dictionary[pW]
        cW←first Character of Dictionary[cW]
        Dictionary[j]←Prefix. cW
        j←n+1
    pW←cW
else
    Prefix←Dictionary[pW]
    cW←first Character of Prefix
    charstream←prefix.cW
    Dictionary[j]←Prefix. C
    pW←cW
    j←n+1
    end
```

[例 3.10] 编码字符串如表 3-15 所示，编码过程如表 3-16 所示，现说明如下。

（1）"步骤"栏表示编码步骤；

表 3-15 被编码的字符串

位置	字符	位置	字符	位置	字符
1	A	4	A	7	B
2	B	5	B	8	A
3	B	6	A	9	C

表 3-16 LZW 的编码过程

步骤	位置	词典		输出
		（1）	A	
		（2）	B	
		（3）	C	
1	1	（4）	AB	（1）
2	2	（5）	BB	（2）
3	3	（6）	BA	（2）
4	4	（7）	ABA	（4）
5	6	（8）	ABAC	（7）
6				（3）

（2）"位置"栏表示在输入数据中的当前位置；

（3）"词典"栏表示添加到词典中的缀—符串，它的索引在括号中；

（4）"输出"栏表示码字输出。

表 3-17 解释了译码过程。每个译码步骤译码器读一个码字，输出相应的缀—符串。并把它添加到词典中。例如，在步骤 4 中，先前码字（2）存储在先前码字（pW）中，当前码字（cW）是（4），当前缀—符串 string.cW 是输出（"AB"），先前缀—符串 string.pW（"B"）是用当前缀—符串 string.cW（"A"）的第一个字符，其结果（"BA"）添加到词典中，它的索引号是（6）。

表 3-17 LZW 的译码过程

步骤	代码	词典		输出
		（1）	A	
		（2）	B	
		（3）	C	
1	（1）	—	—	A
2	（2）	（4）	AB	B
3	（2）	（5）	BB	B
4	（4）	（6）	BA	A B
5	（7）	（7）	ABA	A B A
6	（3）	（8）	ABAC	C

LZW 算法得到普遍采用，它的速度比使用 LZ77 算法的速度快，因为它不需要执行那么多的缀—符串比较操作。对 LZW 算法进一步的改进是增加可变的码字长度，以及在词典中删除老的缀—符串。在 GIF 图像格式和 UNIX 的压缩程序中已经采用了这些改进措施之后的

LZW 算法。

LZW 算法取得了专利，专利权的所有者是美国的一个大型计算机公司 Unisys（优利系统公司）。除厂商业软件生产公司之外，可以免费使用 LZW 算法。

3.3.5 LZW Coding Algorithm

Before introducing the LZW algorithm, let's have a look at some terminologies and signs used in LZ88 algorithm.

(1) Charstream: The character stream waiting for the encoding

(2) Character: The basic unit inside the Charstream.

(3) Prefix: The character string before a particular character.

(4) String: The prefix plus character.

(5) Code word: The basic message unit within the input character stream and represents a string in the dictionary.

(6) Codestream: The encoder output sequence consists of encoded codes and signs.

(7) Dictionary: The string table. Based on the dictionary index, the encoder finds one code word for each string.

(8) Current prefix: Used in the encoding algorithm. It means the currently processing prefix and represented as P.

(9) Current character: Used in the encoding algorithm. It means the character after the current prefix and represented as C.

(10) Current code word: Used in the decoding algorithm. It means the currently processing code word. The W is used to represent the current code word and String.W used to represent the string of current code word.

The terminologies used in LZW alogorithm is the same as in LZ78 except one added terminology, root. It is the string consisted of single characters.

Comparing LZW and LZ78 algorithms, LZW algorithm has the following differences.

① LZW only outputs the code word of the string in the dictionary. This means that the dictionary can't be empty even at the very beginning. It must include every character appearing in the input charstream, which is the root.

② Since every single character possibly appearing in the input charstrem is included in the dictionary and one one-character-prefix will be used at the beginning of each encoding step, the first string searched in the dictionary will contain two characters.

The following are the descriptions of LZW encoding and decoding algorithms.

1. The encoding algorithm

LZW encoding is finished by so called dictionary transform. There is a transform form that contains the character sequence which is called prefix used to finish the transform. It assigns a code word which is also called sequence number for every item inside the form as well. This is shown in Table 3-14.

Table 3-14 Dictionary

Code Word	Prefix	Code Word	Prefix	Code Word	Prefix
1		194	B	⋮	⋮
⋮	⋮	⋮	⋮	1305	a bcdefxy F01234
193	A	255		⋮	⋮

This form actually extends the 8 digits ASCII sign collection. The added signs are used to represent variable ASCII strings appeared in the texts or images. The extended code words could be represented by among 9, 10, 11, 12 or even more digits code. Welch used 12 digits in his paper. The 12 digits could give 4096 different 12 digit codes. In this case, the transform form has 4096 items. The 256 items of them are used to save the already defined signs and the other 3840 items are used to save the prefixes.

LZW encoder (either software or hardware encoder) finishes the input and output transform by managing this dictionary. The input of the encoder is charstream which could be character string consisted of 8 digits ASCII codes. The output of the encoder is the code word string represented in n digits (such as 12 digits). The code words mean the character string consisted of either single or multiple characters.

LZW encoder applied a very practical parsing algorithm which is called greedy parsing algorithm. In this algorithm, it serially checks the character string within the charstream in each analyzing, from which, it finds out the identified longest character string, which is the longest prefix appeared in the dictionary. Add next input character C to the already known prefix will be this prefix's extended character which is the current character. This forms the new extended character string -- string: Prefix.C. Whether add this new string to the dictionary depends on whether there is a same string already in the dictionary. If there is, this new string will become a new prefix and continue to input the next character. Otherwise, write this string to the dictionary, create a new prefix, and assign a new code to it.

The detailed calculation steps of LZW algorithm are as follows.

1. The dictionary includes all the possible roots at the beginning. The current prefix P is empty.

2. Current character (C): = The next character in the Charstream.

3. Judge whether the string P + C is in the dictionary.

    (A) If yes, P = P + C　// (extend P with C)

    (B) If no,

        a. Output the code word representing the current prefix to the Codestream.

        b. Add the string P + C to the dictionary.

        c. Let P: = C　//(The P only contains one character C now.)

4. Judge whether there is still characters in the Charstream need to be encoded.

    (A) If yes, go back to 2;

    (B) If no

        a. Output the code word representing the current prefix P;

        b. Finish.

LZW coding could be described by pseudo code. Suppose there have been some defined single code word in the dictionary at the beginning. For example, there have been 256 characters' code words in the dictionary. The pseudo code could be written as.

```
Dictionary[j]←all n single-character, j=1, 2, ···, n
j←n+1
Prefix←read first Character in Charstream
while((C←next Character)!=NUI.I.)
Begin
     If Prefix. C is in Dictionary
          Prefix←Prefix. C
     else
          Codestream←cW for Prefix
          Dictionary[k]←Prefix.C
          j←n+1
          Prefix←C
     end
Codestream←cW for Prefix
```

Suppose we have data ACDBBCDBACCCCBADBACCBA.

There are only 4 different characters, A, B, C, D in the stream. They could be represented as 0, 1, 2, 3. The original data could be represented as 023112310 2222103102210. Total there are 22 digits.

If we further define 4 represent BA, 5 represent CC, and 6 represent CD, the data after the compression encoding will be 0611645543454 which have total 13 digits, 9 digits less than the original one.

Let's have a look at a couple of examples to understand the calculation process.

[Example 3.7] The input charstream is "abacaba". There are three characters in the charstream. What's charstream's LZW cord word?

Solution: 1. The charstream has three roots, a, b, and c. Put the three roots into the dictionary. Assign the code word 0 to a, 1 to b, and 2 to c. The result is as follows.

0	1	2	3	4	5	6	7
a	b	c					

Current character is (C): = a.

The string P + C = a is in the dictionary. Then P = P + C = a.

2. Read the next character in the charstream and get the current character (C): = b.

The string P + C = ab.

The string is not in the dictionary.

a. Output the cord word 0 of the current prefix a.

b. Add the P + C = ab to the dictionary.

c. Let the P = C = b.

The dictionary and the output stream at this step are as follows.

0	1	2	3	4	5	6	7
a	b	C	ab				

Input Charstream: abacaba	
Output Codestream: 0	

3. Read next character a in the charstream and get (C) = a.

The string P + C = ba is not in the dictionary.

a. Output the cord word 1 of the current prefix b.

b. Add the P + C = ba to the dictionary.

c. Let the P = C = a.

The dictionary and the output stream at this step are as follows.

0	1	2	3	4	5	6	7
a	b	c	ab	ba			

Input Charstream: abacaba	
Output Codestream: 01	

4. Read next character c in the charstream and get (C) = c.

The string P + C = ac is not in the dictionary.

a. Output the cord word 0 of the current prefix a.

b. Add the P + C = ac to the dictionary.

c. Let the P = C = c.

The dictionary and the output stream at this step are as follows.

0	1	2	3	4	5	6	7
a	b	c	ab	ba	ac		

Input Charstream: abacaba	
Output Codestream: 010	

5. Read next character a in the charstream and get (C) = a.

The string P + C = ca is not in the dictionary.

a. Output the cord word 2 of the current prefix c.

b. Add the P + C = ca to the dictionary.

c. Let the P = C = a.

The dictionary and the output stream at this step are as follows.

0	1	2	3	4	5	6	7
a	b	c	ab	ba	ac	ca	

Input Charstream: abacaba	
Output Codestream: 0102	

6. Read next character b in the charstream and get (C) = b.

The string P + C = ab is in the dictionary. Then P = P + C = ab.

7. Read next character a in the charstream and get (C) = a.

The string P + C = aba is not in the dictionary.

a. Output the cord word 3 of the current prefix ab.

b. Add the P + C = aba to the dictionary.

c. Let the P = C = a.

The dictionary and the output stream at this step are as follows.

0	1	2	3	4	5	6	7
a	b	c	ab	ba	ac	ca	aba

Input Charstream: abacaba
Output Codestream: 01023

1. There is no more new character in the input charstream now.

a. Output the code word 0 representing the current prefix P = a.

b. Finish.

The final result is as follows.

Output Codestream: 01023 0

[Example 3.8]　What is the LZW code of the input charstream ababcbaefbabaaa?

Solution: 1. The charstream has five roots, a, b, c, e, and f. Put the five roots into the dictionary. Assign the code word 0 to a, 1 to b, 2 to c, 3 to e, and 4 to f. The result is as follows.

0	1	2	3	4	5	6	7	8	9	A	B	C	D	E
a	b	c	e	f										

Current character is (C): = a.

The string P + C = a is in the dictionary. Then P = P + C = a.

2. Read the next character in the charstream and get the current character (C): = b.

The string P + C = ab.

The string is not in the dictionary.

a. Output the cord word 0 of the current prefix a.

b. Add the P + C = ab to the dictionary.

c. Let the P = C = b.

The dictionary and the output stream at this step are as follows.

0	1	2	3	4	5	6	7	8	9	A	B	C	D	E
a	b	c	e	f	ab									

Input Charstream: ababcbaefbabaaa
Output Codestream: 0

3. Read next character a in the charstream and get (C) = a.

The string P + C = ba is not in the dictionary.

137

a. Output the cord word 1 of the current prefix b.

b. Add the P + C = ba to the dictionary.

c. Let the P = C = a.

The dictionary and the output stream at this step are as follows.

0	1	2	3	4	5	6	7	8	9	A	B	C	D	E
a	b	c	e	f	ab	ba								

Input Charstream: ababcbaefbabaaa
Output Codestream: 01

4. Read next character b in the charstream and get (C) = b.

The string P + C = ab is in the dictionary. Then P = P + C = ab.

Read next character a in the charstream and get (C) = c.

The string P + C = abc is not in the dictionary.

a. Output the cord word 5 of the current prefix ab.

b. Add the P + C = abc to the dictionary.

c. Let the P = C = c.

The dictionary and the output stream at this step are as follows.

0	1	2	3	4	5	6	7	8	9	A	B	C	D	E
a	b	c	e	f	ab	ba	abc							

Input Charstream: ababcbaefbabaaa
Output Codestream: 015

5. Read next character b in the charstream and get (C) = b.

The string P + C = cb is not in the dictionary.

a. Output the cord word 2 of the current prefix c.

b. Add the P + C = cb to the dictionary.

c. Let the P = C = b.

The dictionary and the output stream at this step are as follows.

0	1	2	3	4	5	6	7	8	9	A	B	C	D	E
a	b	c	e	f	ab	ba	abc	cb						

Input Charstream: ababcbaefbabaaa
Output Codestream: 0152

6. Read next character a in the charstream and get (C) = a.

The string P + C = ba is in the dictionary. Then P = P + C = ba.

Read next character e in the charstream and get (C) = e.

The string P + C = bae is not in the dictionary.

a. Output the cord word 6 of the current prefix ba.

b. Add the P + C = bae to the dictionary.

c. Let the P = C = e.

The dictionary and the output stream at this step are as follows.

0	1	2	3	4	5	6	7	8	9	A	B	C	D	E
a	b	c	e	f	ab	ba	abc	cb	bae					

Input Charstream: ababcbaefbabaaa
Output Codestream: 01526

7. Read next character f in the charstream and get (C) = f.

The string P + C = ef is not in the dictionary.

a. Output the cord word 3 of the current prefix e.

b. Add the P + C = ef to the dictionary.

c. Let the P = C = f.

The dictionary and the output stream at this step are as follows.

0	1	2	3	4	5	6	7	8	9	A	B	C	D	E
a	b	c	e	f	ab	ba	abc	cb	bae	ef				

Input Charstream: ababcbaefbabaaa
Output Codestream: 015263

8. Read next character b in the charstream and get (C) = b.

The string P + C = fb is not in the dictionary.

a. Output the cord word 4 of the current prefix f.

b. Add the P + C = fb to the dictionary.

c. Let the P = C = b.

The dictionary and the output stream at this step are as follows.

0	1	2	3	4	5	6	7	8	9	A	B	C	D	E
a	b	c	e	f	ab	ba	abc	cb	bae	ef	fb			

Input Charstream: ababcbaefbabaaa
Output Codestream: 0152634

9. Read next character a in the charstream and get (C) = a.

The string P + C = ba is in the dictionary. Then P = P + C = ba.

Read next character b in the charstream and get (C) = b.

The string P + C = bab is not in the dictionary.

a. Output the cord word 6 of the current prefix ba.

b. Add the P + C = bab to the dictionary.

c. Let the P = C = b.

The dictionary and the output stream at this step are as follows.

0	1	2	3	4	5	6	7	8	9	A	B	C	D	E
a	b	c	e	f	ab	ba	abc	cb	bae	ef	fb	bab		

Input Charstream: ababcbaefbabaaa
Output Codestream: 01526346

10. Read next character a in the charstream and get (C) = a.

The string P + C = ba is in the dictionary. Then P = P + C = ba.

Read next character a in the charstream and get (C) = a.

The string P + C = baa is not in the dictionary.

a. Output the cord word 6 of the current prefix ba.

b. Add the P + C = baa to the dictionary.

c. Let the P = C = a.

The dictionary and the output stream at this step are as follows.

0	1	2	3	4	5	6	7	8	9	A	B	C	D	E
a	b	c	e	f	ab	ba	abc	cb	bae	ef	fb	bab	baa	

Input Charstream: ababcbaefbabaaa
Output Codestream: 015263466

11. Read next character a in the charstream and get (C) = a.

The string P + C = aa is not in the dictionary.

a. Output the cord word 0 of the current prefix a.

b. Add the P + C = aa to the dictionary.

c. Let the P = C = a.

The dictionary and the output stream at this step are as follows.

0	1	2	3	4	5	6	7	8	9	A	B	C	D	E
a	b	c	e	f	ab	ba	abc	cb	bae	ef	fb	bab	baa	aa

Input Charstream: ababcbaefbabaaa
Output Codestream: 0152634660

12. There is no more new character in the input charstream now.

a. Output the code word 0 representing the current prefix P = a.

b. Finish.

The final result is as follows.

Output Codestream: 0152634660 0

The code words in the above two examples started from 0. Actually, code word could start either from 0 or 1. This way, the result code words will be different. But the principle is the smae though. Let's have a look at the following example which starts the code word from 1.

[Example 3.9] Suppose the input charstream is ABABBABCABABBA, what is its LZW code?

Solution: 1. The charstream has three roots, A, B, and C. Put the three roots into the dictionary. Assign the code word 1 to A, 2 to B, and 3 to C. The result is as follows.

1	2	3	4	5	6	7	8	9	A	B
A	B	C								

Current character is (C): = A.

The string P + C = A is in the dictionary. Then P = P + C = A.

2. Read the next character in the charstream and get the current character (C): = B.

The string P + C = AB.

The string is not in the dictionary.

a. Output the cord word 1 of the current prefix A.

b. Add the P + C = AB to the dictionary.

c. Let the P = C = B.

The dictionary and the output stream at this step are as follows.

1	2	3	4	5	6	7	8	9	A	B
A	B	C	AB							

Input Charstream: ABABBABCABABBA
Output Codestream: 1

3. Read the next character in the charstream and get the current character (C): = A.

The string P + C = BA.

The string is not in the dictionary.

a. Output the cord word 2 of the current prefix B.

b. Add the P + C = BA to the dictionary.

c. Let the P = C = A.

The dictionary and the output stream at this step are as follows.

1	2	3	4	5	6	7	8	9	A	B
A	B	C	AB	BA						

Input Charstream: ABABBABCABABBA
Output Codestream: 12

4. Read next character a in the charstream and get (C) = B.

The string P + C = AB is in the dictionary. Then P = P + C = AB.

Read next character B in the charstream and get (C) = B.

The string P + C = ABB is not in the dictionary.

a. Output the cord word 4 of the current prefix AB.

b. Add the P + C = ABB to the dictionary.

c. Let the P = C = B.

The dictionary and the output stream at this step are as follows.

1	2	3	4	5	6	7	8	9	A	B
A	B	C	AB	BA	ABB					

Input Charstream: ABABBABCABABBA		
Output Codestream: 124		

5. Read next character A in the charstream and get (C) = A.

The string P + C = BA is in the dictionary. Then P = P + C = BA.

Read next character B in the charstream and get (C) = B.

The string P + C = BAB is not in the dictionary.

a. Output the cord word 5 of the current prefix BA.

b. Add the P + C = BAB to the dictionary.

c. Let the P = C = B.

The dictionary and the output stream at this step are as follows.

1	2	3	4	5	6	7	8	9	A	B
A	B	C	AB	BA	ABB	BAB				

Input Charstream: ABABBABCABABBA	
Output Codestream: 1245	

6. Read the next character in the charstream and get the current character (C): = C.

The string P + C = BC.

The string is not in the dictionary.

a. Output the cord word 2 of the current prefix B.

b. Add the P + C = BC to the dictionary.

c. Let the P = C = C.

The dictionary and the output stream at this step are as follows.

1	2	3	4	5	6	7	8	9	A	B
A	B	C	AB	BA	ABB	BAB	BC			

Input Charstream: ABABBABCABABBA	
Output Codestream: 12452	

7. Read next character A in the charstream and get (C) = A.

The string P + C = CA.

The string is not in the dictionary.

a. Output the cord word 3 of the current prefix C.

b. Add the P + C = CA to the dictionary.

c. Let the P = C = A.

The dictionary and the output stream at this step are as follows.

1	2	3	4	5	6	7	8	9	A	B
A	B	C	AB	BA	ABB	BAB	BC	CA		

Input Charstream: ABABBABCABABBA	
Output Codestream: 124523	

8. Read next character B in the charstream and get (C) = B.

The string P + C = AB is in the dictionary. Then P = P + C = AB.

Read next character A in the charstream and get (C) = A.

The string P + C = ABA is not in the dictionary.

a. Output the cord word 4 of the current prefix AB.

b. Add the P + C = ABA to the dictionary.

c. Let the P = C = A.

The dictionary and the output stream at this step are as follows.

1	2	3	4	5	6	7	8	9	A	B
A	B	C	AB	BA	ABB	BAB	BC	CA	ABA	

Input Charstream: ABABBABCABABBA
Output Codestream: 1245234

9. Read next character B in the charstream and get (C) = B.

The string P + C = AB is in the dictionary. Then P = P + C = AB.

Read next character B in the charstream and get (C) = B.

The string P + C = ABB is in the dictionary. Then P = P + C = ABB.

Read next character A in the charstream and get (C) = A.

The string P + C = ABBA is not in the dictionary.

a. Output the cord word 6 of the current prefix ABB.

b. Add the P + C = ABBA to the dictionary.

c. Let the P = C = A.

The dictionary and the output stream at this step are as follows.

1	2	3	4	5	6	7	8	9	A	B
A	B	C	AB	BA	ABB	BAB	BC	CA	ABA	ABBA

Input Charstream: ABABBABCABABBA
Output Codestream: 12452346

10. There is no more new character in the input charstream now.

a. Output the code word 1 representing the current prefix P = A.

b. Finish.

The final result is as follows.

Output Codestream: 12452346 1

We give out the LZW algorithm encoding procedures as follows.

```
BEGIN
    s = Next inputting character;
    while not EOF
    {
    c = Next inputting character;
```

```
            If  s + c existing in the dictionary;
                s = s + c;
            Else
                {
                Output the code correspondent to s;
                Add character string s+c to the dictionary and assign a new
                correspondent
                code to it;
                s = c;
                }
        }
        Output the correspondent code of s;
        END
```

2. LZW Decoding Algorithm

There are another two terminologies in LZW decoding algorithm.

One of them is the current code word which means the processing cord word and expressed as cW, meanwhile, the string.cW represents current character string.

The other is the previous code word which means the code word appeared at an earlier time and expressed as pW, at the same time, the string.pW represents the previous charcter string.

At the beginning of the decoding, the decoding dictionary is the same as the encoding dictionary. It has all the possible roots. During the decoding, LZW decoding algorithm will remember the pW, after read the current code word cW, output the current chacracter string string.cW, and then put the privious appeared word code string string.pW extended from the first chacracter of string.cW to the dictionary.

The following shows the detailed calculation steps.

1. Put all the possible roots into the dictionary at the beginning.

2. cW: = The first cord word in the input cord word string.

3. Output the current cord word string string.cW to the character string.

4. Let pW: = cW.

5. Let cW: = next cord word in the input cord word string.

6. Juge whether string.pW is in the dictionary.

(1) If yes,

a. Output the string.pW to the character string.

b. Let current prefix P: = previous appeared string string.pW.

c. Let current character C: = the first character of current string string.cW.

d. Add string P + C to the dictionary.

(2) If no,

a. Let the current prefix P: = string.pW.

b. Let current C: = the first character of string.cW.

c. Output the P + C to the character string and then add it to the dictionary.

7. Whether there is still cord word in the input cord word string.

(1) If yes, go back to step 4.

(2) If no, finish.

The LZW decoding calculation steps could be also written in the following pseudo code.

```
Dictionary[j]←all n single character, j=1, 2, ···, n
j←n+1
cW←first code from Codestream
Charstream←Dictionary[cW]
pW←cW
While((cW←next Code word)!=NUI.I.)
    Begin
        If cW is in Dictionary
        Charstream←Dictionary[cW]
        Prefix←Dictionary[pW]
        cW←first Character of Dictionary[cW]
        Dictionary[j]←Prefix. cW
        j←n+1
    pW←cW
else
    Prefix←Dictionary[pW]
    cW←first Character of Prefix
    charstream←prefix.C
    Dictionary[j]←Prefix. C
    pW←cW
    j←n+1
end
```

[Example 3.10] The encoding character string is showing in Table 3-15. The encoding process is showing in Table 3-16. The columns of the tables are explained as follows.

Table 3-15　Encoding Charstream

Position	Character	Position	Character	Position	Character
1	A	4	A	7	B
2	B	5	B	8	A
3	B	6	A	9	C

Table3-16　LZW Encoding Process

Steps	Position	Dictionary		Ouputs
		（1）	A	
		（2）	B	
		（3）	C	
1	1	（4）	AB	（1）
2	2	（5）	BB	（2）
3	3	（6）	BA	（2）
4	4	（7）	ABA	（4）
5	6	（8）	ABAC	（7）
6				（3）

(1) The "Step" column represents the processing steps.

(2) The "Position" column represents the character's position in the original character string.

(3) The "Dictionary" column represents the strings added to the dictionary. Their indexes are shown in the bracket.

(4) The "Output" column' represents the cord words outputs.

The Table 3-17 explains the decoding process. At each decoding step, the decoder reads one cord word, output the corresponding character string, and add it to the dictionary. For example, in step 4, the previous appeared cord word "2" is saved in pW. The current cord word cW is "4". The output of current character string string.cW is "AB". The previous appeared character string string.pW, "B" is as rhe first character of the current character string string.cW, "A". Add the result "BA" to the dictionary and its index number is "6".

Table 3-17　LZW Decoding Process

Steps	Cord word	Dictionary		Ouputs
		(1)	A	
		(2)	B	
		(3)	C	
1	(1)	—	—	A
2	(2)	(4)	A B	B
3	(2)	(5)	B B	B
4	(4)	(6)	B A	A B
5	(7)	(7)	A B A	A B A
6	(3)	(8)	A B A C	C

LZW algorithm has been broadly accepted. Its calculation speed is faster than LZ77 algorithm's since it doesn't do that many string comparison as LZ77 does. There is some room increasing the length of the variable code and delete the used string in the dictionary to further improve the LZW algorithm. Those improved LZW algorithm has been applied in the GIF image format and UNIX compression programs.

LZW algorithm got the patent. Unisys in US is the owner of the patent. Other than professional software manufacturer, the LZW is free to use.

练　习　题

一、选择题

1. 在无损压缩编码算法中，有一种编码算法将被编码的信息表示成实数轴上 0 和 1 之间的间隔，信息越长，间隔越小，表示这一间隔所需的二进制位数就越多。 这种编码算法是（　　　）。

　　A．哈夫曼编码　　　　B．香农编码　　　C．算术编码　　　D．行程长度编码

2. 有关行程长度编码（RLE）方法，以下说法正确的是（　　　）。

　　A．行程长度编码是一种有损压缩方法

　　B．编码过程中需要根据符号出现的概率来进行编码

C．编码过程中需要建立"词典"

D．行程长度编码方法尤其适用于计算机生成的图像

3．数据压缩编码方法可以分为无损压缩和有损压缩，其中（　　　）属于无损压缩。

A．模型编码　　　B．DCT 编码　　　C．哈夫曼编码　　　D．矢量量化编码

4．常用的统计编码方法包括哈夫曼编码和算术编码，其中（　　　）。

A．算术编码需要传送码表，哈夫曼编码采用 0 到 1 之间的实数进行编码

B．哈夫曼编码需要传送码表，算术编码采用 0 到 1 之间的实数进行编码

C．哈夫曼编码需要传送码表，并且采用 0 到 1 之间的实数进行编码

D．算术编码需要传送码表，并且采用 0 到 1 之间的实数进行编码

二、填空题

1．压缩编码的理论基础是_____，即任何信息都存在冗余，冗余大小与信息中每个符号的出现概率或者说不确定性有关。（参考答案：信息论）

2．数据压缩的对象是_____，而不是_____。（参考答案：数据　信息）

3．_____是指信息所具有的各种性质中多余的无用空间。（参考答案：信息冗余）

4．规则物体的表面具有物理相关性，将其表面数字化后表现的数据冗余称为_____。（参考答案：空间冗余）

5．_____的主导思想是，任何一个字符的编码，都不是另一个字符编码的前缀。反过来说就是，任何一个字符的编码，都不是由另一个字符的编码加上若干位 0 或 1 组成。（参考答案：前缀编码）

6．_____编码的主导思想是：在对文件进行编码时，需要生成特定字符序列的表以及对应的代码，该编码多用于图像数据的压缩。（参考答案：LZW 编码）

三、计算题

1．求字符串"Rissanen"的算术编码。

2．设有输入字符流 bbcabbc，试对其进行 LZW 编码（假定初始字典为 a=1#；b=2#；c=3#）。

3．现有 8 个待编码的符号，$m_1, m_2, \cdots, m_8$，它们的概率如练习—表 1 所列。使用哈夫曼编码算法求出这 8 个符号所分配的代码，并填入表中。（答案不唯一，参考答案：1，000，001，011，0101，01000，010010，010011）。

4．字符流的输入如练习-表 2 所列，使用 LZW 算法计算输出的码字流。如果对本章介绍的 LZW 算法不打算进行改进，并且使用练习-表 3 进行计算，请核对计算的输出码字流是否为

(1) (2) (4) (3) (5) (8) (1) (10) (11)…

请将码字流中的码字填入练习-表 2 对应的位置。

5．LZ78 算法和 LZ77 算法的差别在哪里？

6．LZSS 算法和 LZ77 算法的核心思想是什么？它们之间有什么差别？

练习-表 1

待编码的符号	概率	分配的代码	代码长度（位数）
m_0	0.4		
m_1	0.2		

待编码的符号	概率	分配的代码	代码长度（位数）
m_2	0.15		
m_3	0.10		
m_4	0.07		
m_5	0.04		
m_6	0.03		
m_7	0.01		

练习-表 2

输入位置	1	2	3	4	5	6	7	8	9	10	11	12	13	14	15	16	17	⋯
输入字符流 输出码字	a	b	a	b	c	b	a	b	a	b	a	a	a	a	a	a	a	⋯

练习-表 3

步骤	位置	词典		输出码字
		（1）	a	
		（2）	b	
		（3）	c	
1	1			
2				
⋮				
9				
⋮				

Exercises

一、Multiple Choices

1. Within the lossless compression coding algorithms, there is one algorithm that represents the encoding information as the intervals between the number 0 and 1 along the real number axel. The longer the information is, the smaller the interval is, which means the more binary bits this interval will need. This encoding algorithm is (   ).

    A. Huffman encoding   B. Shannon encoding

    C. Arithmetic encoding   D. Run length encoding

2. About the run length encoding, RLE, the right saying listed below is (   ).

    A. The RLE is one of the lossy algorithms.

    B. The encoding needs the symbols' appearing probabilities to finish the process.

    C. It's necessary to build the "dictionary" for the encoding process.

    D. The RLE is especially efficient for the computer created images.

3. The data compression encoding methods could be classified into two types, the lossy and

the lossless. The (   ) is one of the lossless algorithms.

    A. Model encoding   B. DCT encoding

    C. Huffman encoding   D. Vector quantification encoding

4. The common statistic encoding methods include Huffman encoding and arithmetic encoding. Between those two, (   ) is true.

    A. The arithmetic encoding needs to transmit the code table and Huffman encoding applies the real number between 0 and 1 to encode.

    B. The Huffman encoding needs to transmit the code table and arithmetic encoding applies the real number between 0 and 1 to encode.

    C. The Huffman encoding needs to transmit the code table and applies the real number between 0 and 1 to encode.

    D. The arithmetic encoding needs to transmit the code table and applies the real number between 0 and 1 to encode.

二、Filling out

1. The theory foundation of compression encoding is _____, which means any information data includes redundancy. The quantity of the redundancy is related to the probabilities or uncertainties of the symbols appeared in the information data. (Reference answer: information theory)

2. The data compression object is (   ) and not (   ). (Reference answer: data, information)

3. _____ means the useless data contained inside the normal information data. (Reference answer: redundancy)

4. The regular object's surface has physical relativities. Digitalize the surface will get some data redundancy which is called _____. (Reference answer: space redundancy)

5. _____'s main idea is that any symbol's code is not the prefix of another symbol's code. In another word, any symbol's code can't be formed by adding some "0" or "1" in front of another symbol's code. (Reference answer: prefix encoding)

6. _____'s main idea is that when encoding a file, create the certain symbol series table and their correspondent code. This algorithm is commonly applied in image data compression. (Reference answer: LZW encoding)

三、Calculations

1. Find out the arithmetic codes of character string "Rissanen".

2. Suppose the input character string is bbcabbc, please find out its LZW codes (suppose the initial dictionary is $a=1$, $b=2$, and $c=3$).

3. There are 8 symbols needs to be encoded, m_1, m_2, ..., m_8. Their correspondent probabilities are shown in Exercise Table-1. Please use Huffman encoding algorithm to get those 8 symbols' assigned codes and put them into the table. (The answer is not sole. Reference answer: 1, 000, 001, 011, 0101, 01000, 010010, 010011)

4. The input character stream is shown in Exercise Table-2. Please use LZW algorithm to calculate the output code stream. If not modifying the LZW described in this chapter and using the

Exercise Table-3 to calculate, please check whether the output code stream is as follows

(1) (2) (4) (3) (5) (8) (1) (10) (11) ⋯

Please fill the output code stream in the Exercise Table - 2's correspondent position.

5. What's the difference between LZ78 and LZ77 algorithms?

6. What are the core ideas of LZSS and LZ77 algorithms? What's the difference between them?

Exercise Table-1

Symbols wait for encoding	Probobilities	Assigned codes	Code lengthes(bits)
m_0	0.4		
m_1	0.2		
m_2	0.15		
m_3	0.10		
m_4	0.07		
m_5	0.04		
m_6	0.03		
m_7	0.01		

Exercise Table-2

Input position	1	2	3	4	5	6	7	8	9	10	11	12	13	14	15	16	17	⋯
Input stream Output codes	A	b	a	b	c	b	a	b	a	B	a	a	a	a	a	a	a	⋯

Exercise Table-3

Steps	Position	Dictionary		Output codes
		(1)	a	
		(2)	b	
		(3)	c	
1	1			
2				
⋮				
9				
⋮				

150

第四章　音频压缩编码

数字电话、数据通信和多媒体数据容量日益增长，在不希望明显降低传送话音信号的质量的前提下，除了提高通信带宽之外，对话音信号进行压缩是提高通信容量的重要措施。涉及到多媒体通信的一个可说明音频数据压缩的重要性的例子是，用户无法使用 28.8kb/s 的调制解调器来接收互联网上的 64kb/s 话音数据流，这是一种单声道、8 位/样本、采样频率为 8kHz 的话音数据流。

ITU-TSS 为此制定了并且继续制定一系列话音数据编译码标准。其中 G.711 使用 μ 律和 A 律压缩算法，信号带宽为 3.4kHz，压缩后的数据率为 64kb/s；G.721 使用 ADPCM 压缩算法，信号带宽为 3.4kHz，压缩后的数据率为 32kb/s；G.722 使用 ADPCM 压缩算法，信号带宽为 7kHz，压缩后的数据率为 64kb/s。在这些标准基础上还制定了许多话音数据压缩标准，例如 G.723，G.723.1，G.728. G.729 和 G.729.A 等。这里将重点介绍音频编码的基本思想，而详细计算则留给那些开发和具体设计编译码器软硬件的读者去研究。具体内容有音频与听觉，声音数字化和数字音频的文件格式等。

Chapter 4　Audio Signal Compression Coding

Nowadays, the data capacity of digital phones, data communications, and multimedia has been rapidly increased. This has been putting the pressure on the communication channel capacity. Under the pressure, without losing much quality on the transmitted audio signals, besides adding more bandwidth, audio signal compress coding will be the key to increase the communication capacity. A good example in multimedia communication scope showing the significance of audio data compression is that users can't use the 28.8kb/s modem to receive the 64kb/s speech data stream from the internet. This kind of 64kb/s speech data stream is a typical speech data stream with single channel, 8 bits/sample, and 8kHz sampling frequency.

ITU-TSS has proposed and is still proposing a serious speech data encoding and decoding standards. Among those, G.711 applied μ regulating and A regulating compression algorithms. The signal bandwidth is 3.4kHz and the data rate after the compression is 64kb/s. G721 applied ADPOM compression algorithm. The signal bandwidth is 3.4kHz and the data rate after the compression is 32kb/s. G.722 applied ADPOM compression algorithm. The signal bandwidth is 7kHz and data rate after the compression is 64kb/s. Besides those standards, ITU-TSS proposed many speech data compression standards as well, such as G.723, G.723.1, G.728, G.729, and G.729A. We're going to put the most of our attention to the basic idea of audio signal coding and leave the very detailed calculation procedure to those who develop and design the hardware and software of the encoder and decoder. Those basic ideas include audio and hearing, audio signal digitalizing, and the file formats of the digital audio signal.

4.1 音频与听觉

4.1.1 话音波形的特性

了解话音波形的基本特性对声音数据的压缩编码、声音的识别和文本—声音的转换等都有很重要的意义。

1. 声音是什么

人们的周围充满着各种各样的声音，如人发出的声音、动物发出的声音、乐器发出的声音、机械发出的声音，以及自然界产生的各种声音，像风声、雷声和雨声。那么，声音是如何产生的呢?就人的听觉来说，声音是空气压力发生快速变化对人的听觉系统产生影响的现象。若空气的大气压力保持着恒定不变的某种状态，就听不见声音了。大气压力变化传播到耳朵时，人们就听到了声音。当大气压力发生百万分之一的改变时人的耳朵就能感知到，听到声音。人的耳朵是否听到声音，还与空气的振荡快慢有关，振荡速度快于每秒 20 次且慢于每秒 2 万次时，人耳可以感知其振动，听到声音。

我们可以归纳为以下的一些要点。

- 声音是由于物体的__振动__而产生。
- 声音的传播需要__介质__。
- 它可以在__固体、液体、气体__中传播，__真空__不能传声。
- 声音以__纵波__的形式传播，它具有__能量__。
- 通常情况下，声音在空气中的传播速度约为__340m/s__。
- 声音在金属中和液体中的传播速度比它在空气中的要__快__，约为__5000m/s__和__1000m/s__。
- 人耳能听到的声波的频率范围通常在__20~20000Hz__之间。
- 频率比可听声高的声波叫__超声波__。它具有__定向性好__、__穿透能力强__等特点，可用于__测距、成像、测速、清洗、焊接、碎石__等。
- 频率比可听声低的声波叫__次声波__，监测与控制它有助于减少它的危害，并可用来__预报地震、预测台风和监测核爆炸__。
- 乐音通常是指那些__动听的、令人愉快__的声音。乐音的波形是__有规律__的。
- 噪声通常是指那些__难听的、令人厌烦__的声音。噪声的波形是__杂乱无章__的。从环保角度看，凡__影响人们正常学习、工作和休息__的声音都属于噪声。

2．听觉系统

1）听觉器官

简单地说，声音是由物体的振动产生的。由振动产生的声波传入人耳，到达鼓膜，使鼓膜发生振动。鼓膜的振动通过耳小骨和淋巴液传递到"基底膜"，最终引起有毛细胞的纤毛振动，激起神经细胞信号。这种神经细胞信号传递到大脑，人们便感知到了声音信号。大脑对声音信号进行解读，人就获得了听觉信息，人耳结构如下所述，可参照图 4-1～图 4-5 的耳结构图。

耳是听觉的外周感觉器官，它由如下的各部分组成。

- 外耳：耳廓、外耳道。
- 中耳：鼓膜、听小骨、咽鼓管和听小肌。
- 内耳：半规管、耳蜗。

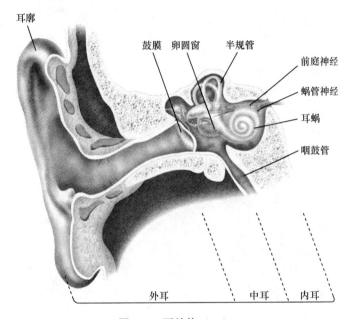

图 4-1　耳结构（一）

2）声音的传递（外耳、中耳和内耳的功能）

（1）耳廓。

它具有如下特点：

① 利于集音；

② 判断声源，依据声波到达两耳的强弱和时间差判断声源。

（2）外耳道。

它的特点如下：

① 传音的通路；

② 增加声强，与 4 倍于外耳道长的声波长（正常语言交流的波长）发生共振，从而增加声强。

（3）中耳的功能。

① 鼓膜。

结构特点：

是一个具有一定紧张度、动作灵敏、斗笠状的半透明膜，它对声波的频率响应较好，失真度较小。

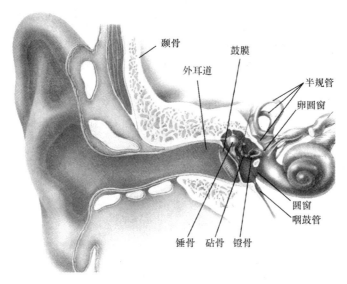

图 4-2 耳结构（二）

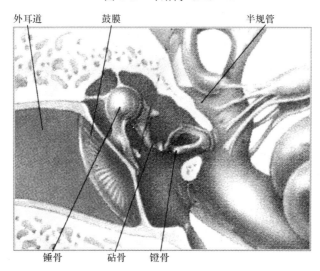

图 4-3 耳结构（三）

功能作用：

能如实地把声波振动传递给听小骨。

② 听小骨。

结构特点：

由锤骨—砧骨—镫骨依次连接成呈弯曲杠杆状的听骨链。

功能作用：

传递振动，增强振压（1.3 倍），减小振幅（约 1/4），防止卵圆窗膜因振幅过大造成损伤。

（4）咽鼓管。

结构特点：

是鼓室与咽腔相通的管道,其鼻咽部的开口通常呈闭合状态,当吞咽、打呵欠或喷嚏时则开放。

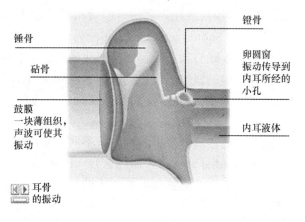

图 4-4 耳结构（四）

功能作用：

① 调节鼓膜两侧气压平衡、维持鼓膜正常位置、 形状和振动性能。

② 咽鼓管黏膜上的纤毛运动可排泄中耳内的分泌物。

3）声波在内耳耳蜗转变为动作电位

内耳耳蜗形似蜗牛壳，蜗管腔被前庭膜和基膜分隔为三个腔：前庭阶、蜗管和鼓阶。

基膜上有螺旋器：由内、外毛细胞、支持细胞及盖膜等构成。

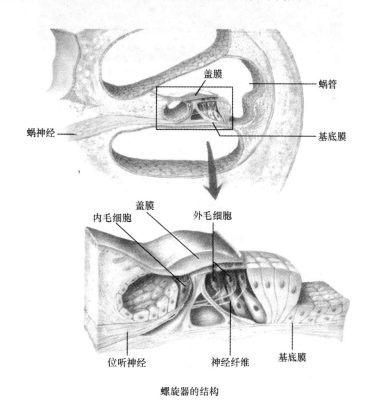

螺旋器的结构

图 4-5 耳结构（五）

当肺部中的受压空气沿着声道通过声门发出时就产生了话音。普通男人的声道从声门到嘴的平均长度约为 17cm，这个事实反映在声音信号中就相当于在 1ms 数量级内的数据具有相关性，这种相关称为短期相关（Short—term Correlation）。声道也被认为是一个滤波器，这个滤波器有许多共振峰，这些共振峰的频率受随时间变化的声道形状所控制。例如舌的移动就会改变声道的形状。许多话音编码器用一个短期滤波器（Short Term Filter）来模拟声道。庆幸的是声道形状的变化比较慢，模拟滤波器的传递函数的修改没有那么频繁，典型值在 20ms 左右。

压缩空气通过声门激励声道滤波器，根据激励方式不同，发出的话音分成三种类型：浊音（Voiced Sounds）、清音（Unvoiced Sounds）和爆破音（Plosive Sounds）。

声音是一种机械波，它具有普通波的物理特征。声波可以发生折射、反射、衍射、**掩蔽**等物理现象。声音的传播需要介质，气体、固体和液体都能传播声音，它们是不同的介质，这些不同的介质传播声音的速度不同。声音的这些特性使得人们可以感知到不同的声音效果。例如，声音在空间的来回反射，造成声音的空间效果，使得人们在剧场中听到的声音和在公园中听到的声音效果是不一样的。

声音是通过介质传播的一种连续的波，如图 4-6 所示，这种连续性包括时间上的连续和幅度上的连续。发声的物体称为声源，声源产生声音，只含有一种频率的声音称为单音。

图 4-6　某声波的一段波形

在多数情况下，许多不同频率、不同幅度的声波同时存在，发生叠加，这样的声音称为复合音。图 4-7 说明了多个不同频率、不同振幅的振波单音信号可以叠加起来，产生复杂的声波信号。反之，复杂的声波信号也可以分解为许多个不同频率、不同幅度的单音振波。傅里叶变换为声波的合成与分解提供了数学理论的支持，这种合成与分解的变换，为研究和处理声音信号提供了基本的和有效的方法。

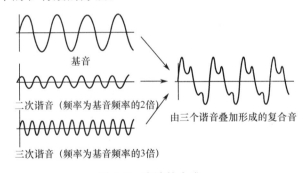

图 4-7　声波的合成

3．声音的分类

根据声音的不同特征，可以将声音分为不同的种类。

根据声音的内容可以分为语音和乐音，根据声音信号的类型可以分为模拟声音和数字化声音。计算机处理的声音信号是经过离散化了的信号，因此又称为音频数据。

根据声音信息学特性可以将声音简单地分为两类：不规则声音和规则声音。不规则声音一般是指不包含任何信息的噪声。规则声音通常分为语音、乐音和音效。语音是指具有语言内涵和人类约定俗成的特殊声音。乐音是规范的、符号化的声音。音效则是指人类熟悉的其他声音，如动物和机器产生的声音、自然界中风雨雷电的声音等。

多媒体技术中主要研究的是规则声音中的语音和乐音信号。

4.1 Audio Signal and Hearing

4.1.1 The Audio Waveform's Characteristics

Getting knowing the basic characters of audio waveform is critical to understand the audio signal data compression, speech identification, and text-speech converting.

1. What the Sound is?

There are kinds of sounds around people daily, such as the sounds created by human being including speaking, the sounds created by animals, the sounds created by instruments, the sounds created by mechanics, and kinds of natural sounds including wind's, thunder's, and rain's. Then, how those sounds originally created? To human being's hearing ability, the sound is the affection influence of the fast changed air pressure to the human being's hearing system. If the air pressure keeps no changing all the time, we won't hear anything anymore. When this kind of changing propagated to our ear, we can hear the sounds. This kind of air pressure changing doesn't have to be big for us the feel and hear the sound. Actually, we could feel and hear the sound over only one over one millionth air pressure changing. Another factor also affects the hearing, which is the speed of the air vibration. When the speed of air vibration is faster than 20 times/second and slower than 20,000 times/second, we could feel the vibration and hear the sound.

We could conclude the above and other sound related topics as following.

- The sound is created from the object's <u>vibration</u>.
- The sound needs <u>media</u> to propagate.
- The sound could propagate in <u>solid, fluid, and air</u>. But not in <u>vacuum</u>.
- The sound is propagating in <u>vertical wave format</u> and it has <u>energy</u>.
- Usually, the speed of the sound propagation in air is <u>340m/s</u>.
- The sound propagation speeds in metal and fluid are <u>faster</u> than in air, which are about <u>5000m/s</u> and <u>1000m/s</u>.
- The sound wave frequency range human being could hear is <u>between 20 and 20kHz</u>.
- The sound whose wave frequency is higher than 20kHz is called <u>super-sound wave</u>. It has the characters of <u>good directional, strong penetrating ability</u>, and etc. It could be applied in <u>distance measurement, image creation, speed detection, cleaning, fire joining, stone splashing,</u> and etc.
- The sound whose wave frequency is lower than 20Hz is called infer-sound wave. Monitoring and controlling it is better to the reducing the damage it may cause. It could also be used <u>to predict the earthquake, to predict the tornado, and to monitor nuclear bomb explosion.</u>

- Musical sound means the sound which is <u>touching, pleasant and making you feel happy.</u> It has <u>regulated wave form.</u>
- Noise sound means the sound which is <u>hard to listen to and unpleasant.</u> Its wave from is <u>unregulated.</u> From the view of environment, any sound <u>disturbing people's normal studying, working, and resting</u> is noise sound.

2. Hearing System

1) Hearing Organs

Simply speaking, the sound is produced from the object's vibration. This sound wave propagated to the ear, get to the eardrum, and cause the eardrum vibrating. This eardrum vibrating transfers to the oval window via ossicles and fluid. This will cause a vibration of the vestibulo-cochlear nerve which will create a nerve cell signal. After the created nerve cell signal transfers to the brain, one will sense the sound signal. The brain will read the sound signal and get the hearing information. The human's ear structure is showing as follows. The lore spondent structures are shown from Fig.4-1 to Fig.4-5.

Ear is an outer sensing organ of the hearing system. It consists of the following arts.

- Outer ear: Pinna and Ear canal
- Middle ear: Eardrum, Ossicles, and Eustachian tube
- Inner ear: semicircular canals and cochlear.

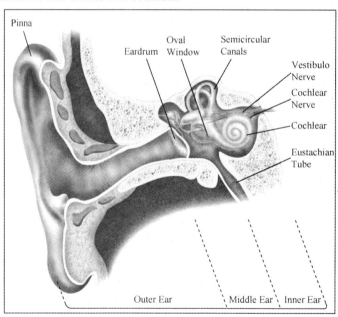

Fig. 4-1   Ear Sturcture 1

2) The sound transmission: The functions of outer ear, middle ear, and inner ear.

(1) Pinna.

It has the following characters.

① Better for the sound collection;

② Tell the sound source. Based on the strengthens and weakness of the sound wave and time difference the sound wave reaching the ears, judge where the sound comes from.

(2) Ear canal.

It has the following characters.

① It is the tunnel of the sound transmission;

② Strengthen the sound. Strengthen the sound by resonant with the wave frequency whose wave length is 4 times of the ear canal length (The normal talking sound wave length).

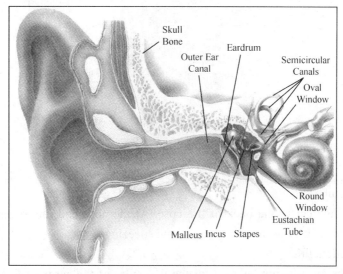

Fig. 4-2   Ear Sturcture 2

(3) Middle ear.

① Eardrum.

It has certain tension. Moves fast with flexibility. It is a half-penetrating form with cone shape. It has a very good frequency response with miner distortion to the sound signal.

Its functionality is to transfer the sound wave vibration to the ossicles without distortion.

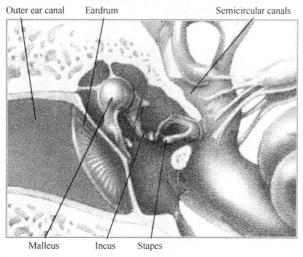

Fig. 4-3   Ear Sturcture 3

② Ossicles.

They are movable bones and formed by three parts, the malleus, incus, and stapes.

Its functionality is to transmit the vibration, strengthen the vibration pressure (about 1.3 times), and reduce the vibration amplitude (roughly 1/4) to prevent the possible over amplitude damage to the oval window.

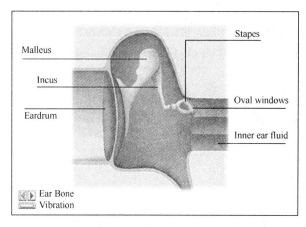

Fig. 4-4   Ear Sturcture 4

(4) Eustachian tube.

It links the middle ear to the back of the nose. At the nose end, the opening part usually is closed and only opens when one swelling, yawn, and sneezing.

It has two functionalities.

① It adjusts the air pressure balance for the two sides of the eardrum to keep the eardrum's normal position, shape, and vibration function.

② The movement of the hair grown on the custachian tube could expel the substances created inside the middle ear.

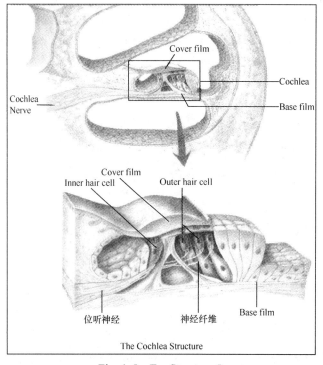

Fig. 4-5   Ear Sturcture 5

160

3) Sound wave transformed to moving voltage inside the inner ear cochlea

The inner cochlea looks like a snail shell. The cochlea tube is divided into three parts by front room membrane and base membrane and they are front cell, tube, and drum cell.

There are spiral cell in the base membrane and it is consisted of inner and outer hire cell, supporting cell, and cover membrane.

The voice will be created when air inside the lung compressed, along the voice tunnel travels, and reaches the vocal cords. The average length from vocal cords to month of a normal male's is about 17cm. This actually tells that the audio signal within time length 1ms is correlated. This kind of correlation is particularly called short-term correlation. Voice tunnel could be taken as a filter which has multiple co-vibrating points. Those co-vibrating points' frequencies are controlled by the voice tunnel's shape which changes with time. For example, the voice tunnel's shape will change when tone moving with time. There are many speech encoders use a short term filter to simulate the voice tunnel. Fortunately, the voice tunnel's shape changing is slow, it's not necessary to change the filter's transform function very frequently. The typical change time is 20ms.

The compressed air ignites the voice tunnel filter via vocal cords. There are three types of voices will be produced depends on the different igniting, voiced sounds, unvoiced sounds, and plosive sounds.

The sound wave is a kind of mechanical wave. It has a normal mechanical wave's characters, such as refraction, reflection, diffraction, shelter, and etc. It needs media to propagate. Air, solid, and fluid are able to propagate sound. But different media could propagate sound in different speed. The characters mentioned made it possible for people to tell different sound affects. For example, the sound back and forth reflection in the air creates the sound space affection. This makes the difference for people to hear the sound in the theatre and in the park.

Furthermore, the sound is a continuously wave propagated via medias. (as shown in Fig. 4-6) This continuation is both in time zone and in amplitude. We call the object which produces the sound as sound source. In another word, the sound source produces sound. The sound only including one frequency is called single frequency sound.

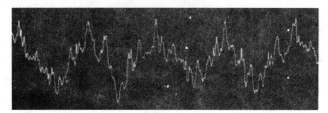

Fig. 4-6   One Section of a Sound Wave

Under most situations, the sound waves with many different frequencies and amplitudes are coexisting and integrated together. This kind of sound is called compound sound. In Fig. 4-7, one example shows that multiple frequencies and multiple amplitudes single frequency sound could be integrated to form a complicated compound sound. On the other hand, the complicated compound sound could be separated into several or many single frequency sound with its own different frequency and amplitude. The Fourier transform known provided the mathematical base for this kind of bidirectional converting. At the same time, this kind of integration and separating gives the basic and efficient way for researching and processing sound/audio signals.

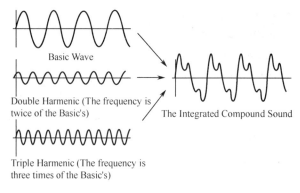

Fig. 4-7　The Integraton of Sound Waves

3. The Categories of the Sounds

The sounds could be classified into different types based on their particular characters.

Based on their contents, the sounds could be classified into speaking sound and music sound. Based on their different signal types, the sounds could be classified into analogue and digital sounds. The audio signals processed in the computer are sampled signals and so called sound/audio data as well.

Simply speaking, the sounds/audios could be classified into two types, regulated and unregulated. The unregulated sounds mean the noises which contain none of useful information. The regulated sounds/audios could be further divided into three kinds, speaking sounds, musical sounds, and affection sounds. The speaking sounds mean the special sounds used between the human being who has the particular ability to understand them. The musical sounds mean the regularized, marked sounds. |Sounds affection means some other sounds that people familiar with such as the sounds produced by animals, machines, winds, thunders, and etc.

The sounds interested in the multimedia area are the speaking sounds and musical sounds within the regular sound limits.

4.1.2　声音的三要素

1．声音的物理特征

声音有三个要素：响度（Ioudness）、音调（Pitch）、音色（Music quality）。

响度又称为音强，是人们主观上感觉到的声音大小，它由声波的振幅决定，与振幅成正比，振幅越大，响度就越大。人耳的听觉与响度成对数关系，人耳对于声音细节的分辨只有在响度适中时才最灵敏。度量响度的单位是分贝（dB）。一般人耳能察觉到 3dB 的音强变化，下面的数据有助于进一步理解响度的单位——分贝（dB）。微风的响度是 10dB，轻声耳语的响度是 20dB，图书馆内噪音的响度是 40dB，讲课时嗓音的响度是 60～70dB，大炮开火时的响度是 120～150dB，正常生活环境下的噪音响度不应超过 7dB。

在物理上，声音的响度使用客观测量单位声压（达因/平方厘米）来度量。在心理上，主观感觉的声音强弱使用响度级 phon 或者 sone 来度量。当声音弱到人耳刚刚可以听见时，称此时的音强为"听阈"。另一种使人耳感到疼痛的声音音强称为"痛阈"。在听阈和痛阈之间的区域就是人耳的听觉范围，人类能够感知的声音幅度范围为 0～120dB。

音调是指声音的高低（高音、低音），是人对声音频率的感觉。频率快，则声音高；频率慢，则声音低。音调虽然主要由频率决定，但同时也与声音强度有关。人耳听觉的频率范围是 20Hz～20kHz，人的发声频率约为 85Hz～11kHz，话音频率一般为 300～3400Hz。对一定强度的纯音，音调随频率的升降而升降；低频纯音，音调随声强增加而下降，高频纯音的音调随强度增加而上升。大体上低于 2kHz 纯音的音调随强度的增加而下降，高于 3kHz 纯音的音调随强度的增加而上升。

音调的单位是美（mel）。取频率为 1kHz、声压级为 40dB 的纯音的音调作标准，称为 1000美。如果一个纯音，听起来音调高一倍，则称为 2000 美，音调低一倍的称为 500 美，以此类推，可建立起整个可听频率内的音调标度。

音调还与声音持续时间长短有关。非常短促的纯音，只能听到像打击或弹指那样的"喀嚓"一响，感觉不出音调。持续时间 10～50ms 时，听起来觉得音调是由低到高连续变化的。超过 50ms 时，音调就稳定不变了。乐音（复音）的音调更复杂些，一般可认为主要由基音的频率决定。音调控制就是人为地改变信号中高、低频成分的比重，以满足听者的爱好、渲染某种气氛、达到某种效果、补偿扬声器系统及放音场所的音响不足等目标。

发音体振荡的基本频率称为基音，此外，还伴随有许多更高频率的、振幅较小的声音谐波，称为泛音。音色是由混入基音的泛音所决定的，高频谐波越丰富，音色就越有明亮感和穿透力。不同谐波具有不同的幅值和相位偏移，由此产生出不同的音色效果。音色是音乐中最为吸引人、能直接触动感官的重要表现手段。人耳具有很好的区分音色的能力，人耳通过音色能区别出男声和女声，能区别出各种不同的乐器的声音等。只要音色不同，即使在同一音高和同一声音强度的情况下，人耳也能区分出是不同的声音。

在听音过程中，音色的作用很大。可以认为声音包含乐音和噪声，乐音由源音和节拍构成，源音的特征包括音调、响度和音色，音色由纯音、变换和混合方式等决定。波形和音色密切相关，确定的波形具有确定的音色。而同一种音色可能有多种波形。两个截然不同的波形，但频谱却是一样的，因为频谱关系不表示波形的相位。人的听觉系统对声波的相位没有感觉，所以两种不同波形声音听上去可能是一样的。

声音的这三个要素可简单归纳如下。

- 响度表示声音的　强弱　，是由声源振动的　振幅　决定的。
- 音调表示声音的　高低　，是由声源振动的　频率　决定的。
- 音色与发声体的材料和结构有关。音色不同，声波的　波形　也不同。

听觉的心理特征变量与声波的物理特征变量有关，但不直接对应。表 4-1 给出听觉的心理特征变量与声波的物理特征变量间的关系。

表 4-1　听觉的心理特征变量与声波的特理特征变量间的关系

听觉的心理特征变量	首要物理变量	听觉的心理特征变量	首要物理变量
响度	声强度	音量	频率和强度
音调	声波频率	密度	频率和强度
音色	声波复合		

2．声音质量的度量

声音质量通常取决于声音信号的带宽，等级由高到低依次是 DAT—CD—FM—AM 一电话。声音质量的评价是一个很困难的问题。声音质量有两种基本的度量方法：客观质量度量和主观质量度量。

声音客观质量主要用信噪比（Signal to Noise Ratio，SNR）来度量。其特点是计算简单，但是不能完全反映出人对语音质量的感觉。

主观质量评价最常用的方法是平均意见得分（Mean Opinion Score，MOS）。MOS 得分采用 5 级评分标准，如表 4-2 所列。这种方法是通过采集若干实验者对声音质量的好坏评分进行统计计算的，得分最高的声音质量最好。由于主观和客观的原因，每次实验者给出的评价分数会有变化，因此 MOS 得分也会有所波动。

表 4-2　MOS 评分标准和声音质量描述

分数	质量级别	失真级别	分数	质量级别	失真级别
5	优	无	2	差	有失真，讨厌，不反感
4	良	有失真，不讨厌	1	劣	有失真，讨厌，反感
3	中	有失真，令人讨厌			

在数字语音及多媒体通信中，语音质量分为四类：广播质量、网络质量、通信质量和合成质量。广播质量语音通常在数据率为 64kb/s 以上获得，MOS 得分为 5 分；网络质量语音在数据率为 16kb/s 以上获得，MOS 得分为 4～4.5 分；通信质量语音在数据率为 4.0kb/s 以上获得，MOS 得分为 3.5 分左右，这时能感觉重建语音质量有所下降，但不妨碍正常通话，可以满足多数语音通信系统的使用要求；合成质量语音的 MOS 得分在 3.0 以下，主要是指一些声码器合成的语音所能够达到的质量，一般具有足够高的可懂度，但自然度和讲话人的确认度等方面不够好。

4.1.2　The Three Major Characteristics of the Sound

1. Sound's physical Attributes

The three major characteristics of the sound are loudness, pitch, and music quality.

Loudness is also taken as strength. It's how we feel the strength of the sound. It varies with the amplitude of the sound and proportional to the amplitude of the sound. The bigger the amplitude, the stronger the loudness is. People's hearing has a logarithm relationship with the loudness. People only can tell the sound details when the loudness is in medium level, not too loud and not too weak. The loudness unit is dB. Generally, people can tell the 3dB loudness change. The following data will give a better understanding of dB. The loudness of the gentle wind is roughly 10dB. The loudness of whisper is roughly 20dB. The loudness of the noise inside the library is roughly 40dB, The loudness of lecturing is about 60 to 70dB. The loudness of the cannon firing is about 120 to 150dB. The loudness of the noise within the people's living area shouldn't exceed 7dB.

In physics, the loudness is measured by dyne/cm². Psychologically, feeling about the loudness is measured by phon or sone. When the sound is getting weaker and weaker until is bared heard, at this point, the loudness is called hearing threshold. On the other hand, when the sound is getting stronger and stronger until it hurts the ear, at this point, the loudness is known as pain threshold. The loudness between those two points is the range people can hear. The sound amplitude range people can hear is from 0dB to 120dB.

The pitch indicates the low and high of the sound. It's the hearing reaction to the frequency. The faster the frequency, the higher the sound is. The slower the frequency, the lower the sound is. Although the pitch is mainly decided by frequency, it's also related to the loudness. The sound frequency range that people can hear is from 20Hz to 20kHz. The sound frequency human being produced is roughly from 85Hz to 11kHz. The speaking sound frequency is roughly from 300Hz to 3400Hz. To certain loudness pure sound, pitch goes alone with the frequency changing. The higher the frequency, the higher the pitch is. To lower frequency pure sound, pitch goes down with frequency increasing. To high frequency pure sound, pitch goes up with frequency increasing. Roughly speaking, for the pure sound whose frequency is lower than 2kHz, pitch goes down with frequency increasing; for the pure sound whose frequency is higher than 3kHz, pitch goes up along with frequency increasing.

The unit of pitch is mel. We take pure sound whose frequency is 1Hz and loudness is 40dB as a standard pitch of 1000mel. If a pure sound sounds double the standard pitch, we call it 2000mel. On the other hand, if a pure sound sounds half the standard pitch, we call it 500mel. In this way, we could build a pitch measurement within the hearing frequency range.

Pitch is also related to the existing length of the sound. For very short pure sound, people only can hear something like a click and wouldn't be able to feel the pitch. When existing time goes up to 10 to 50ms, people can hear the pitch changing from low to high. When the existing time exceeds 50ms, the pitch won't change anymore.

The musical sound (compound sound) is more complicated. In most situations, we think the pitch is determined by the frequency of base sound. When we say pitch control, we mean manually change the ratio of high and low frequency partition within the musical sound to reach the goal like satisfying the listener's preference, evoking some emotional atmosphere, reaching certain affection, compensating some musical lack of speaker system and HiFi system of under certain environment, and etc.

The sound producing object's base vibrating frequency is called base pitch. Besides this base pitch, there are many higher frequencies and smaller amplitudes harmonics called as overtones. Music quality is mainly depends on the overtones mixed inside the base pitch. The more the high frequency overtones, the more the music quality has bright and penetrating feeling. Different overtones have different amplitudes and phase shift and then produce different musical effect. The music quality is the most attractive part inside the music and directly touches the people's sense organ. People's ear has the ability to tell the different musical qualities. Based on the musical quality, people could tell whether the sound comes from male or female and identify the different musical instrument and sounds. As long as the music quality is different, people can tell different sounds even the pitches and the frequencies are the same.

Pitch is playing a critical role in the listening process. We could believe that sound includes music and noise. The music consists of source sound and beat. The characteristics of the source sound include pitch, loudness, and music quality. The music quality is determined by pure sound, transformed sound, and mixed sound. The sound's wave form is closely related to the music quality. Certain wave form has certain music quality. But one music quality may have several waves. Two total different waves could have same spectrum since the spectrum won't reflect the different phase relation. People's hearing system is not sensitive to the wave's phase. This makes that the two different waves sound same to human beings.

The sound's three major characteristics described above could be concluded as follows.

- The loudness represents the sound's <u>strength and weakness</u> and it is determined by the source sound's <u>vibrating amplitude</u>.
- The pitch represents the sound's <u>high and low</u> and it is determined by the source sound's <u>vibrating frequency</u>.
- Music quality is related to the material and structure of the object that produced the source sound. Different music quality is correspondent to the different sound waves.

The hearing's psychological characteristics are related to the sound wave's physical characteristics. But they are not one to one correspondent to each other. The Table 4-1 shows their relationships.

Table 4-1　The Relationship Between Hearing's Psychological Characteristical Variables and Sound Wave's Physical Characteristical Variables

Hearing's Psychological Characteristical Variables	Main Physical Variables	Hearing's Psychological Characteristical Variables	Main Physical Variables
Loudness	Wave Amplitude	Volume	Frequency and Amplitude
Pitch	Wave Frequency	Density	Frequency and Amplitude
Music Quality	Intigerated Wave		

2. The Measurement of Sound Quality

The sound quality usually decided by the bandwidth of the sound wave. The levels are normally sequenced from higher to lower as DAT - CD - FM - AM - Phone. It's not easy to valuate a sound quality. There are two basic measurement methods available and they are objective quality measurement and subjective measurement.

The objective measurement is normally done by checking the Signal to Noise Ratio, SNR. The advantage of this measurement is its simplicity. The disadvantage is that it can't reflect people's feeling about the sound quality.

The subjective measurement is usually done by taking the Mean Opinion Score, MOS. There are 5 levels within the MOS as shown in Table 4-2. The result is achieved by statistically collecting the valuation marks from different dedicated listeners. The higher the marks, the better the sound quality is. Each time, the same dedicated listeners may give different valuation marks because of both objective and subjective reasons. The MOS varies each time accordingly.

Table 4-2　MOS Score Standards and Sound Quality Description

Marks	Quality Levels	Distortion Levels	Marks	Quality Levels	Distortion Levels
5	Excellent	None	2	Poor	Distortion, Nasty, Not strongly Dislike
4	Good	Distortion, Not Nasty	1	Bad	Distortion, Nasty, strongly Dislike
3	Fare	Distortion, Nasty			

In digital speech and multimedia communication, the sound qualities are classified into 4 classes, broadcasting, network, communication, and integrated. The broadcasting sound quality usually could be got at the data rates over 64kbps and MOS score is 5. The network sound quality could be got at the data rate over 16kbps and MOS score is from 4 to 4.5. The communication sound quality could be got at the data rate over 4kbps and MOS score is about 3.5. At this point, people can feel the rebuilt sound's quality is not that good. But it still not affects the normal oral communication. It could meet the requirements of most speech communication systems'. The integrated sound quality could get the MOS score of 3 or lower. It mainly indicates the sound quality of some sound encoder integrated. It contains enough info for people to understand the contents with lacking of speaking naturalization and identification.

4.1.3 听觉感知特征

人的听觉系统非常复杂，迄今人们对听觉系统的生理结构、生理学原理等并不完全清楚，或对听觉感知实验结果不能做到完全合理解释。人耳听觉感知还受到心理因素的很大影响。人耳听觉特征的研究限于心理声学和语言声学。其中，人耳对响度、音高的感知特征和掩蔽效应可以直接应用于声音数据的压缩编码。

1. 对响度的感知

声音的响度就是声音的强弱。在物理上，声音的响度使用达因/平方厘米（声压）或瓦特/平方厘米（声强）的单位进行客观度量。在心理上，主观感觉声音强弱使用响度级 phon 或 sone 来度量。以上这两种感知声音强弱的度量概念完全不同，但又有一定的联系。

当声音弱到人的耳朵刚刚可闻，称此时的声音强度为"听阈"。

听阈：在安静环境中，能被人耳听到的纯音的最小值（当声音弱到人的耳朵刚刚可以听见时的声音强度）。对应于不同的频率有其不同的听阈，对应于各个频率 f 可绘出标准听阈曲线。

在听阈曲线以下的各种声音将不能被人耳察觉。

例如，1kHz 纯音的声强达到 10^{-16}W/cm² （定义成 0dB 声强级）时，人耳刚能听到，此时的主观响度级定为零方。据实验结果，听阈是随频率变化的。由此得出的"听阈一频率"曲线如图 4-8 所示。图中最下面的曲线称为"零方等响度级"曲线，也称为"q 绝对听阈"曲线，表示在安静环境中，人耳听到的纯音的最小值。

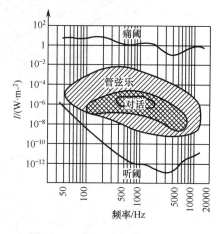

图 4-8　听阈、痛阈与听觉区域

另一种极端的情况是声音强到使人耳感到疼痛。

相对于人们感受给定各频率的正弦式纯音，开始使人耳感到疼痛的阈值，称为相应频率的 "痛阈"。图 4-9 给出了根据各频率 f 绘出的标准痛阈曲线。

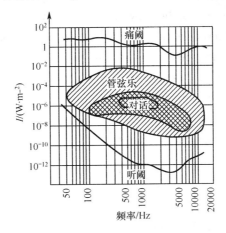

图 4-9　听阈、痛阈与听觉区域

从已有的实验结果可以得知，如果频率为 1kHz 的纯音的声强级达到 120dB 左右，人的耳朵就感到疼痛，这个阈值称为 "痛阈"。通过实验测量，可得到 "痛阈一频率" 曲线，即图 4-9 最上方的曲线，同理，它也就是 120 方等响度级曲线。

在 "听阈—频率" 曲线和 "痛阈一频率" 曲线之间的区域是人耳的听觉范围。这个范围内的等响度级曲线也可通过实验结果绘出。由图 4-10 可知，1kHz 的 10dB 的声音和 200Hz 的 30dB 的声音，在人耳听起来具有大致相同的响度。而在低频区和高频区，能被人耳听到的声音的信号幅度要高得多。

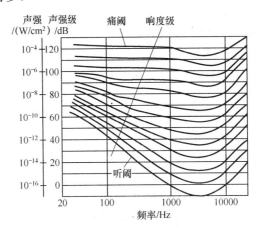

图 4-10　"听阈—频率" 曲线

响度是表示人耳对声音的强弱的主观感觉量。

具有相等响度的不同频率的点连接起来构成的一条条曲线被称为等响度曲线。又称为 Fletcher-Munson（弗莱彻—芒森曲线）曲线，即图 4-10 所示的曲线。

等响曲线是反映人们对一定的声音振幅范围，心理和生理因素反应的曲线，每条曲线上对应于不同频率的声压级是不相同的，但人耳感觉到的响应却一样，因此称为等响曲线，每条曲线上对应一个响度值，由等响曲线族可以得知，当音量较小时，人耳对高低音感觉不足，而音量较大时，高低音感觉充分，人对 2Hz~4kHz 之间的声音量最为敏感。

2．对音高的感知

音频是指声音信号的频率。

音高也称音调，表示人耳对声音调子高低的主观感受。主要与声音的频率有关。但不与频率成正比，类似于响度，音调的感觉与频率成对数关系，因此通常用频率的倍数或对数关系来表示音调。

用频率来表示声音的音高，其单位是 Hz。主观感觉的音高单位则是 Mel。主观音高与客观音高的关系是：

$$Mel=1000\log_2(1+f)$$

类似于测量响度时是以 1kHz 纯音为基准，测量音高时，以 40dB 声强为基准，而且同样由主观感觉来确定。

测量时，让实验者听两个声强级为 40dB 的纯音，然后固定其中一个纯音的频率，调节另一个纯音的频率，直到实验者感到后者的音高为前者的两倍，就标定这两个声音的音高差为两倍。从实验结果我们可得出音高与频率之间的关系如"音高频率"曲线（图 4-11）所示。

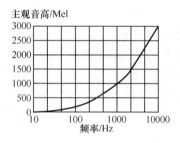

图 4-11 "音高—频率"曲线

3．声音音色的感知

音色又称音品（Musical Quality），是指音的感觉特性。

不同的物体发出的声音不同。正因为如此，在听同一音高和同一声音强度下的钢琴和小提琴演奏时，能够区分出是不同乐器发出的。

人们通过音色可以分辨出不同的材料、物体包括人。

通过以下几个有趣的小百科问题来概括一下音色的概念。

（1）为什么会有不同的"音色"？

答：声音是由发声的物体震动产生的，当其整体震动时发出基音，但同时其各部分也有复合的震动，这些各部分震动产生的声音组合成泛音。由于部分小于整体，所有不同的泛音都比基音的频率高，但强度都相当弱。

音色的不同取决于不同的泛音，每一种乐器、不同的人以及所有能发声的物体发出的声音，除了一个基音外，还有许多不同频率的泛音伴随，正是这些泛音决定了其不同的音色，使人能辨别出是不同的乐器甚至不同的人发出的声音。

（2）"音色"有哪些特征？

答：人耳对音色的听觉反应非常灵敏，并具有以下特征。

① 记忆力。

当熟人跟你谈话时，即使你未见到他（她）也会知道是谁在跟你谈话。甚至连熟人的走路声，你都可以辨认出。这说明人耳对经常听到的音色具有很强的记忆力。

② 分辨力。

熟知乐器者，只要听到音乐声就能迅速指出是何种乐器演奏的。即使在同一频段内演奏，你仍能分辨出是那一种弦乐器演奏的。这说明每种乐器都有其独特的音色，人耳对各种音色的分辨能力非常强。

（3）什么是"音色感"？

答：音色感是指人耳对音色所具有的一种特殊的听觉上的综合性感受。即使选用世界上最先进的电子合成器模拟出各种乐器，如小号、钢琴或其他乐器，虽然频谱、音色可以做到完全一样，但对于音乐师或资深的发烧友来讲，仍可清晰地分辨出它们是合成出来的，而不是自然演奏出来的。这说明频谱、音色虽然一样，但复杂的音色感却不相同，以至人耳听到的音乐效果不同。这也说明音色感是人耳特有的一种复杂的听觉上的综合性感受，是无法模拟的。

4．声音的掩蔽效应

在聆听一个声音的同时，由于被另一个声音（称为隐蔽声）所掩盖而听不见的现象称为掩蔽现象。一个声音的听阈因另一个声音的掩蔽作用而提高的效应，称为掩蔽效应，如图4-12所示。

掩蔽效应有如下特点：被掩蔽声的频率越接近掩蔽声的频率时，掩蔽量越大；掩蔽声的声音越强，掩蔽量越大；低频声容易掩蔽高频声，而高频声较难掩蔽低频声。典型的例子是我们在听音箱放送音乐时，在音乐播放的过程中，人们感觉不到噪声的存在，但当音乐停止或间歇过程中，人们就可以感觉到音箱发出的本底噪声。

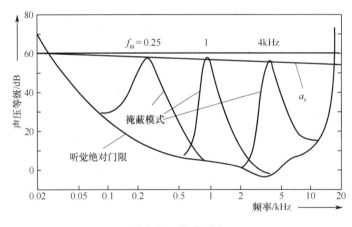

图 4-12　掩蔽效应

进一步讲，声音的隐蔽效应可主要分为频域掩蔽和时域掩蔽两种。

（1）频域掩蔽。

对于频率相近的声音，响度高的阻碍另一个响度较低声音的听觉感知的现象，称为频域掩蔽效应。响度高者称为掩蔽声音，响度低者称为被掩蔽声音。例如，一群学生在教室里正在大声地相互讨论着一个问题，这时老师大声说"请大家暂停讨论，请安静！"，可是学生们听不清或听不见老师的话。老师又提高嗓门喊了一遍，学生们杂乱的话音即刻停止了。在这个场景中，老师和学生们的话音频率相近，当师生的话音响度也相近时，老师的话音被淹没了；当老师提高嗓门喊了一声时，其话音响度高于学生们，老师的声音盖过了学生们的声音，学生们便听到老师的声音了。

一个强纯音会掩蔽在其附近同时发声的弱纯音,如图 4-13 所示。例如,一个声强为 60dB、频率为 1kHz 的纯音,另外还有一个 1.1kHz 的纯音,前者比后者高 18dB,这时人耳只能听到 1kHz 的强音。如有一个 1kHz 的纯音和一个声强比它低 18dB 的 2kHz 的纯音,由于频率相差较大,此时人耳会同时听到这两个声音。要使 2kHz 的纯音也听不到,则需要把它降到比 1kHz 的纯音低 45dB。一般来说,弱纯音频率离强纯音频率越近就越容易被掩蔽。

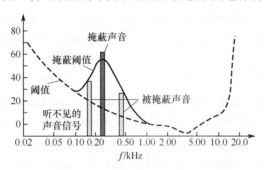

图 4-13 频域掩蔽

(2)时域掩蔽。

从物理过程来看,当声波遇到物体势必会发生反射,产生回音。但有时能听到声音的回音,有时听不到。这取决于人听到声音那一刻与听到回音的那一刻之间的时间差。当时间差很小时,人们听不见回音。时间差主要是由声音传播的距离所决定的。所以在大山里能听见回音,而在一般的房间里听不见回音。

当声音与其回声的时间差很小时,回声是听不到的。这说明,在时间上相邻的声音之间存在掩蔽现象,称为时域掩蔽。时域掩蔽又分为超前掩蔽和滞后掩蔽,如图 4-14 所示。一般超前掩蔽大约只有 5~20ms,而滞后掩蔽可持续 50~200ms。

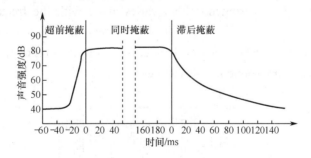

图 4-14 时域掩蔽

(3)声音感知特性与 MPEG 声音编码。

MPEG 声音编码是指 MPEG-1 Audio、MPEG-2 Audio 和 MPEG-2AAC 声音编码。MPEG 声音编码进行数据压缩编码的主要依据是人耳的听觉特性,在此听觉特性的基础上建立"心理声学模型",从而控制、实现音频数据压缩。

心理声学模型的一个基本依据是听觉系统中的听觉阈值电平,低于这个强度的声音信号,人耳就听不见了,所以可以直接把这部分声音数据去掉。

心理声学模型的另一个基本依据是听觉掩饰特性,据此可以对听觉阈值进行自适应调节。声音压缩算法建立了这种特性的模型,用以消除声音数据的冗余,实现数据压缩。

4.1.3　The Major Characteristics of Hearing

The human being's hearing system is complicated. Up to now, people still don't clearly understand the hearing system's psychological structure and principle. In another word, people can't give a really reasonable explanation for the haring system's experiment result. Furthermore, people's hearing is often greatly affected by the psychological factors. Hearing characteristics research is limited to psychological acoustics and language acoustics. Among those, people's sensitivity characteristics to loudness and pitch and shelter effect could be directly applied to sound data compression encoding.

1. Loudness Sensitivity

The sound loudness represents the sound's strength or weakness. In physics, the loudness is measured by either dyne/cm² (sound pressure) or watt/cm² (sound strength). Psychologically, subjectively feel the sound strength and weakness measured either by phon or sone. Those two measurement methods are in totally different ideas and related each other at the same time.

When sound weaken and weaken until just be able to be heard, the loudness of the sound at this point is called hearing threshold.

Hearing Threshold: Under the quiet surroundings, the smallest pure sound people can hear. (The correspondent loudness with the weakest sound people can barely hear.) At different frequency point, there are different hearing threshold. The standard hearing threshold curve could be drawn correspondent to different frequency point f.

After we get the hearing threshold curve, any sound below the curve can't be heard.

For example, when 1kHz pure sound's strength reaches 10^{-16}W/cm² (defined as 0dB strength), people can just barely hear, we define the objective loudness at this point as zero phon. Based on the experiment's results, the hearing threshold varies along with the frequency changing. This relationship is showing in Fig. 4-8. The curve the lowest one in the figure is called "zero phon even loudness curve" and "absolute q hearing threshold curve" as well. It represents that under the quiet circumstances, the lowest pure sound people can hear.

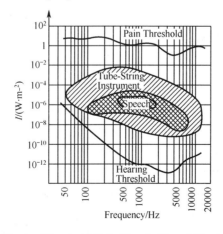

Fig. 4-8　Hearing Threshold, Pain Threshold, and Hearing Regions

The other extreme situation is that the sound is too strong that it hurts people's ears.

To the correspondent sine shape pure sound with different frequencies, the threshold people's ear starting to get hurt is called the "pain threshold" to the related frequencies. Fig. 4-9 shows the "pain threshold" curve correspondent to frequency f.

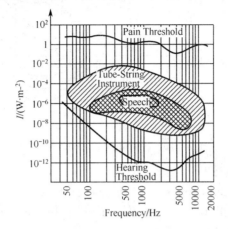

Fig. 4-9　Hearing Threshold, Pain Threshold, and Hearing Regions

From the experiment results done, we know that when the loudness of pure sound with 1kHz reaches about 120dB, people who are hearing the sound will feel ear hurt. This threshold is called "pain threshold". From the experimental measurements, we could get the pain threshold - frequency curve on top of Fig. 4-9. For the same reason as above, it is also called 120 phon even loudness curve.

The region between the hearing threshold and pain threshold is the area people can normally hear. The even loudness curves within this region could be got from related experiments as well as shown in Fig. 4-10. From the Fig. 4-10, we could see that the sound with 10dB amplitude at the frequency of 1kHz has similar loudness to the sound with 30dB amplitude at the frequency of 200Hz to people's hearing. In the lower frequency and higher frequency area, the sound could be heard has much higher amplitude.

The loudness represents the objective feeling of human being's ear to sound's strength and weakness.

Link the points with same loudness and different frequencies, we could get many curves and we call them equivalent loudness curves. They are also called Fletcher-Munson curves as shown in Fig.4-10.

The equivalent loudness curves reflect people's reaction to different sound wave amplitudes psychologically and physiologically. On each curve, the sound wave amplitudes /sound strengths vary along with the different frequencies. But at the same time, the loudness people's ear felt is the same. This is so called equivalent loudness curve. Each curve is correspondent to one loudness value. From a set of equivalent loudness curves showing in Fig. 4-10, we know that people are not that sensitive to the high or low tone sounds when the sound strength is low. On the other hand, when the sound strength is getting stronger, people are getting more sensitive to the high or low tone sounds. The most sensitive areas are the regions where the frequencies are between 2 to 4kHz.

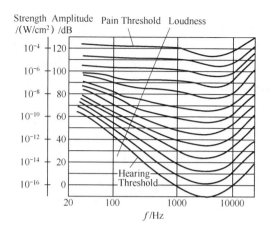

Fig. 4-10　Hearing-Frequency Curves

2. Sensitivity to Pitch/Tone

Sound frequency means the sound signal's frequency.

Pitch is called tone as well. It represents the people's objective hearing feeling to sound's high or low tone. It majorly related to the sound frequency. It's not proportional to sound frequency though. Similar to loudness, the sensitivity of tone has a logarithmic relationship with the sound frequency. That's why the tone is usually represented as multiple or logarithm of frequency.

The unit of using frequency to represent the sound's tone is Hz. The unit of subjective feeling of the tone is mel. The relationship between objective and subjective tone is indicated in the formulae below.

$$Mel = 1000\log_2(1+f)$$

Simile to the measurement of loudness taking 1kHz pure sound as base reference, the measurement of tone takes 40dB strength as a base reference and decide the correspondent tone subjectively.

When measuring, ask the listener who is in the experiment to listen to two 40dB strength pure sound. Fix the frequency of one pure sound and raise up the frequency of the other pure sound until the listener feels the tone of this pure sound is the two times higher than the fixed one. Then mark down the tone of this pure sound is the two times of the fixed one's. From the results of the experiment, we could get the tone-frequency curve indicating the relationship between tone and frequency as shown in Fig. 4-11.

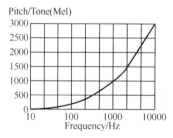

Fig.4-11　Tone -Frequncy Curve

3. Feeling of the Sound Musical Quality

Timber is also called musical quality which is the sound's feeling characteristic.

Different objects could produce different sounds. For this reason, we could tell which one is piano and which one is violin when they are playing with the same pitch and loudness at the same time.

We could identify the different materials, objects, and even people by musical quality.

We'll finish this part by introducing several interesting normal living musical quality example questions.

(1) Why there is different musical quality?

Answer: The sound is produced by the vibrating object. The whole object's vibrating produce the base tone, at the same time, the other parts of the object has compound vibrating as well. The sound produced by those compound vibrating forms the harmonic tones. Since they are parts of the object, the frequencies produced by them are higher than the base's with much weaker loudness.

The different tones depend on the different harmonic tones. The sound produced by every musical instrument, different people, and all other objects that can create the sound has base tone. Based the base tone, there are many different frequency harmonic tones existing. It's just those harmonic tones that made the different musical qualities which provided the chance for people to identify the sound produced by different musical instrument and even the different people.

(2) What kind of characteristics does the musical quality have?

Answer: People are very sensitive to the musical qualities. Some of the characteristics are as follows.

① Remembering.

When you talk to somebody you know, you know whom are you talking to without seeing him/her. When you know him/her well, you can identify he/she is walking closely. This tells us that our ear can well remember the musical quality of the sounds we often hear.

② Identity.

People who are familiar with the musical instrument could quickly tell which sound comes from which instrument as soon as they hear the sound. Even they are playing in the same frequency region, one still could tell the sound comes out form which string instrument. This means that each instrument has its own particular musical quality. People's ear is doing very well to identify the different musical qualities.

(3) What is musical quality sensitivity?

Answer: The musical quality sensitivity means a particular integrated musical hearing feeling characteristic people's ears have. Even you use the best music integrator in the world to simulate kinds of musical instrument, such as horn, piano, and other instrument and get the exact frequencies and musical quality, the musicians and skilled music listeners still can clear tell the music is produced by the integrator instead of natural playing. From this we know that even the frequencies and musical quality are the same, the human being's complicated musical quality sensitivity is still not the same. This gives the result that music hearing feeling is different. This also shows that the musical quality sensitivity is the human being's own particular complicated integrated feeling and can't be simulated.

4. The Sound's Masking Effect.

When we are listening to one sound and can't really hear it since another sound around sheltered it. This is called masking effect and the overriding sound is called masking sound. as shown in Fig 4-12. The sound's hearing threshold would rise because of another overriding sound's masking. This phenomena is called masking effect.

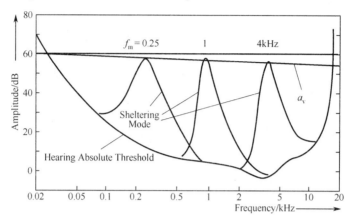

Fig. 4-12   Masking Effect

The masking effect has following characteristics. The closer the frequency of the sheltered sound to the masking sound's, the bigger the shelter is. The stronger the masking sound, the bigger the shelter is. It's easy for the sound that has low frequency to shelter the sound with high frequency. It's not that easy the opposite way. The typical example is that when we are listening the HiFi playing music, we won't hear noise when the music is playing and we do hear the HiFi's background noise when the music stops or pauses.

Furthermore, the masking effect mainly could be classified into two kinds. One is frequency domain masking and the other is time domain masking.

(1) Frequency Zone Masking.

For the sounds which have similar frequencies, the one which has bigger loudness will shelter the one which has smaller loudness. This is called frequency zone masking. The one that has bigger loudness is called masking sound and the one that has smaller loudness is called sheltered sound. For example, when the students inside the classroom are talking about one question loudly, the teacher may say in a high voice: "Please stop talking and be quiet!". But the students may not be able to hear clearly or absolutely not hear what the teacher said at this point. The teacher may have to rise up his voice again and repeat the words. This time, the students heard what the teacher said and stopped talking immediately. In this scenario, the talking sound frequencies of the teacher's and the students' are close each other. The teacher's talking sound was sheltered at the beginning. The second time, the teacher rose up his voice which increased the sound loudness and made it higher than students'. This way, the teacher's sound masking the students' and the students heard.

One strong pure sound could shelter its nearby weak pure sounds as shown in Fig 4-13. One of the scenarios is that one pure sound has loudness 60dB and frequency 1kHz and another pure sound close by has loudness 42dB and frequency 1.1kHz. The first one's loudness is 18dB higher than the second one. In this situation, people only can hear the first pure sound and not the second one. Another scenario is that one pure sound has frequency 1kHz and 18dB stronger than another pure sound close by with frequency 2kHz. This time, people can still hear both pure sounds since their frequencies are apart from each other. If one particular don't want to hear the pure sound with 2kHz, then the loudness of this pure sound's loudness has to be reduced to 45dB lower than the one's with 1kHz. General speaking, the closer the frequencies of the stronger and the weaker pure sounds, the easier the weaker one could be sheltered.

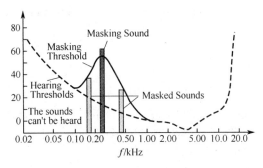

Fig. 4-13   Frequency Zone Masking

(2) Time Zone Masking.

From the principle of physics, when sound wave meets objects, it will reflect from the object. But sometimes we can hear the reflected sound and sometimes can't. It depends on the time difference between the original sound and the reflected sound. When the time difference is very small, we won't be able to hear the reflected sound. This time difference is mainly decided by the distance the sound propagated. For this reason, we could hear the reflected round in mountains and can't in normal size rooms.

The phenomena that we can't hear the reflected sound when the time difference between the original sound and the reflected one is very small indicates that the sounds in time zone being close each other have masking effect. This masking effect is called time zone masking. The time zone masking is classified into two kinds, ahead masking and behind masking as shown in Fig. 4-14. Usually the ahead masking only lasts about 5 to 20ms and the behind masking lasts much longer in about 50 to 200ms.

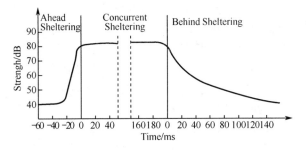

Fig. 4-14   Time Zone Sheltering

(3) Sounds' Perceiving Characteristics and MPEG Audio Encoding.

MPEG audio encoding means MPEG-1 Audio, MPEG-2 Audio, and MPEG-2AAC audio encodings. The foundation of the MPEG audio encoding data compression is the people's hearing characteristics. The "psychological audio model" is created based on this base to control and realize the audio data compression.

A base of the psychological audio model is the hearing threshold. Any sound lower than the hearing threshold won't be heard. This way, we could directly get rid of those audio data.

Another base of the psychological audio model is the hearing masking characteristic. The hearing thresholds could be auto-adjusted accordingly. Audio compression algorithm creates the model with the characteristic as above to remove the redundancy within the audio data and compress the audio data.

4.1.4　多声道声音

空间听觉又称为立体声，是具有空间感的声音。

由于人类的耳朵能够判别出声波到达左右耳的相对时差和音强差，所以人耳能够判别声音的来源。同时人的双耳能够在同一时刻听到两个或两个以上的不同声源发出的声音，因此可以产生出复杂多变的听音效果。

立体声效就是利用人耳的方位感特性提高音响效果的。

立体声技术是利用立体声效，在放音时重现各种声源的方向及相对位置的技术。立体声重放给人的感觉不像单道声那样从一个"点"发出，而是感到声源分布到了一个较宽的范围。

电话、中波广播的声音由一个声源发出，称为单声道声音。单声道声音不能很好地表现出声音的内涵质量，例如不能表现出声音的时空变换效果。使用两个发声装置，发出两个不同的声音，分别通过不同的路径送到人的左右耳，就称为双声道声音。

如果在人的周边有许多声源同时发声，声音分别从不同的位置、以不同的方式、不同的路径传播到人的双耳，这就是多声道声音。对于多声道声音，人耳不仅能产生更多的声音细节，还能感知到音源的位置、位置移动、变化、音源距离等情况，有关声音在声场空间中产生的效果，称为三维音效。由此人的大脑可以在眼睛没有看到画面的情况下，设想出与实际情形十分符合的画面场景；也可以把听音效果与所看到的画面场景精细地结合，使声、画互动，声音促进对画面的理解，画面则促进精细地听音，从而产生出与现实、自然场景一致的声场效果。声源本身的信号特征、声源的空间位置、声源所处的环境这三个因素描述了声源信息。多声道系统之所以能够产生如此的听音效果是由人耳的听音特征和耳轮结构所决定的。在前边已经提到过，人的双耳在听到声音的同时，还能识别出某个声源与人的左右耳距离的差异造成的时间差异，识别出声音到达左右耳时的强度差别，图 4-15 和图 4-16 描述了这两种关系。这种人类感知声源位置变化的理论称为双工理论，它是根据实验得到的结果。

耳间时间差和耳间强度差是人耳声音定位的主要线索。强度差是由于信号在传输过程中发生衰减造成的。声音信号的衰减是因为距离或物体遮挡而产生的。人的头部遮挡了声音，使声音衰减，在人的左右耳之间产生了强度的差别，靠近声源一侧的耳朵听到的声音强度要大于另一耳（图 4-15）。时间差是由声音传输距离差造成的。当声音从正面传来时，到达人耳距离相等，时间差为零。若音源偏左或偏右 30°，则声音到达双耳的时间就有约 30μm 差异，正是这点时间差，使得人耳辨别出了声源的位置（图 4-16）。

在三维音效产生过程中，耳廓也发挥了重要作用。耳廓结构复杂，对声波产生的作用也复杂。人们意识到耳廓对声音的定位发挥着重要作用，提出了声源定位的耳廓效应。人们还不能很好地构造出耳廓的数学应用模型。实际上，到达人耳的声信号会受到人的外耳、肩部以及躯干的滤波作用，这些声学滤波器效果目前采用头部相关传递函数 HRTF（Head—Related Transfer Function）来表示，用于解释人耳对声信号的三维定位问题。

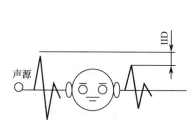
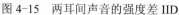

图 4-15　两耳间声音的强度差 IID

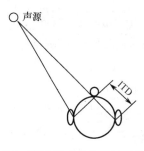

图 4-16　两耳间声音的到达时间差 IID

以上的声学原理被用于实现多媒体计算机系统中的一些音效处理过程。

根据需求的不同，现有的立体声效系统有 2/0、3/0、3/1、3/2、5.1 声道、7.1 声道等。5.1、7.1 中的 ".1" 就是指 LFE（Low Frequency Effects，低频音效。频率为 15～120Hz，又称为 Woofer）声道，如图 4-17 所示。

在实际应用中，常见的多声道音频系统是 5.1 声道和 7.1 声道系统。

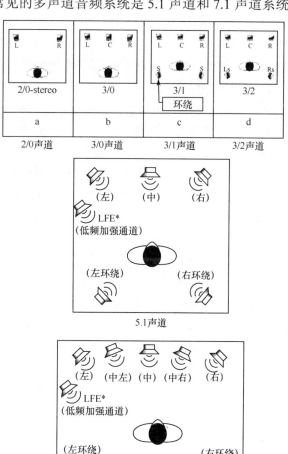

图 4-17　多声道音频系统

4.1.4 Multi-Channel Sounds

Space hearing - Space hearing is also called stereo. It is the sound has space feeling.

Since people's ears could tell the time and strength differences the sound reach the left ear and right ear, the ear could tell where the sound come from. At the same time, ears could listen to the sounds comes from two or more sources simultaneously. For this reason, the complicated variable hearing effect could be produced.

The stereo is applying the ear's direction oriental characteristic to improve the HiFi effect.

The stereo technology is applying the stereo sound effect to represent the sound sources' direction and relative position when broadcasts the sound. The stereo playing is not like single channel the sound feels like coming out from one point. In stereo playing, the sounds feel like coming from serious distributed sound sources.

The sound of telephone and AM comes from one source and so is called single channel sound. The single channel sound can't represent the sound inner deep affection. It can't represent the sound time space exchange effect. It can't produce two different sounds using two different speakers at the same time to send them to people's ears. This kind of sound is called dual channel sounds.

Let's say there are several sound sources producing the sounds at the same time and the sounds come out from different locations. The produced sounds get to the people's ears in different formats and via different paths. This type of sound is called multiple channel sound. To this kind of multiple channel sound, people's ears not only could create many sound details, but also could sense the sound source's position, moving, changing, distance, and etc. Those related to the effects created by the sound in the hearing range are called three dimension sound effects. From those effects, people could imagine the real picture which is close to the real one without seeing it. What people also could do is to thoroughly combine the sound with the seen picture to make them compensate each other. The sound could help people to better understand the picture and the picture could help people to better listen to the details of the sound. This way could produce the real and natural pictures with the vivid sounds. Three factors, sound source's own characteristics, its position, and its surrounding condition give out the sound source's information people need to know. The reason for the multi channel sound system could give out the effects as described is for the ears' hearing characteristics and its structure. As we mentioned, when people's ears are listening, they also could identify the time difference for the sound get to the left and right ear from one source because of the distance difference from the source to the left and right ear. They also could identify the strength difference for the sound get to the left and right ear. Those two types of relationship are shown in Fig. 4-15 and Fig. 4-16. This ears sense sound source position changing theory is called two way working theory. It's an experimental result.

The strength and time differences between the two ears are the major clue for people's ears to identify the sound source. The strength difference comes from the sound signal's attenuation during the propagation. The sound signal's attenuation is caused by either distance or object obstructing.

One's head blocks the sound and created the attenuation and so the strength difference between the two ears. The ear closer to the sound source would receive a stronger sound and the other one would receive a weaker sound as shown in Fig. 4-15. Time difference is created by the propagation distance difference. When the sound comes from the opposite, the distances to the one's two ears are the same, the time difference will be zero. If the sound source leans 30° either to the left or to the right, there will be about 30μs time difference for the sound reach two ears. It is just because of this small time difference, the ears could identify where the sound source is as shown in Fig. 4-16.

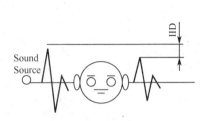

Fig. 4-15   The Sound Strength
Difference Between Two Ears

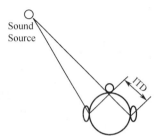

Fig. 4-16   The Sound Reaching Time
Difference Between Two Ears

During the process of three-dimension sound effects creating, the outer ear plays a crucial role. The structure of the outer ear is complicated and has quite difficult to understand functions to the received sound. People have realized its crucial position for the ear to locate the sound source and proposed the outer ear effect on sound source locating. Up to now, there is not an ideal outer ear mathematical model available. Actually, the sound reaching the one's ear would filtering affected by one's outer ears, shoulders, and body. Those sound filtering functions are represented by HRTH, Head-Related Transfer Function. It is used to explain how the ears do the three dimension sound source locating.

The acoustic principles stated above are applied in the multimedia computer dealing with the sound effects processing.

Based on the different sound environmental requirements, there are different stereo systems available, such as 2/0, 3/0, 3/1, 3/2, 5.1, and 7.1, where the ".1" means the LFE, Low Frequency Effect. Its frequency is from 15 to 120Hz. It is also called "woofer" as we familiar with. This is shown in Fig. 4-17.

In reality, the most often seen systems are 5.1 and 7.1 channels systems.

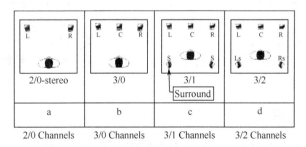

a	b	c	d
2/0 Channels	3/0 Channels	3/1 Channels	3/2 Channels

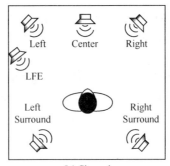

5.1 Channels

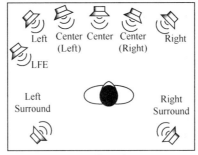

7.1 Channels

Fig. 4-17　Multiple Channels Audio Systems

4.2　声音数字化

在这一节里，将讲述以下几部分内容。

（1）什么是数字音频；

（2）数字音频的特点；

（3）采样和量化；

（4）影响数字音频质量的技术参数；

（5）数字音频文件的存储量。

4.2.1　数字音频

有史以来，大多数电信号的处理一直是用模拟元器件（如晶体管、变压器、电阻、电容等，对模拟信号进行处理。但是，开发一个具有相当精度且几乎不受环境变化影响的模拟信号处理元部件是相当困难的，而且成本也很高。

如果把模拟信号转变成数字信号，用数字来表示模拟量，对数字信号做计算，那么就避开了难点。把开发模拟运算部件的问题转变成开发数字运算部件的问题，这就出现了数字信号处理器（Digital Signal Processor，DSP）。DSP 与通用微处理器（Processor）相比，除了它们的结构不同外，其基本差别是 DSP 有能力响应和处理采样模拟信号得到的数据流，如做乘法和累加求和运算。

4.2.2　数字音频的特点

在数字域而不在模拟域中做信号处理的主要优点如下。首先，数字信号计算是一种精确的运算方法，它不受时间和环境变化的影响；其次，所涉及的数学运算不是物理上实现的功能部件，而是仅用数学运算去实现，所以相对容易；此外，可以对数字运算部件进行编程，如改变算法或改变其某些功能，可仅对数字部件进行再编程即可。

语音信号是典型的连续信号，不宜在时间上是连续的，而且在幅度上也是连续的。在时间上"连续"是指在一个指定的时间范围里，声音信号有无穷多个值。在幅度上"连续"是指每一个特定的时间点上，幅度的数值有无穷多个。在时间和幅度上都连续的信号称为模拟信号。

模拟音频：在时间和幅度上都是连续变化的音频信号。

4.2.3　采样和量化

1. 采样

在某些特定的时刻对模拟信号进行取值称为采样（Sampling），由这些特定时刻采样得到的信号称为离散时间信号。采样得到的幅值仍然是无穷多个实数值中的一个，因此幅度还是连续的。

如果把信号幅度取值的数目加以限定，这种由有限个数值组成的信号就称为离散幅度信号。这种过程就是量化（Quantization）。简略地说，量化的过程是一个四舍五入的过程，量化误差是存在的。一般来说，这限定的有限个数值越多，或说表达信号幅度的值，其小数点后的位数越多，误差越小，但数据量也随之越大。所以需要根据实际的情况的需要来均衡和取舍。

假设某音频信号输入电压的范围是 0.0～0.7V，限定它的取值在 0，0.1，0.2，…，0.7 共 8 个值的范围内，取值长度限定在小数点后一位。如果在某个采样点得到的实际幅度值是 0.123V，则它的取值就应算作 0.1V，如果得到的实际幅度值是 0.26V，则它的取值就算作 0.3，以此类推，就把可取值无限的幅度值以这种方式以有限的数值取代来表示，这些取代的数值就称为离散数值。

总结一下，通过采样把无限的实数集里的时间值用有限的数字值来表示，通过量化把无限的实数集里的幅度值用有限的数字值来表示，这样就得到了把信号在时间和幅度上都用离散的数字表示的数字信号。对于音频信号，通过类似的处理，就可得到我们所要的数字化了的音频——在时间和幅度上都是离散的、不连续的音频信号。

现在的多媒体计算机功能很强大，但再强大其本质上也是计算机，其内部只能处理数字信息。各种命令是由不同的数字表达的，各种幅度的物理量也是由不同的数字表达的。而人们通常听到的声音都是模拟信号，怎样才能让计算机也能处理这些声音数据呢？唯一的方法是将音频信号用如上的办法数字化，用一系列有限的数字来表示音频信号，即数字音频。

那么模拟音频与数字音频有什么主要不同？数字音频有哪些优点呢？

简单地说，数字音频的特点是保真度好，便于存储、传输和处理。

人们熟悉的模拟音频播放，通过扩大器、扬声器，就可以听到模拟音频的声音。那么数字音频又是如何播放的呢？

其实最后的步骤和模拟音频是很相似的，但到那之前，需要做几件事情。首先，将这些由大量数字描述而成的音乐送到数/模转换器（Digital to Analog Converter，DAC）里，它将数字回变成一系列相应的模拟电压（电流）值，然后通过有助于稳定的保持线路，最后将信号由低通滤波器输出。这样，比较平缓的具有脉动电压的模拟信号可持续不断地发送至放大器和扬声器，伴随的电流经过放大再转变成人们听到的声音，这就是数字音频播放的一个非常简单的描述。

如上所述，声音进入计算机的第一步就是数字化，数字化实际上就是采样和量化再加上编码，如图 4-18 和图 4-19 所示。

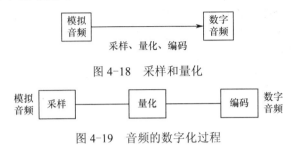

图 4-18　采样和量化

图 4-19　音频的数字化过程

如前所述，连续时间的离散化通过采样来实现，就是每隔相等的一小段时间采样一次，这种采样称为均匀采样（Uniform Sampling）；连续幅度的离散化通过量化来实现，如我们前面所做的，把信号的强度划分成一小段一小段，如果幅度的划分是等间隔的，就称为线性量化，否则就称为非线性量化。图 4-20 表示了声音数字化的概念。

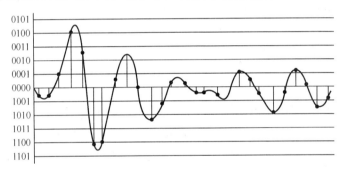

图 4-20　声音的采样和量化

进一步来看一下这个过程。

图 4-21 中，横轴代表时间轴，纵轴代表幅度轴。

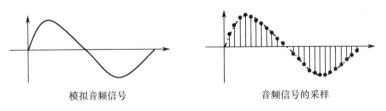

模拟音频信号　　　　　　　音频信号的采样

图 4-21　采样过程

分别来看。

采样：将时间上连续的取值变为有限个离散取值的过程。

显而易见，采样点是有限的、离散的，那么每秒钟需要采集多少个声音样本，也就是采样频率为多少才算合理呢？

来看下面几幅有趣的图形。

从数据存储和传输的角度来看，采样点越少，所得到的数据就越少，那么对同样一个音频信号而言，所需要存储和传输的数据就越少，就越经济。但是如前所述，当播放数字音频

184

的时候，是需要把数字音频转换为模拟音频的，换句话说，采样之后的数字音频在一定阶段是需要还原成模拟音频的。这时，问题就出来了，从数据存储和传输的角度来看，数据越少越好，反推回去，就是采样点越少越好。但是，从数字音频还原回原来的模拟音频的角度来看，是个什么情形呢？直观的想，数据如果少到一定程度后，还能否还能还原成所要的效果呢？来看看图4-22所示的图形。

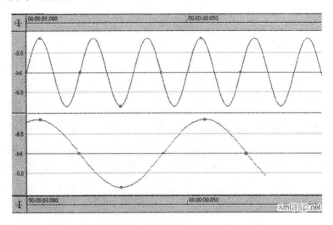

图 4-22　采样频率为多少才算合理呢（一）
图中，上半部分表示原始音频的波形；
下半部分表示恢复后的波形；红色的点表示采样点。

显然，恢复后的波形与原始波形相差很远，这说明这样采样后的信号不能够代表原始的信号。这是不能接受的，造成的原因是采样点太少了。

模拟声音在时间上是连续的，而数字音频是一个数字序列，在时间上只能是断续的。因此当把模拟声音变成数字声音时，需要每隔一个时间间隔在模拟声音波形上取一个幅度值，称为采样，采样的时间间隔称为采样周期。每秒的采样次数称为采样频率。上面的情况是采样周期过长或采样频率过低形成的。

那么采样周期多长或采样频率多高才是合适的，才能够从采样后的信号里恢复出原始的信号呢？奈奎斯特采样定理回答了这一问题。

先来看图4-23所示的图形。

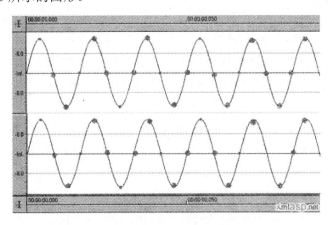

图 4-23　采样频率为多少才算合理呢（二）
图中，上半部分表示原始音频的波形；
下半部分表示恢复后的波形；红色的点表示采样点。

图 4-23 中，从采样后的音频信号中可以完整的恢复原信号，图中所使用的采样频率刚好是音频信号频率的两倍，这其实就是奈奎斯特采样定理的内容。

奈奎斯特定理指出：采样频率不应低于声音信号最高频率的两倍，这样就能把以数字表达的声音还原成原来的声音，这称为无损数字化（Lossless Digitization）。采样定律用公式表示为

$$f_s \geq 2f \text{ 或者 } T_s \leq T/2$$

式中：f 为被采样信号的最高频率。

结合上面的采样图形里的典型正弦波波形来这样理解奈奎斯特定理：声音信号可以看成由许许多多正弦波组成的，这一点从典型的傅里叶变换当中也可以得到。一个振幅为 A、频率为 1 的正弦波每秒至少需要两个采样样本来表示，因此，如果一个信号中的最高频率为 f_{max}，则采样频率最低要选择 $2f_{max}$。例如，电话话音的信号频率为 3.4kHz，采样频率通常选为 8kHz。

或说设连续信号 $X(t)$ 的最高频率分量为 F_m，以等间隔 T_s（T_s 称为采样间隔，$F_s=1/T_s$ 称为采样频率）对 $X(t)$ 进行采样，得到 $X_s(t)$。如果 $F_s \geq 2F_m$，则 $X_s(t)$ 保留了 $X(t)$ 的全部信息（从 $X_s(t)$ 可以不失真地恢复出 $X(t)$）。

也即只要采样频率高于信号中最高频率的 2 倍，就可以从采样中完全恢复原始信号的波形。

4.2　Sound Signal Digitalizing

In this section, we'll discuss the following contents.

（1）What is digital audio;

（2）The characteristics of the digital audio;

（3）Sampling and quantifying;

（4）The technical parameters affecting digital audio quality;

（5）The storage quantity of the digital audio files.

4.2.1　Digital audio

It has been for a long time that most signal processing was using analog devices such as transistor, transformer, resistor, and capacitor to deal with the analog signals. But it's hard to develop a high accuracy analog signal processing component or device which is not affected by the surrounding conditions. The cost is very high as well.

If we transfer the analog signal to digital signal and represent the analog values in digital format, then deal with the digital values instead of analog ones, we convert the hard ones to much easier ones. We turned the issue of developing analog signal processing device to develop digital signal processing device. This is where the digital signal processor, DSP devices come from. Comparing the DSP and the microprocessor, besides the structure difference, the DSP is capable of responding to and processing the data stream produced from the analogue signal sampling, such as dealing with the multiple and summarizing accumulate adding.

4.2.2　The Characteristics of the Digital Audio Signals

The advantages of digital signal processing via the analogue signal processing are as follows. First of all, the digital signal calculation is a precise algorithm. It is not affected by time and the surrounding condition changing. Secondly, the mathematical calculation is not the physically realized device. But only need mathematical calculations. Plus the mathematical calculation is easy to put into the practice. Lastly, one could program the digital calculation device. Whenever one needs to change the algorithm or some of its functions, all he/she needs to do is to reprogram the device.

The audio signal is a typical continuous analog signal. It's not only continuous in time axis, but also in amplitude. Continuous in time axis means that within a time frame, there are unlimited audio signal values available. Continuous in amplitude means that at certain time point, there are unlimited amplitude values available. The signal which is continuous both in time axis and amplitude is called analog signal.

Analog audio signal - the audio signal continuous both in time axis and amplitude.

4.2.3　Sampling and Quantifying

1．Sampling

Taking amplitude values at certain time points is called sampling. The signal formed by those sampled values is called time discrete signal. The amplitude value got from the sampling is still one of the unlimited values. So, at this point, the signal is still continuous in amplitude.

If we limit the amplitude values within a certain number of values, then the signal will contain certain limited number of values. This kind of signal is call amplitude discrete signal. This process is quantization. Simply speaking, the quantization is a round up process. The quantization error exists. General speaking, the more the limited amplitude values picking from or the more digits you take after the decimal points, the smaller the quantization error is. But the necessary storage data will be getting bigger at the same time. That's why we have to make the choice based on particular applications.

Let's say we there is an audio signal whose input voltage is between 0.0 to 0.7V. We limit its taking value within the range of 8 serious limited values as 0.1, 0.2, ⋯ 0.7 and each value has one digit after the decimal point. In this case, if one actual sampled value is 0.123V, we take it as 0.1V. If another actual sampled value is 0.26V, we take it as 0.3V. Repeat the process, we could convert the unlimited amplitude values between 0.0～0.7V to a limited values as one of the eight values, 0.1 to 0.7 which are called discrete amplitude values.

We put the above together, we represent the signal's unlimited time zone values inside the real number collection with limited digital values via sampling and represent the signal's unlimited amplitude values within the real number collection as well with limited digital ones via quantization. This way we get the digital signal we wanted which is represented in limited digital values both in time zone and amplitude. Similar to audio signal, we process the audio signal same way and could get the digitalized audio signal both represented with limited digital values in time zone and amplitude. In another words, both discrete in time zone and amplitude.

As we know, multimedia computers are strong enough to handle many things at the same time. But no matter how good it is, it is actually still a computer which can only deal with digital values. Inside the computer, the commands are passed in digital format and different actual physical values are presented in digital format as well. The sounds we normally heard are analog signals. What should we do to deal with them? What we have indicated above actually answered this question already, digitalization. We could get the digital audio. after the digitalization process shown above.

Then what is the main difference between the digital and analog audio signals? What is the benefit of the digital audio signal?

The simple answer to the above question is that the digital audio signal has high audio quality, easier to store, transmit, and process.

We are familiar with how the analog audio signals broadcasted. We used amplifier and speakers to listen to the analog audios. Then how the digital audio signal is broadcasted?

Actually, the final steps are very similar to the analog ones. But before we get to there, we need to do several things. Firstly, we need to send those digitalized audio to a device, DAC, digital analog converter. This device will convert back the input digital audio to analog audio, a serious of correspondent voltages/currents. Secondly, send those analog signal values to a stabilizing circuit. Finally, send the signal go through a low band filter. This way, the slowly varying pulse analog voltage continuously sent to the amplifier and speakers. The accompanied current via an amplifier converted to the sound we heard. This is a very simple description of the digital audio signal broadcasting.

As we talked above, the first step for the audio signals get into the computer is the digitalizing. Simply speaking, the digitalizing is the process of sampling and quantifying, in reality, plus the encoding as shown in Fig. 4-18 and Fig. 4-19.

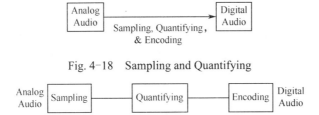

Fig. 4-18   Sampling and Quantifying

Fig. 4-19   Audio Digitalizing Process

As indicated earlier, we could get the time zone discrete signal by sampling. That is taking one sample for every certain time interval. This is called uniform sampling. Furthermore, we could get the discrete amplitude by quantization. As we did, divide the signal's amplitude into one after one small section. If every section is evenly from each other, the quantization is called linear quantization; otherwise, it is called non-linear quantization. Fig. 4-20 shows the idea of quantization.

Let's have a close look at the process.

In Fig. 4-21, the horizontal axis represents the time and the vertical axis represents the amplitude.

Let's have a look at the sampling and the quantifying separately.

Sampling: the process of converting the analog signal which is continuous in time zone to time discrete signal with limited values.

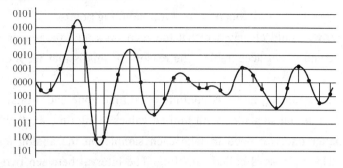

Fig. 4-20   Audio Sampling and Quantifying

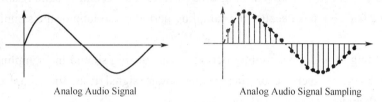

Analog Audio Signal   Analog Audio Signal Sampling

Fig. 4-21   The Process of Sampling

It's easy to see that the sampling points are discrete and limited. Then how many samples should we take within one second will be reasonable? In another words, what is the reasonable sampling frequency?

Let's have a look at several interesting graphics as follows.

If we think about data storage and transmission, for a certain audio signal, the fewer samples we take, the less data we have. As a result, the less data needed to be stored and transmitted is and it's more economical. It's a good thing. As we talked, when we broadcast the digital audio, we need to convert the digital audio back to analog audio. In another words, we need to convert the digital audio back to analog audio at certain point. Now we have an issue. For data storage and transmission, the less data, the better. Therefore, the less sampling points, the better. For digital audio converting back to analog audio, what would it be? We could imagine that when the data quantity is getting less and less, to certain point, could we still get what we want? Please refer to the following fig. 4-22.

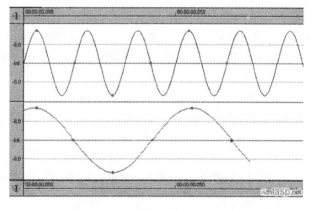

Fig. 4-22   Reasonable Sampling Frequency (1)

189

In this figure, the above part is an original analog audio signal and the lower part represents the recovered audio signal. The red dots represent the sampling points.

From the figure, we could clearly see that the recovered audio signal is distorted too much from the original one and is not acceptable. The reason for this is that we took too less sampling points.

As we already know, analog audio is continuous in time zone and digital audio is discrete in time zone. The latter is digital sequence with limited values along time axel. When we convert the former one to the latter one, we need to take each sample on the analog audio signal at each discrete time point. This process is called sampling. The interval between two sampling points is called sampling period. The total sampling times within one second is called sampling frequency. The situation we had above is because the sampling period is too long or the sampling frequency is too low.

Then how long should the sampling period be or how big should the sampling frequency be? Thereafter, we could recover the original analog audio signal from the sampled discrete audio signal. Naquist theorem has an answer for this question.

Let's have a look at the Fig.4-23.

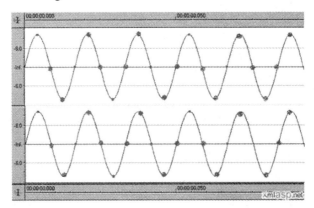

Fig. 4-23   Reasonable Sampling Frequency (2)

Within the figure, the top half represents the original audio signal, the lower half represents the audio signal recovered from the sampled samples. The red dots represent the sampling points.

In this figure, the recovered audio signal is the same as the original one. This means that from the sampled audio signal we could completely recover the original audio signal. The sampling frequency used in this example is just the two times of the original audio signal's. This is actually what the Naquist theorem will say.

Nyquist theorem: Whenever the sampling frequency is not lower than two times of the highest frequency of the original signal, the original signal could be completely recovered from the sampled discrete signal.

This kind of digitalization is called lossless digitalization.

The Naquist theorem could be represented in mathematical format as follows.

$$f_s \geqslant 2f \text{ or } T_s \leqslant T/2$$

The f is the original signal's highest frequency.

Looking at the typical sinewave graphics in the sampling figures above, we could understand the Naquist theorem this way: signal, in our situation, audio signal composes many sinewaves with different frequencies which has been approved in the signal Fourier transform. A sinewave with amplitude A and frequency 1 needs at least two samples to represent within one second. So, if the audio signal's highest frequency is f_{max}, the sampling frequency is at least $2f_{max}$. A real example is the telephone signal's frequency is 3.4kHz and the correspondent sampling frequency is 8kHz.

In another words, suppose the continuous signal $X(t)$'s highest frequency is F_m, use the even time interval T_s sampling the signal, where T_s is called sampling interval and $F_s = 1/ T_s$ is called sampling frequency). The sampled signal we got is $X_s(t)$. If $F_s \geqslant= 2F_m$, then $X_s(t)$ contains all the information $X(t)$ has. In another words, from $X_s(t)$, could without distortion completely recover the original signal $X(t)$.

This is to say that as soon as the sampling frequency is equal to or more than two times the highest frequency of the original signal's, one could completely recover the original signal from the sampled signal.

2. 量化

音频量化：将经采样后幅度上无限多个连续的样值变为有限个离散值的过程。

在数字音频技术中，经采样后，时间轴上变成了离散信号，但模拟电压的幅值仍然是连续的，而用数字表示音频幅度时，只能把无穷多个电压幅度用有限个数字表示，即把某一幅度范围内的电压用一个数字表示，这就称为量化。图4-24形象地表示出了这一过程。

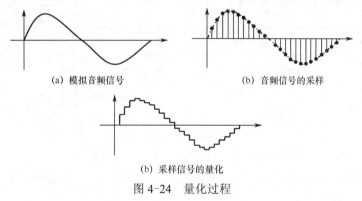

(a) 模拟音频信号　　　　　　　　(b) 音频信号的采样

(b) 采样信号的量化

图4-24　量化过程

其具体的过程可简述为：先将整个幅度划分成为有限个幅度（量化阶距）的集合，把落入某个阶距内的样值归为一类，并赋予相同的量化值，如图4-25所示。

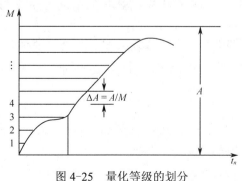

图4-25　量化等级的划分

量化精度或说样本大小是用每个声音样本的位数(bit/s 或 b/s)表示的，它反映度量声音波形幅度的精度。例如，每个声音样本用 16 位（2 字节）表示。测得的声音样本值是在 0～65535 的范围里，它的精度就是输入信号的 1/65536。

样本位数的多少影响到声音的质量，位数越多，声音的质量越高，位数越少，声音的质量越低；但反过来，位数越多，而需要的存储空间也越多，这是不理想的地方。位数越少，需要的存储空间越少，这是人们希望的。所以，声音的质量和位数的多少需要根据具体应用情况折中考虑。

再来看一下前面提到过的一个量化例子。假设输入电压的范围是 0.0V～0.7V，并假设它的取值只限定在 0，0.1，0.2，…，0.7 共 8 个值。如果采样得到的幅度值是 0.123V，它的取值就应算作 0.1V，如果采样得到的幅度值是 0.26V，它的取值就算作 0.3，这种数值就称为离散数值。

图 4-26 中表示出了不同的三种量化精度的量化数值。

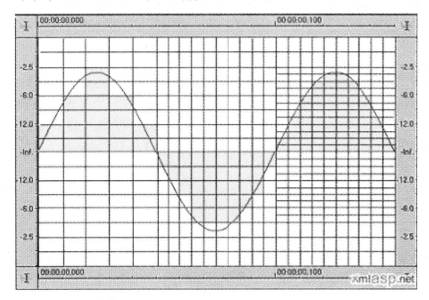

图 4-26　不同的三种量化精度的量化数值

在此，另外值得一提的是，量化精度的另一种表示方法是信号噪声比（信噪比），并用下式计算：

$$\text{SNR} = 10\lg[(V_{\text{signal}})^2/(V_{\text{noise}})^2] = 20\lg(V_{\text{signal}}/V_{\text{noise}})$$

式中：V_{signal} 为信号电压；V_{noise} 为噪声电压；SNR 的单位为分贝（dB）。

下面的两个列子解释了此公式的计算方式。

例 1：假设 $V_{\text{noise}} = 1$，量化精度为 1 位表示 $V_{\text{signal}} = 2^1$，它的信噪比 SNR＝6dB。

例 2：假设 $V_{\text{noise}} = 1$，量化精度为 16 位表示 $V_{\text{signal}} = 2$ 的 16 次方，它的信噪比 SNR＝96dB。

2．Quantifying

Audio signal quantifying: It is the process to convert the sampled unlimited continuous amplitude values into the limited discrete values.

In digital audio technology, the continuous analog audio signal becomes the discrete audio signal along time axel after the sampling. At this point, the amplitude is still continuous with unlimited values. When we use numbers to represent the discrete audio signal's amplitude, we're trying to use limited digits to represent the unlimited amplitude values. The only way to make this happen is to divide those unlimited values into separate regions where put all the values inside the region into one integrated digital value. This process is called quantifying. Fig. 4-24 shows the process.

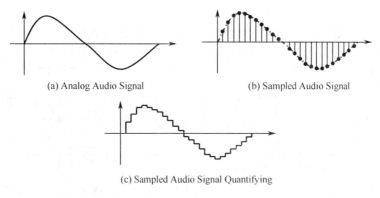

(a) Analog Audio Signal   (b) Sampled Audio Signal

(c) Sampled Audio Signal Quantifying

Fig. 4-24   Quantifying Process

The detailed process could be simplified as follows. Divide the whole amplitude range into limited divisions which is called quantifying stages. Put all the values inside each divided division into one and represent it with one digital value as shown in the following Fig.4-25.

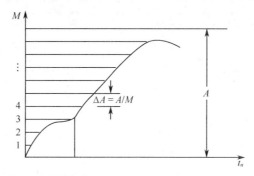

Fig. 4-25   Get the Quantifying Stages

The quantifying accuracy or quantifying sample size is represented by the sample's digits, bit/s or b/s. It reflects the accuracy of representing the quantified amplitude. For example, if each amplitude sample is represented by a 16 digits binary number which in size is two bytes, the amplitude range could be represented by those 16 digits is $0\sim65535$. The quantifying accuracy in this quantization is $1/65535$ of the input time discrete signal.

The size of the quantifying sample affects the sound quality. The bigger the size or the more the digits, the better the sound quality is. The smaller the size or the less the digits, the poorer the sound quality is. On the other hand, the more the digits, the more storage space will be needed for data storage. This is the downside. The less the digits, the less storage space will be needed for data storage. This is what we want. In this case, we have to compensate the sound quality and storage space each other and get what we need.

Let's have another look at the quantifying example we mentioned. Suppose the input signal's amplitude range is 0.0~0.7V and its quantified values will be one of the group of 8, 0.0, 0.1, 0.2, ··· 0.7. If time sampled amplitude is 0.123V, its amplitude quantifying value shall be 0.1V. If time sampled amplitude is 0.26V, its amplitude quantifying value shall be 0.3V. The values, 0.1, 0.3 and etc. are called discrete values.

Fig. 4-26 gives three quantifying examples with three different quantifying accuracies (three different digits sizes).

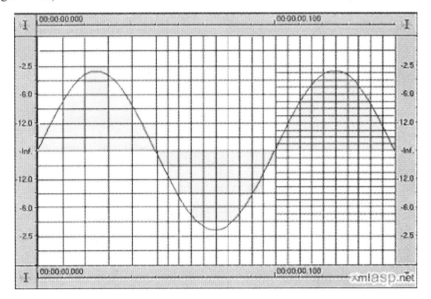

Fig. 4-26　Three Different Quantifying Accuracies and Quantifying Values

Another thing worth to mention is that another way to represent the quantifying accuracy is the signal noise ratio. Simply expressed as SNR. It could be calculated from the following formula.

$$SNR = 10\lg[(V_{signal})^2/(V_{noise})^2] = 20\lg(V_{signal}/V_{noise})$$

Where the V_{signal} represents signal voltage and the V_{noise} represents the noise voltage. The unit of the SNR is dB.

The following two examples give the calculation process of the formula.

Example 1. Suppose $V_{noise} = 1$, the quantifying accuracy is 1 digit and represented as $V_{signal} = 2^1$, its SNR = 6dB.

Example 2. Suppose $V_{noise} = 1$, the quantifying accuracy is 16 digit and represented as $V_{signal} = 2$ to the power of 16, its SNR = 96dB.

4.2.4　影响数字音频质量的技术参数

对模拟音频信号进行采样量化编码后，得到数字音频。数字音频的质量取决于采样频率、量化位数和声道数三个因素。

1．采样频率

如前所述，采样频率是指一秒钟时间内采样的次数。在实际的计算机多媒体音频处理中，采样频率通常采用三种：11.025 kHz（语音效果）、22.05kHz（音乐效果）、44.1kHz（高保真效果）。常见的 CD 唱盘的采样频率即为 44.1kHz。

2．量化位数

量化位数也称"量化精度"，是描述每个采样点样值的二进制位数。例如，8 位量化位数表示每个采样值可以用 2 的 8 次方即 256 个不同的量化值之一来表示，而 16 位量化位数表示每个采样值可以用 2 的 16 次方即 65536 个不同的量化值之一来表示。在实际应用中，常见的量化位数有 8 位、12 位、16 位。

如前所述的例子中，每个声音样本用 16 位（2 字节）表示，测得的声音样本值是在 2 的 16 次方，0～65536 的范围里，它的量化精度就是输入信号的 1/65536。

3．声道数

声音通道的个数称为声道数，是指一次采样所记录产生的声音波形个数。记录声音时，如果每次生成一个声波数据，称为单声道；每次生成两个声波数据，称为双声道（立体声）。随着声道数的增加，所占用的存储容量也相应地成倍增加。

4.2.5　数字音频文件的存储量

存储量指的是数据在计算机及存储介质中的存储量，通常以字节为单位。模拟波形声音被数字化后音频文件的存储量（这里假定是未经压缩的数据）为

$$存储量=采样频率×量化位数/8×声道数×时间$$

计算出的结果的单位为字节。

例如，用 44.1kHz 的采样频率进行采样，量化位数选用 16 位，如果录制 1s 的立体声节目，则其波形文件所需的存储量为多少？

解：根据以上计算公式，可得出结果如下：

$$存储量=44100×16/8×2×1=176400(B)$$

其存储量为 176.4kB（176.4k 字节）。

这里讲的是数字音频文件的存储量，鉴于上述原因，其单位用的是字节。

下面来看一下数字音频文件的传输率。在多媒体通信中，实际传输的是经过处理后的二进制的 0 或 1 码，是按位（bit，比特）来传输的。所以当谈及数字音频文件的传输率时，所用的单位是数据位（bit，比特）。

以数据位为单位，模拟波形声音被数字化后音频文件的数据传输率（这里假定是未经压缩的数据）为

$$传输率=采样频率×量化位数×声道数$$

例如，用 44.1kHz 的采样频率进行采样，量化位数选用 16 位，则传输该立体声节目所需的数据传输率为多少？

解：根据以上计算公式，可得出结果如下：

$$传输率=44100×16×2=1411200(b/s)$$

4.2.4　The Technical Factors Affect the Audio Quality

After we finish the analog audio signal's sampling and quantifying, we got the digital audio signal. The quality of the digital audio signal is mainly decided by the following three technical factors, sampling frequency, quantifying digits, and audio signal channels.

1. Sampling Frequency

As indicated, the sampling frequency means how many samples have been taken within one second. In real computer multimedia audio signal processing, three common sampling frequencies are applied, which are 11.025kHz for speaking effect, 22.05kHz for music effect, and 44.1kHz for HiFi (High Fidelity) effect. Most music CDs on the market used 44.1kHz sampling frequency.

2. Quantifying Digits

As mentioned, quantifying digits are also called quantifying accuracy. It is the digits of the binary number representing the quantified amplitude. For example, 8 quantifying digits means that each quantified amplitude will be represented by one of the power 8 of 2, 256 binary numbers. The 16 quantifying digits means that each quantified amplitude will be represented by one of the power 16 of 2, 65536 binary numbers. In reality, the most common quantifying digits are 8, 12, and 16.

As indicated, if each amplitude sample represented in 16 quantifying digits, 2 bytes, the quantified values we got would be in the range of 0~65536, 2 to power 16. Its quantifying accuracy is the 1/65536 of the input signal.

3. Audio Channels

The number of the sound channels is called audio channels. It means how many audio wave samples will be recorded at one sampling time point in the sampling stage. When recording the audio signal, if only one audio wave sample will be recorded at each sampling time point, this is called single channel. If two audio wave samples will be recorded at each sampling time point, it is called two channels, stereo channels. Along with the channels increasing, the required storage capacity will be doubled increasing as well.

4.2.5　Required Digital Audio Files Storage Capacity

As soon as talking about the storage capacity, we mean the data storage capacity in computer system and its related storage media. Its common unit is byte. The required storage capacity for the digitalized analog audio signal which has not been compressed yet could be calculated as follows

Storage capacity = Sampling frequency×Quantifying digits/8×Channels×Time

The results unit is byte.

Example: If we apply the 44.1kHz sampling frequency to sample the sampling process, use 16 quantifying digits to quantify the amplitude, and record 1 second stereo audio, What will be the storage capacity we need for the digitalized audio wave files?

Solution: Based on the given formula, we could get the following result.

Storage capacity = 44100×16/8×2×1 = 176400(B)

Therefore, its storage capacity is 176.4kB, 176.4k bytes.

Since we talked about is the storage of digital audio data files, we used the byte as the storage unit.

Let's have a look at the digital audio data files communication transmission rate now. As we know, the eventually actually transmitted in the communication channel is the processed binary data, either 1 or 0. It is transmitted bit by bit. For this reason, when we talk about the communication transmission rate, we use the unit bit.

Take the bit as the unit, the data transmission rate of the digital audio file coming from the digitalized analog audio signal could be calculated as follows. Suppose the digitalized data hasn't been compressed yet

$$\text{Transmission rate} = \text{Sampling frequency} \times \text{Quantifying digits} \times \text{Channels}$$

Example: If we apply 44.1kHz sampling frequency to the sampling process and use 16 quantifying digits to quantify the amplitude. What will be the communication transmission rate we need to transmit this stereo program?

Solution: Based on the formula above, we could get the result as follows.

$$\text{Transmission rate} = 44100 \times 16 \times 2 = 1411200(\text{b/s})$$

Where the bps stands for bits per second.

4.2.6 声音质量与数据率

在不同的应用场合，对音频的质量有不同的要求，也即对采样频率、样本精度和声道数有不同的要求。

根据声音的频带，通常把声音的质量分成 5 个等级，由低到高分别是电话（Telephone）、调幅（Amplitude modulation，AM）广播、调频（Frequency modulation，FM）、光盘（Compact disk，CD）和数字录音带（Digital audio tape，DAT）的声音。在这 5 个等级中，使用的采样频率、样本精度、通道数和数据率不同。

表 4-3 中列出了常见的不同质量声音的数字化性能指标

表 4-3 不同质量声音的数字化性能指标

质量	采样频率/kHz	样本精度/bit	声道	数据率/（kb/s）	频率范围/Hz
电话	8	8	单声道	64.0	200～3400
AM	11.025	8	单声道	88.2	50～5000
FM	22.050	16	立体声	705.6	20～10000
CD	44.1	16	立体声	1411.2	20～20000
DAT	48	16	立体声	1536.0	20～20000

4.2.6 Sound Quality and Data Rates

There are different sound quality requirements to variety applications. That is actually saying that there are different sampling frequency, quantifying accuracy, and channel quantities to meet the requirements of different applications.

The sound qualities are classified into 5 levels based on the audio signal's bandwidths. From low to high, the five levels are Telephone, Amplitude modulation (AM), Frequency modulation (FM), Compact disk (CD), and Digital audio tape (DAT). Among those five levels, each one of them uses different sampling frequency, quantifying accuracy, channel quantities, and data rates.

The Table 4-3 below gives the different digitalizing technical characters for several common sound qualities.

Table 4-3　Different Sound Quality's Digitalizing Technical Characters

Sound Quality	Sampling Frequency /kHz	Quantifying Accuracy /bit	Channel Quantity	Data Rates /(kb/s)	Frequency Range /Hz
Telephone	8	8	Single Channel	64.0	200～3400
AM	11.025	8	Single Channel	88.2	50～5000
FM	22.050	16	Stereo	705.6	20～10000
CD	44.1	16	Stereo	1411.2	20～20000
DAT	48	16	Stereo	1536.0	20～20000

4.3　数字音频的文件格式

对音频数据进行编码处理后，经常要以文件的形式将其数据保存在磁盘上。音频编码数据在文件中的存储形式、排列顺序、相关参数等称为文件格式，因各种应用需求不同，存在着多种多样的音频文件格式，有些文件格式可以存储多种不同的音频编码数据，也有些文件格式是为某一种音频编码特制的。这些格式既有流行的也有不那么流行的，有存在长久的也有昙花一现的，正是它们构成了五彩缤纷的数字音频世界，这里一一介绍。

4.3.1　主流音频格式

目前的主流音频格式不少，如图 4-27 所示，不同的格式有自己不同的用途。

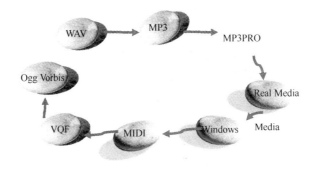

图 4-27　主流音频格式

1．WAV 文件格式

用.wav 为扩展名的文件格式称为波形文件格式（Wave File Format），它在多媒体编程接口和数据规范 1.0（Multimedia Programming Interface and Data Specifications 1.0）文档中有详细的描述。该文档是由 IBM 和微软公司于 1991 年 8 月联合开发的，它是一种为交换多媒体资源而开发的资源交换文件格式（Resource Interchange File Format，RIFF）。

波形文件格式支持存储各种采样频率和样本精度的声音数据，并支持声音数据的压缩。波形文件由许多不同类型的文件构造块组成，其中最主要的两个文件构造块是 format chunk（格式块）和 sound data chunk（声音数据块）。格式块包含有描述波形的重要参数，例如采样频率和样本精度等，声音数据块则包含有实际的波形声音数据。RIFF 中的其他文件块是可选择的。

这种由微软和 IBM 开发的 WAV 文件格式广泛应用于 Windows 系统中，是 Microsoft

Windows 本身提供的音频格式，由于 Windows 本身的影响力，这个格式已经成为了事实上的通用音频格式。WAV 格式可以存储多种不同编码的声音数据，但常用于存放 1～2 声道的 PCM 编码声音数据，不进行压缩编码，可以保持原始数据的最好音质。

通常使用 WAV 格式都是用来保存一些没有压缩的音频。

2．MP3 文件格式

MP3 文件格式是由 Fraunhofer-IIS 研究所开发，第一个实用的有损音频压缩编码。

MP3 利用了知觉音频编码技术，也就是利用了人耳的特性，削减音乐中人耳听不到的成分，同时尝试尽可能地维持原来的声音质量。

MP3 文件格式是现今应用最多的文件格式，是专门用于存储 MP3 编码声音数据的文件格式。在 MP3 文件格式中加入了 ID3V1 和 ID3V2 标签，以提供版权声明。但标签未加密，可被任意修改。标签还可用于存储歌曲名、专辑名等信息。

MP3 歌曲文件内不带有歌词。可以在外部配合一个文本格式的歌词文件，两个文件配合，可以使音频播放软件边唱边同步显示歌词内容，歌词文件常见的格式是 Lrc 格式。

歌与词能即时配合播放是因为在歌词文件中的每一行歌词前面都标记了显示时间。新的技术方法也可以将歌词嵌入到 MP3 歌曲文件内。

3．MP3PRO 文件格式

MP3PRO 文件格式由德国 Fraunhofer-IIS 研究所、瑞典 Coding Technologies 公司、法国 Thomson multimedia 公司共同推出。

在原来 MP3 技术的基础上专门针对原来技术中损失了的音频细节进行独立编码处理并捆绑在原来的 MP3 数据上，在播放的时候通过再合成而达到良好的音质效果。

4．Real Media——网络流媒体鼻祖

Real Media 文件，格式是 RA、RMA，由 Real Networks 公司发明，特点是可以在非常低的带宽（28.8kb/s）下，提供足够好的音质让用户能在线聆听。

用途主要是在线聆听，并不适于编辑，所以相应的处理软件并不多。

5．Windows Media——霸气十足

Windows Media 由 Microsoft 公司推出，一种网络流媒体技术，唯一一个能提供全部种类音频压缩技术（无失真、有失真、语音）的解决方案 。在 64kb/s 的码率情况下，WMA 可以达到接近 CD 的音质。

由于是微软的杰作，故而具有微软的一切特征。

ASF（Advanced Streaming Format）高级数据流格式是 Windows98 中的流媒体文件格式，ASF 编码及存储格式是微软 Windows Media 的核心技术。2000 年，随着 Windows Media Player7 的发布，微软将 ASF 改造为 WMA（Windows Media Audio）和 WMV 格式。WMA 成为微软使用的自有音频文件格式，它支持流媒体应用方式、可变码流率技术、有损和无损数据压缩编码技术，以及微软的 DRM（Digital Rights Management）内容数字版权加密保护技术。

WMA 数据的压缩比为 18：1，高于 MP3。WMA 编码在 128kb/s 码流率时音质超过 MP3。在 64～128kb/s 低码流率时具有优势，这时在兼顾数据压缩的同时，声音的细节特征也得以保留，能基本满足对容量有一定要求的听音者对音质的需要。192～320kb/s 码流率时适用于对音质要求较高的应用过程。

使用 Windows Media Player 程序可以进行 WMA 格式声音数据的压缩编码和播放，也可以在其他应用软件中进行压缩编码和播放。WMA 格式是 MP3 播放机所支持的除 MP3 格式

外的最主要格式，应用非常广泛。

WMA 是微软的私有技术，属于听觉感知类的数据压缩编码技术。

6．MIDI 文件格式

MIDI（Music Instrument Digital Interface）格式文件是计算机与电子乐器的桥梁

MIDI 文件格式是专用于存储 MIDI 音乐乐谱编码的文件格式。其数据可以经过代码转换后，用于手机铃声。

MIDI 是数码音乐文件，由曲谱、时序、乐器编号、音高等信息组成，告诉一个 MIDI 播放器何时用何种音高去演奏何种乐器，附带演奏一些效果比如颤音、混响等。

电子乐器数字接口（Musical Instrument Digital Interface，MIDI）是用于在音乐合成器（Music Synthesizers）、乐器（Musical Instruments）和计算机之间交换音乐信息的一种标准协议。从 20 世纪 80 年代初期开始，MIDI 已经逐步被音乐家和作曲家广泛接受和使用。MIDI 是乐器和计算机使用的标准语言，是一套指令（即命令）的约定，它指示乐器（即 MIDI 设备）要做什么，怎么做，如演奏音符、加大音量、生成音响效果等。MIDI 不是声音信号，在 MIDI 电缆上传送的不是声音，而是发给 MIDI 设备或其他装置让它产生声音或执行某个动作的指令。

MIDI 标准之所以受到欢迎，主要是它有下列几个优点：生成的文件比较小，因为 MIDI 文件存储的是命令，而不是声音波形；容易编辑，因为编辑命令比编辑声音波形要容易得多；可以作背景音乐，因为 MIDI 音乐可以和其他的媒体，如数字电视、图形、动画、话音等一起播放，这样可以加强演示效果。

产生 MIDI 乐音的方法很多，现在用得较多的方法有两种：一种是频率调制（Frequency Modulation，FM）合成法，另一种是乐音样本合成法，也称为波形表（Wave Table）合成法，这两种方法目前主要用来生成音乐。

国际 MIDI 协会出版了标准 MIDI 文件（Standard MIDI Files）规范，该标准说明了处理定时标记 MIDI 数据的一种标准化方法。这种方法适合各种应用软件共享 MIDI 数据文件，这些软件包括音序器、乐谱软件包和多媒体演示软件。

标准 MIDI 文件规范定义了 3 种 MIDI 文件格式，MIDI 音序器能够管理文件标准规定的多个 MIDI 数据流，即声轨（Tracks）。MIDI 文件格式 0（Format 0）规定所有 MIDI 音序数据（MIDI Sequence Data）必须存储在单个声轨上，它仅用于简单的单声轨设备；MIDI 文件格式 1（Format 1）规定数据以一个声轨集的方式存储；MIDI 文件格式 2（Format 2）可用几个独立模式存储数据。

7．VQF——生不逢时

VQF 实际指的是日本 Nippon Telegraph and Telephone（NTT）与 YAMAHA 公司开发的一种比较先进的音频压缩技术，通常认为 96kb/s VQF 与 128kb/s MP3 质量相同。

VQF 在 YAMAHA 公司的大力推动下也曾有相当的市场份额。不过时至今日，VQF 已经在逐步淡出舞台。

8．Ogg Vorbis——开放、免费

Ogg Vorbis 是一种音频压缩格式，类似于 MP3 等现有的通过有损压缩算法进行音频压缩的音乐格式。但有一点不同的是，Ogg Vorbis 格式是完全免费、开放源码且没有专利限制的。但到目前为止，其应用仍很有限。

4.3 The Digital Audio File Formats

After done the encoding process to the digital audio signal data, we usually need to save the data as data file to the disk. The saving format, sequence order, and correlated coefficients of the encoded audio data existing inside a file is called file format. There exist kinds of audio file formats to meet the different application requirements. Some file formats could be used to save a variety of digital audio data and some only could be used to save one particular kind of digital audio data. It depends. Some file formats are common and some are not. Some of them stay for a long time and some wouldn't. All of those give us a variety of digital audio formats. Let's have closer look at them.

4.3.1 Main Class Digital Audio Formats

There are many kinds of formats in this class as shown in the following Fig.4-27. They are applied in different applications.

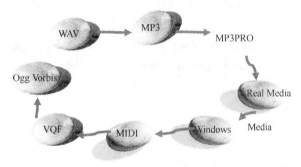

Fig. 4-27   Main Class Digital Audio Formats

1. WAV File Format

The file format with extension .wav is called wave file format. Its detailed description could be found in Multimedia Programming Interface and Data Specification 1.0. The document is proposed by IBM and Microsoft in August, 1991. It is a resource interchange file format, RIFF and used for multimedia resource interchanging.

The RIFF supports saving kinds of digital audio data with variety of sampling frequencies and quantifying accuracies. The RIFF consists of different file construction chucks. The two major ones are format chuck and sound data chuck. The format chuck contains the important coefficients describing the audio wave, such as sampling frequency, quantifying accuracy, and etc. The sound data chuck contains the actual digital audio wave data. The other chucks inside the RIFF are optional.

This WAV file format developed by Microsoft and IBM are widely applied in the Windows systems. It is the digital audio format provided by the Microsoft Windows itself. Because of the broadly acceptance of the Microsoft Windows, this format has been the real major digital audio format. WAV format could be used to save variety of encoded digital audio data. But it is usually used to save either single or two channels PDM encoded digital audio data. Without compressed encoding, this format could maintain the best sound quality of the original digital audio data.

We usually take the WAV format to save some uncompressed digital audio file data.

2. MP3 File Format

It is the first practical lossy digital audio signal data compression encoding method proposed by Fraunhofer-IIS.

MP3 takes the advantage of hearing audio encoding technology. It uses the characteristics of the human being's ears to takes out the components that human being's ear can't hear anyways. Meanwhile, maintain the original sound quality as high as possible.

MP3 file format is the most common format currently. It is particularly designed to save the MP3 encoded digital audio data. The labels ID3V1 and ID3V2 used show the version and copyright. But the labels are not encrypted and could be changed without authority. The labels could be also used to save the titles of music and episodes.

The song lyric is not included in the MP3 audio music files. It could be attached as an extra file in txt format. The two files cooperate each other could make the digital audio playing software plays the music and shows the song lyric synchronically. The common song lyric file format is lrc format.

The reason for the music and the song lyric could be played synchronically is that at the beginning of each line of the song lyric had been marked along time in advance. The newer technology could also insert the song lyric into the MP3 music files.

3. MP3 PRO File Format

It is proposed by Fraunhofer-IIS, German, Coding Technologies, Swiss, and Thomson Multimedia, France.

It is based on the MP3 technology particularly to find musical details lost in the original MP3 encoding, independently encode them, and bond them with the original MP3 data. Combine them together when playing and reach a very good sound quality.

4. Real Media——The Ancients of the Network Multimedia Streaming

Its file formats are RA, and RMA. It is proposed by Real Networks. The best of this format is that it allows listeners to hear the good enough quality sound on line with relatively very low bandwidth data, 28.8kbps.

It is mainly applied at multimedia streaming which allows people to on line listen. It's not good for editing. Therefore, there is not that many related processing software available.

5. Windows Media——Appears Aggressive Full

It is a network streaming technology and proposed by Microsoft. It is the only one solution that could provide all kinds of compression for digital audio data, lossless, lossy, and speech. With 64kbps code rate, WMA could get to very close to CD's sound quality.

Proposed by Microsoft, it naturally has all Microsoft's characteristics.

ASF, Advanced Streaming Format is the streaming file format in Windows 98. The ASF encoding and storage format is the core technology of Microsoft Media Technology. Along with the Windows Media Player7 proposed in 2000, Microsoft changed ASF to WMA and WMV formats.

WMA became the Microsoft's its own audio data file format. It supports data streaming application, variable code flow rate technology, lossless and loosy data compression encoding technology, and Microsoft DRM, Digital Right Management, content digital copyright encryption protection technology.

WMA's data compression rate is 18 : 1 which is higher than MP3's. WMA encoding's sound quality is over MP3's when the code rate reaches 128kb/s. It is dominant in lower code rate, 64~128kb/s. At this range, it could still keep the music details while finishing the data compression. It meets the sound quality requirement of the normal music listener's. When its code rate reaches 192~320kb/s, it is suitable for the higher sound quality requiring applications.

The Windows Media Player program could be used for WMA format digital audio data encoding and broadcasting. It could also be used to encode and broadcast within other practical software. WMA format is the main format that the MP3 player supports besides the MP3 format. It is widely accepted.

WMA is the Microsoft's own technology. It is a type of hearing sensitive data compression technology.

6. MIDI file format

It is a bridge between the computer and the music.

MIDI file format is particularly applied in storing encoded MIDI sheet music. Its data could be used as cellphone's ring music after the code converting.

MIDI is the abbreviation of Music Instrument Digital Interface. Obviously, it is an interface between music instrument and digital devices.

MIDI is data code musical file. It contains the information like musical score, time sequence, numbered musical instrument, musical tone, and etc. The information tells a MIDI player how to play music such as at what time point using which tone with some side affections like shivering sound, mix sound, and etc.

The musical instrument digital interface, MIDI is also a standard protocol for musical information exchange among music synthesizers, musical instruments, and computer. Since early 1980, 20th century, MIDI has been broadly accepted by musicians and composers. The MIDI is also a language used by both musical instrument and computer. It has a set of commands which tell the instrument, that is MIDI instrument what to do and how to so it, such as playing musical notes, turn up the sound, and create sound effects. The MIDI itself is not an audio signal, but a command sent to the MIDI instrument or other devices to tell the instrument what to do. What transmitted along the MIDI cable is not the sound itself, but those commands.

The reason for MIDI becoming popular is that it has the follows good characteristics. Its file size is relatively small since it contains the commands instead of the audio signal itself. It is easy to be edited since to edit the commands is much easier than to edit the audio signal itself. It could be also used as background music since the MIDI music could be concurrently played with other media such as digital TV, graphics, animation, speech, and etc to get a better musical effect.

There are many ways to produce the MIDI music. Two of them are most common. One of them is the frequency modulation, FM mixture. The other one is the musical sample mixture which is also called wave label mixture. Those two are majorly used to produce music.

International MIDI association published the MIDI standard files which recommended a standard process to deal with the timing marked MIDI data. This method allows various application software to share MIDI data. Those software include time sequence device, sheet music software package, and multimedia demonstration software.

The Standard MIDI files give three types of MIDI file formats. MIDI time sequence device could manage several data streams, tracks standardized by the MIDI standard files. The MIDI file format 0 requires that all MIDI sequence data has to be saved on one single track which is suitable for simple single track devices. The MIDI file format 1 requires that the MIDI sequence data shall be saved as the collection of tracks. The MIDI file format 2 allows that the MIDI sequence data could be saved in several independent formats.

7. VQF——Not a Good Timing to Survive

VQF actually means the relatively advanced digital audio data compression technology developed by Japan Nippon Telegraph and Telephone, NTT and YAMAHA. It's commonly accepted that the 96kbps VQF has the same sound quality as the 128kbps MP3.

Under the hard promoting of YAMAHA, the VQF occupied the certain percent market for a while. But up to now, it is slowly out of the market.

8. Ogg Vorbis——Open and Free

Ogg Vorbis is a digital audio data compression format. It is similar to MP3 and other existing musical formats which compress the digital audio data via lossy data compression algorithms. One major thing is different though. That is the Ogg Vorbis format is completely free. Its source codes are open and have no patent limit. Unfortunately, it's still not that popular.

4.3.2 非主流音频格式

常见的非主流音频格式如图 4-28 所示。

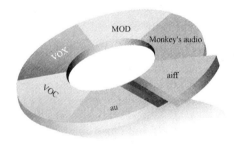

图 4-28 非主流音频格式

1. MOD——最热门的非主流

与 MIDI 有点相似,Module(MOD)是数码音乐文件,由一组 Samples(乐器的声音采样)、曲谱和时序信息组成,告诉一个 MOD 播放器何时以何种音高去演奏在某条音轨的某个样本,附带演奏一些效果等。这使得 MOD 成为一种介乎于像 WAV 或 VOC 那样的纯正样本数据文件和像 MIDI 那样的纯正时序信息文件之间的混合体,成为一种比较灵活的音频格式。

2．Monkey's audio——一个也不能少

Monkey's audio 是一种无损压缩技术，也就是说对压缩数据进行还原之后得到的数据与原来的数据是完全相同的。

这种格式的压缩比远低于其他格式，但能够做到真正无损，因此获得了不少发烧用户的青睐。

3．aiff——苹果专用

aiff 是 Apple 苹果计算机的标准音频格式，属于 QuickTime 技术的一部分。

aiff 虽然是一种很优秀的文件格式，但由于它是苹果计算机上的格式，因此在 PC 平台上并没有得到很大的流行。不过，Microsoft 公司的 WAV 格式就是据此产生的。AIFF 是一种很优秀的文件格式，只要苹果计算机还在，AIFF 就始终还占有一席之地。

4．au——聊胜于无

au 是 unix 下一种常用的音频格式，起源于 Sun 公司。

出现多年，本身也支持多种压缩方式，但文件结构的灵活性就比不上 aiff 和 WAV。对于目前许多新出现的音频技术它都无法提供支持，所以起不到类似于 WAV 和 aiff 那种通用性音频存储平台的作用。目前可能唯一必须使用 au 格式来保存音频文件的就是 java 平台。

5．VOC——曾经的辉煌

VOC 由创新公司（Creative）公司开发，曾经是 DOS 系统下面的音频文件格式标准。但该格式带有浓厚的与硬件相关色彩，变成了明显的缺点。

出现多年，所以现在的很多播放器都仍然支持这种格式。

6．VOX——记住你的声音

VOX 引申是 voice 的意思，表明了该格式专门面向语音音频。由 Dialogic 公司发明，使用 ADPCM 压缩技术进行压缩，主要应用于语音通信方面。

这个文件格式最常见于一些利用互联网进行语音通信的软件，比如 PC2Phone。

在介绍了上述诸多音频媒体格式之后，总结特点概括列于表 4-4 中。

表 4-4　音频媒体格式特点概括表

媒体格式	扩展名	相关公司或组织	主要优点	主要缺点	适用领域
WAV	WAV	Microsoft	可通过增加驱动程序而支持各种各样的编码技术	不适于传播和用作聆听。支持的编码技术大部分只能在 Windows 平台下使用	音频原始素材保存
MP3（MPEG 音频）	MP3（包括 MP2 MP1 MPa 等）	Fraunhofe r-IIS	在低至 128kb/s 的比特率下提供接近 CD 音质的音频质量。广泛的支持	出现得比较早，因此音质不是很好	一般聆听和高保真聆听
MP 3 PRO	MP3	Fraunhofe r-IIS Coding Te chnologies Thomson Multimedia	在低至 64kbps 的比特率下提供接近 CD 音质的音频质量	专利费用较高，支持的软件和硬件不多	一般聆听和高保真聆听
Real Media	ra, rma	RealNet works	在极低的比特率环境下提供可听的音频质量	不适于除网络传播之外的用途。音质不是很好	网络音频流传输

媒体格式	扩展名	相关公司或组织	主要优点	主要缺点	适用领域
Wind ows Media	wma, asf	Microsoft	功能齐全，使用方便。同时支持无失真、有失真、语音压缩方式	失真压缩方式下音质不高。必须在 Windows 平台下才能使用	音频档案级别保存，一般聆听，网络音频流传输
MIDI	MID MIDI RMI XMI 等	MIDI Association	音频数据为乐器的演奏控制，通常不带有音频采样	没有波表硬件或软件配合时播放效果不佳	与电子乐器的数据交互，乐曲创作等
Ogg Vorb is	OGG	Xiph Foundati on	在低至64kbps的比特率下提供接近 CD 音质的音频质量。开放源代码，不需要支付使用许可费用，跨平台	发展较慢。推广力度不足	一般聆听和高保真聆听
VQF	Vqf tvq	NTT Human Interface Laborato ries	在低至96kbps的比特率下提供接近 CD 音质的音频质量	相关软件太少	一般聆听
MOD(Module)	Mods3m it xm mtm ult 669 等	Amiga 和 mod	音频数据由乐器采样和乐谱、演奏控制信息组成。	具体的文件格式太多影响推广和使用	一般聆听
Monkey's Audio	ape	Matthew T.Ashland	无失真压缩。部分开放代码。	由于是个人作品，使用上存在一定风险	高保真聆听和音频档案级别保存
aiff	aiff	Apple	可通过增加驱动程序而支持各种各样的编码技术	一般限于苹果计算机平台使用	苹果计算机平台下音频原始素材保存
au	au	Sun	Unix 和 Java平台下的标准文件格式	支持的压缩技术太少且音频数据格式受文件格式本身局限	Unix 和 Java 平台下音频原始素材保存
VOC	VOC	Creative	对于目前的音频技术来讲，该格式已经没有什么优点了	与具体的硬件相结合因此没有延续性	淘汰
VOX	VOX	Dialogic	面向语言的编码	文件格式缺乏足够的信息，因此不适应作存档用途。技术比较早期	淘汰

4.3.3 Real Networks 公司简介

RealNetworks 总部位于西雅图，是全球领先的数字媒体技术提供公司，主要从事软件产品和服务的开发和销售业务，旨在使个人计算机及其他电子设备用户通过 Web 发送和接收音频、视频及其他多媒体服务。

4.3.2 Non-main Class Digital Audio Formats

The common non-main Class Digital Audio Formats are shown in Fig. 4-28.

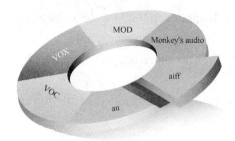

Fig. 4-28   Non-main Class Digital Audio Formats

1. MOD——The Most Common Non-main Class

Similar to MIDI, Module, MOD is a set of digital musical files. It consists of a group of musical instrument audio samples, lyrics, and time sequence information which will tell a MOD player when, how, and in what tone to play a sample on a certain track and some associated side effects. This makes MOD become one of the medium mixtures between the pure sample data file format like WAV and VOC and the pure time sequence information file format like MIDI. It is a relatively flexible audio file format.

2. Monkey's Audio——Can't Without It

It is a kind of lossless data compression technology. That is to say that the audio data recovered from the compressed one is exactly the same as the original one.

This format's data compression ratio is much lower than the others'. But it is a real lossless compression technology. For this reason, many passionate music lovers like it.

3. aiff——Apple's Own

It is Apple computers' standard audio data format. It's a part of QuickTime technology.

Aiff is an outstanding audio data file format. But it is limited to the Apple computer only and was not popular on PC. Meanwhile, the Microsoft's WAV file format was proposed based on Aiff. As we indicated, Aiff is an excellent audio file format as soon as the Apple computer is there, it will be there as well.

4. AU——Better than Nothing

AU is a audio file format under the system, UNIX and is proposed by Sun.

It has been there for years and supports many different data compression algorithms. But it's not that flexible like Aiff and WAV. It doesn't support the nowadays appeared many new audio process technologies. This makes it impossible for AU to act as a general storage platform like WAV and Aiff do. Presently, the Java platform is the only one which must use the audio data file saved as AU format.

5. VOC——Shining in The Past

The VOC is proposed by Creative firm. It was the standard audio data file format under DOS system. It is closely related to the hardware which becomes its apparent shortcoming.

Since it has been there for years, many players still support this format.

6. VOX——Remember The Sound

The VOX's expanding is voice which shows that is particularly designed for speaking audio data. It is proposed by Dialogic. It applies ADPCM technology to compress the data. As mentioned, it is majorly applied in speaking audio data communication.

This format is most often seen in some internet speech audio data communication software such as PC2Phone.

After we talked a lot about those digital audio data file formats, we put them together in the Table 4-4.

Table 4-4　Abstract of the Various Audio Data File Formats

Media Format	Extension Name	Related Firms	Major Benefits	Major Shortcomings	Suitable For
WAV	wav	Microsoft	By adding some driver software, could support various audio encoding technologies	Not very good for transfer and real listening. Most audio encoding technologies it supported could only be used under the Windows	Keep the original audio data files
MP3 (MPEG Audio)	MP3 (including MP2, MP1, MPa, and etc.)	Fraunfer - IIS	Could provide close to CD`s sound quality with as low as 128kbps data rate. It's popular	Early populated and the quality is not that good	General and HIFI listening
MP3PRO	MP3	Fraunfer - IIS; Coding Technologies; Thomson Multimedia.	Could provide close to CD`s sound quality with as low as 64kbps data rate	Patent cost is relatively high with rare hardware and software support.	General and HIFI listening
Real Media	ra, rma	Real Networks	Provide audible sound quality with very low data rate	The sound quality is not that good and not good for the applications besides the network propagation.	Network audio data streaming
Windows Media	wma, asf	Microsoft	Complete functions, easier to use, and support lossless, lossy, and speech data compression formats	The sound quality is poor under the lossy compression format. Only could be used under windows	Archived audio data files storage. General listening and network audio streaming
MIDI	mid, midi, rmi, xmi, etc.	MIDI Association	The data is actually the musical equipment control code and not the real audio sampled data	The effects won't be good without wave table either hardware or software's cooperating	Interchanging audio data with musical instrument. Music creation

Media Format	Extension Name	Related Firms	Major Benefits	Major Shortcomings	Suitable For
Ogg Vorbis	ogg	Xiph Foundation	Could provide close to the CD's sound quality with 64kbps low data rate. Source codes are open to public. Free to use. Support various platforms	Developed slowly. Lack of strong promotion	General and HIFI listening
VQF	vqf, tvq	NTT Human Interface Laboratories	Could provide close to the CD's sound quality with 96kbps low data rate	Lack of relative software	General listening
MOD (Module)	mod, s3m, it, xm, mtm, ult 669	Amiga and mod community	Its audio data consists of musical instrument samples, lyrics, and playing control codes	Too many detailed audio data file formats to spread and apply	General listening
Monkey's Audio	ape	Matthew, T., Ashland	Lossless data compression. Partially open	It is a private product. May have risk to apply	HIFI listening and archiving audio data files
AIFF	aiff	Apple	Could support various audio data compression encoding technologies by adding related software	Usually used in Apple computer platforms	The original audio data keeping under the Apple computer platforms
AU	au	Sun	The standard audio file format under the UNIX and Java platforms	The audio data compression technologies supported is rare. The audio data file format is limited by the format itself	The original audio data keeping under the UNIX and Java platforms
VOC	voc	Creative	There is no benefit anymore	Related closely to hardware and hard to extend	Abandoned
VOX	vox	Dialogic	Particularly for speech encoding	There is not enough information included in the format and is not good to be used to keep the original audio data. It's a older technology	Abandoned

4.3.3　Real Network Introduction

The RealNetwork's headquarters is in Seattle. It is one of the core firms which provide digital media technologies. Its main business is in software product development and services. It has put lots of efforts to let the computer and other electronic equipments transmit and receive audio, video, and other media via Web.

209

4.4 脉冲编码调制（PCM）

4.4.1 PCM 的概念

脉冲编码调制（Pulse Code Modulation，PCM），它是概念上最简单、理论上最完善的编码方法；是最早研制成功、使用最为广泛的编码方法；在实际应用中也是数据量最大的一种编码系统。

PCM 的编码原理比较直观和简单，如图 4-29 所示。在这个编码框图中，它的输入是模拟声音信号，它的输出是 PCM 样本。图中的"防失真滤波器"是一个低通滤波器，用来滤除声音频带以外的信号；"波形编码器"可暂时理解为"采样器"，"量化器"可理解为"量化阶大小（Step—size）"生成器或者称为"量化间隔"生成器。

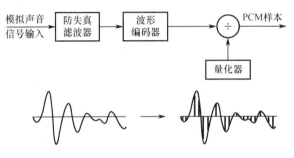

图 4-29　PCM 编码框图

在前面介绍声音数字化的时候，讲到声音数字化有两个步骤：第一步是采样，就是每隔一段时间间隔读一次声音的幅度；第二步是量化，就是把采样得到的声音信号幅度转换成数字值。但那时没有涉及很具体的量化方法，实际上量化有好几种方法，但可归纳成两类：一类称为均匀量化，另一类称为非均匀量化。采用的量化方法不同，量化后的数据量也就不同。因此，从某种意义上来说，量化本身也就是一种压缩数据的方法。

4.4.2 均匀量化

前面已经讲到，如果采用相等的量化间隔对采样得到的信号作量化，那么这种量化称为均匀量化。均匀量化就是采用相同的"等分尺"来度量采样得到的幅度，也称为线性量化，如图 4-30 所示。量化后的样本值 Y 和原始值 X 的差 $E = Y–X$ 称为量化误差或量化噪声。

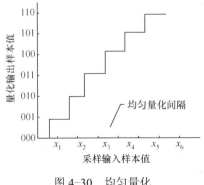

图 4-30　均匀量化

用这种方法量化输入信号时，无论对大的输入信号还是小的输入信号一律都采用相同的量化间隔。为了适应幅度大的输入信号，同时又要一定的满足精度要求，就需要增加样本的位数。但通常对话音信号来说，大信号出现的机会并不多，这时增加的样本位数就是一种浪费，为了克服这个不足，就出现了非均匀量化的方法，这种方法也称为非线性量化。

4.4.3 非均匀量化

为了克服均匀量化的不足，非均匀量化的基本想法是，当对输入信号进行量化时，大的输入信号采用大的量化间隔，小的输入信号采用小的量化间隔，如图 4-31 所示。这样就可以在满足一定精度要求的情况下用总的较少的位数来表示采样后的样本信号。声音数据还原时，采用相同的规则。

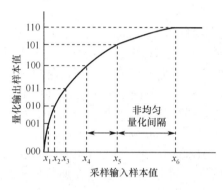

图 4-31 非均匀量化

在非线性量化中，采样输入信号幅度和量化输出数据之间定义了两种对应关系：一种称为 μ 律压扩算法，另一种称为 A 律压扩算法。

4.4.4 μ 律压扩

μ 律（μ—law）压扩（G.711）主要用在北美和日本等地区的数字电话通信中，按下面的式子确定量化输入和输出的关系：

$$F_\mu(x) = \operatorname{sgn}(x)\frac{\ln(1+\mu\,|\,x\,|)}{\ln(1+\mu)}$$

式中：x 为输入信号幅度，规格化成一 $1 \leqslant x \leqslant 1$；$\operatorname{sgn}(x)$ 为 x 的正负极性；μ 为确定压缩量的参数，它反映最大量化间隔和最小量化间隔之比，取 $100 \leqslant \mu \leqslant 500$。

由于 μ 律压扩的输入和输出关系是对数关系，所以这种编码又称为对数 PCM。具体计算时，用 $\mu=255$，把对数曲线变成 8 条折线以简化计算过程。详细计算请看有关参考文献。

4.4.5 A 律压扩

A 律（A—law）压扩（G.711）主要用在欧洲和中国大陆等地区的数字电话通信中，按下面的式子确定量化输入和输出的关系：

$$F_A(x) = \operatorname{sgn}(x)\frac{A\,|\,x\,|}{1+\ln A\mu}, \qquad 0 \leqslant |\,x\,| \leqslant 1/A$$

$$F_A(x) = \operatorname{sgn}(x)\frac{1+\ln(A\,|\,x\,|)}{1+\ln A}, \quad 1/A < |\,x\,| \leqslant 1$$

式中：x 为输入信号幅度，规格化成 $-1 \leqslant x \leqslant 1$；$sgn(x)$ 为 x 的极性；A 为确定压缩量的参数，它反映最大量化间隔和最小量化间隔之比。

A 律压扩的前一部分是线性的，其余部分与 μ 律压扩相同。具体计算时，$A=87.56$，为简化计算，同样把对数曲线部分变成折线。详细计算请看相关参考文献。

对于采样频率为 8kHz，样本精度为 13 位、14 位或者 16 位的输入信号，使用 μ 律压扩编码或者使用 A 律压扩编码，经过 PCM 编码器之后每个样本的精度为 8 位，输出的数据率为 64kb/s。这个数据就是 CCITT 推荐的 G.711 标准：话音频率脉冲编码调制（Pulse Code Modulation（PCM）of Voice Frequencies）。

4.5 PCM 在通信中的应用

PCM 编码早期主要用于话音通信中的多路复用。

通常在电信网中传输媒体费用约占总成本的 65%，设备费用约占成本的 35%，因此提高线路利用率是一个通信传输的重要课题，通常有如下两种方法。

1. 频分多路复用（frequency-division multiplexing，FDM）

频分多路复用的基本思想是把传输信道的频带分成好几个窄带，每个窄带传送一路信号。

例如，一个信道的频带带宽为 1400Hz，把这个信道分成 4 个子信道（Sub-channels）：（820~990）Hz，（1230~1400）Hz，（1640~1810）Hz 和（2050~2220）Hz，相邻子信道间相距 240Hz，用于确保两个子信道之间不相互干扰。在通信两端的每对用户仅占用其中的一个子信道。这是模拟载波通信的主要手段。

2. 时分多路复用（Time-Division Multiplexing，TDM）

时分多路复用的基本思想是把传输信道按时间来分割，为每个用户指定一个时间间隔，每个间隔里传输信号的一部分，这样就可以使许多用户同时使用一条传输线路。这是数字通信的主要手段。

例如，话音信号的采样频率 $f=8000$Hz，它的采样周期 $=125\mu s$，这个时间称为 1 帧。在这个时间里可容纳的话的路数有两种规格：24 路制和 30 路制。图 4-32 表示了 24 路制的结构。

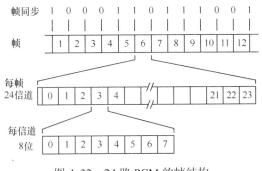

图 4-32 24 路 PCM 的帧结构

24 路制的重要参数如下：

① 每秒钟传送 8000 帧，每帧 125μs。

② 12 帧组成 1 复帧（用于同步）。

③ 每帧由 24 个时间片（信道）和 1 位同步位组成。

④ 每个信道每次传送 8 位代码，1 帧有 24×8＋1＝193 位。

⑤ 数据传输率 R＝8000×193＝1544kb/s。

⑥ 每一个话路的数据传输率＝8000×8＝64kb/s。

30 路制的重要参数如下：

① 每秒钟传送 8000 帧，每帧 125μs。

② 16 帧组成 1 复帧（用于同步）。

③ 每帧由 32 个时间片（信道）组成。

④ 每个信道每次传送 8 位代码。

⑤ 数据传输率：R＝8000×32×8＝2048kb/s。

⑥ 每一个话路的数据传输率＝8000×8＝64kb/s。

时分多路复用（TDM）技术已广泛用在数字电话网中，为反映 PCM 信号复用的复杂程度，通常用"群（Group）"这个术语来表示，也称为数字网络的等级。PCM 通信方式发展很快，传输容量已由一次群（基群）的 30 路（或 24 路），增加到二次群的 120 路（或 96 路），三次群的 480 路（或 384 路），……。图 4-33 表示二次复用的示意图。图中的 N 表示话路数，无论 N＝30 还是 N＝24，每个信道的数据率都是 64kb/s，经过一次复用后的数据率就变成 2048kb/s（N＝30）或者 1544kb/s（N＝24）。在数字通信中，具有这种数据率的线路在北美被称为 T1 远距离数字通信线，提供这种数据率服务的级别称为 T1 等级，在欧洲被称为 E1 远距离数字通信线和 E1 等级。T1/E1，T2/E2，T3/E3，T4/E4 和 T5/E5 的数据率如表 4-5 所列。

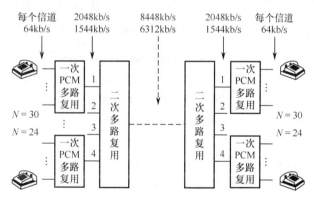

图 4-33　二次复用示意图

注意：上述基本概念都是在多媒体通信中经常用到的。

表 4-5　多次复用的数据传输率

国家和地区	数字网络等级	T1/E1	T2/E2	T3/E3	T4/E4	T5/E5
美国	64kb/s 话路数	24	96	672	4 032	
	总传输率/（Mb/s）	1.544	6.312	44.736	274.176	
	数字网络等级	1	2	3	4	5
欧洲	64kb/s 话路数	30	120	480	1 920	7 680
	总传输率/（Mb/s）	2.048	8.448	34.368	139.264	560.000
日本	64kb/s 话路数	24	96	480	1 440	
	总传输率/（Mb/s）	1.544	6.312	32.064	97.728	

4.4 Pulse Code Modulation

4.4.1 The PCM Idea

The pulse code modulation, PCM is one of the encoding algorithms which have the simplest idea and most completed theory. It is one of the earliest proposed and widest applied encoding algorithms. In the application, it's one of the algorithms which have the biggest quantity of data.

The PCM's encoding principle is simple and easy to understand as shown in Fig. 4 - 29. In this encoding block diagram, the input is an analog audio signal and its output is the PCM sample. The "Distortion Preventive Filter" is a low pass filter which is used to filter the noise except the audio signal's spectrum. The "Wave Encoder" could be temporarily thought as a "Sampler". The "Quantifier" could be thought as a quantifying step-size creator or quantifying interval creator.

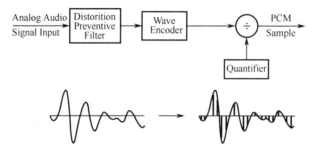

Fig. 4-29   PCM Encoding Block Diagram

When we were talking about the audio digitalization, we mentioned that there are two major steps in audio digitalization. The first one is the sampling which is to take one value of the amplitude of the analog audio signal at each certain interval. The second step is the quantifying which is to covert the sampled continuous amplitude into discrete digital values. We didn't get into the detailed quantifying methods at that time. There are actually a few ways available. They could be classified into two types. One of them is the even quantifying and the other is the non-even quantifying. With different quantifying method used, we'll get different data quantities. That's why to certain point we say that the quantifying itself is one of the data compression algorithms.

4.4.2 Even Quantifying

As we talked, if we use the equal quantifying steps to quantify the sampled discrete signal, we're doing the even quantifying. In another words, the even quantifying is to use the equal "step-size" to measure the sampled discrete signal's amplitude. It is also called liner quantifying as shown in Fig. 4-30. The difference E between the quantified value Y and the original value X is called the quantifying error or quantifying noise.

When using this algorithm to quantify the input discrete sampled signals, we use the same step-size to each input value no matter whether it's big or small. To quantify the bigger input value and meet the requirement of the certain quantifying error, will need to increase the size/digits of the quantifying samples. On the other hand, since the speech signal usually doesn't have that many big values, the increased digits will be a waste. To compensate the situation, the non-even quantification appeared which is also called non-liner quantification.

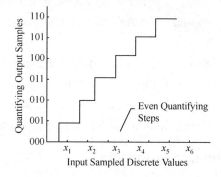

Fig. 4-30   Even Quantification

4.4.3   Non-even Quantification

To compensate the disadvantage of the even quantification, the basic thinking of the non-even quantification is to use bigger quantification step-size or say more digits for the bigger input value and smaller quantification step-size or say fewer digits for the smaller input value as shown in Fig. 4-31. This way could promise to keep certain quantification error and use total smaller mount of digits meanwhile. Use the same rule when converting back to the original data.

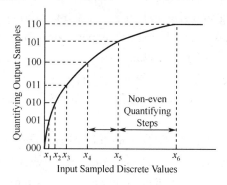

Fig. 4-31   Non-even Quantification

Within the non-liner quantification, there are two types of correspondent relationships between the input sampled discrete signal amplitude and the output quantified data. One of them is called μ-law compressing and expanding algorithm and the other is called A-law compressing and expanding algorithm.

4.4.4 μ-Law Compressing and Expanding

μ-law compressing and expanding is mainly used in digital telephone communications in North America and Japan. The correspondent relationship between the quantification input and out is decided by the following formula.

$$F_\mu(x) = \mathrm{sgn}(x)\frac{\ln(1 + \mu\,|x|)}{\ln(1 + \mu)}$$

Where the x is the amplitude of the input discrete signal and formulated as $-1 < x < 1$. The sgn(x) is the x's positive or negative polarity. μ is the compression or expansion coefficient. It reflects the ratio of the biggest quantifying step-size and the smallest quantifying step-size. Its value is taken as $100 \leqslant \mu \leqslant 500$.

Since the input and output of the μ-law compressing and expanding is logarithmic relationship, this encoding method is also called logarithmic PCM. When you really use this algorithm, take the μ as 255 and convert the logarithmic curve into 8 connected straight lines to simplify the calculation process. The detailed calculation procedure please refers to the related references.

4.4.5　A–Law Compressing and Expanding

A-law compressing and expanding is mainly used in digital telephone communications in Europe and China, main land. The correspondent relationship of the quantification input and output could be calculated by the following formulas.

$$F_A(x) = \mathrm{sgn}(x)\frac{A\,|x|}{1 + \ln A}, \qquad 0 \leqslant |x| \leqslant 1/A$$

$$F_A(x) = \mathrm{sgn}(x)\frac{1 + \ln(A\,|x|)}{1 + \ln A}, \quad 1/A < |x| \leqslant 1$$

Where the x is the amplitude of the input discrete signal and formulated as $-1 < x < 1$. The sgn(x) is the x's positive or negative polarity. A is the compression or expansion coefficient. It reflects the ratio of the biggest quantifying step-size and the smallest quantifying step-size.

The first part of the A-law compressing and expanding is liner and the others is the same as the μ-law compressing and expanding. While really doing the calculation, take the $A = 87.56$. Similarly, to simplify the calculation, replace the logarithmic curve with the close enough several sections of straight lines. Again, for the detailed calculation process, please refer to the related references.

For the typical input signal with 8kHz sampling frequency, 13, 14, or 15 digits quantifying accuracy, use either μ-law compressing and expanding or A-law compressing and expanding. After the PCM encoder, each output sample's quantifying accuracy is 8 digits and the data rate is 64kb/s. This kind of data is the G711 standard, pulse code modulation (PCM) of voice frequencies, recommended by CCITT.

4.5　The Applications of PCM in Communications

The PCM encoding early was used in multiplexing in speech communications.

In telecommunication networks, roughly 65% of the cost comes from the transmission media and 35% comes from the communication equipments. Therefore, increasing the line usage efficiency is a very important object in telecommunications. Majorly, there are two methods shown below available to do so.

1. Frequency-Division Multiplexing, FDM

Its idea is to divide the total available transmission channel bandwidth into smaller subsections with narrower bandwidths. On each one of those narrower bandwidths, transmit one independent signal.

For example, one transmission channel's bandwidth is 1400Hz. Divide this channel into 4 sub-channels as 820~990Hz, 1230~1400Hz, 1640~1810Hz, and 2050~2220Hz. Each interval between the sub-channels is 240Hz used to prevent the interference between the sub-channels. This gives total 170Hz×4+240Hz×3=1400Hz bandwidth. Each pair of users at the ends of the transmission channel occupies one sub-channel instead of the whole one. This is the common way applied in analog carrier wave communications.

2. Time-Division Multiplexing, TDM

Its idea is to divide the total available transmission channel into smaller subsections along with time. Each user will be assigned a certain time section. Within each of such sections, transmit part of the signal. This way, many user could share one transmission line. This is the common way applied in digital communications.

For example, the speech signal's sampling frequency is $f = 8000$Hz. Obviously, its sampling period is 125μs. This time period is call one frame. Within each such a frame, there are usually two standards related to two different numbers of speech sub-channels, 24 and 36. The Fig. 4-32 shows the structure of 24 speech sub-channels.

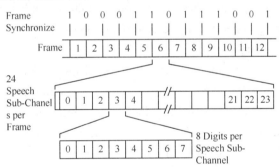

Fig. 4-32   The Structure of 24 Sub-Channels

The important coefficients of the 24 speech paths are as follows.

① Transmit 8000 frames per second. Each frame takes 125μs to be transmitted.

② Every 12 frames forms one dual frame used for synchronization.

③ Every frame consists of 24 time sections (sub-channels) and 1 digit synchronizing code.

④ Each sub-channel transmits 8 digits each time. This way each frame has 24×8+1=193 digits.

⑤ The data transmit rate is R=8000×193=1544kb/s.

⑥ The data transmit rate of each sub-channel is 8000×8 = 64kb/s.

The important coefficients of the 30 speech paths are as follows.

① Transmit 8000 frames per second. Each frame takes 125μs to be transmitted.

② Every 16 frames forms one dual frame used for synchronization.

③ Every frame consists of 32 time sections (sub-channels).

④ Each sub-channel transmits 8 digits each time.

⑤ The data transmit rate is $R=8000\times32\times8=2048$kb/s.

⑥ The data transmit rate of each sub-channel is $8000\times8=64$kb/s.

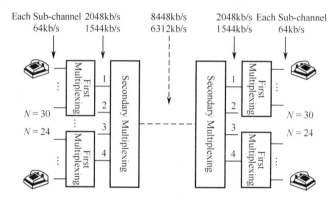

Fig. 4-33   Secondary Multiplexing Rough Drawing

The Time-Division Multiplexing，TDM has been widely applied in digital telephone networks. To reflect the complicated level of the PCM signal multiplexing, the terminology, group has been used which is called digital networks grades as well. The PCM communication style has been developed very fast. The transmitting capacity started from the first/base group which has either 30 or 24 sub-channels, increased to the secondary group which has either 120 or 96 sub-channels, increased again to the third group which has either 480 or 384 sub-channels, and etc. The Fig. 4-33 shows the situation of secondary group.

The N in the figure represents the number of the sub-channels. Each sub-channel's data rate is 64kb/s no matter $N=30$ or $N=24$. After the first multiplexing, the data rate will become either 2048kb/s (for $N=30$) or 1544kb/s (for $N=24$). In digital communication, in North America, the transmission line with this data rate is called T1 long distance digital communication line. The grade providing the service is called T1 grade. In Europe, the same transmission line is called E1 long distance digital communication line and the corresponding service is called E1 grade. The data rates of T1/E1, T2/E2, T3/E3, T4/E4, and T5/E5 are shown in the Table 4-5.

Note: The basic concepts described are seen in multimedia communications from time to time.

Table 4-5   Multiple Multiplexing Data Transmission Rate

Countries/Territories	Digital Network Grades	T1/E1	T2/E2	T3/E3	T4/E4	T5/E5
US	64kb/sSub-channels	24	96	672	4032	
	Overall Transmission Rate/（Mb/s）	1.544	6.312	44.736	274.176	
	Digital Network Grade	1	2	3	4	5
Europe	64kb/sSub-channels	30	120	480	1 920	7 680
	Overall Transmission Rate/（Mb/s）	2.048	8.448	34.368	139.264	560.000
Japan	64kb/sSub-channels	24	96	480	1 440	
	Overall Transmission Rate/（Mb/s）	1.544	6.312	32.064	97.728	

第五章　图像压缩编码—静止图像压缩

无论中文或英文都有一个类似的说法："一幅图胜过千言万语"。图像在人类的信息获取中的重要作用可见一斑。

万千世界里，人们所感知信息的五官中，眼睛接受的信息占 70%，声音占 20%，触觉和味觉、嗅觉大概占 10%。

图像是多媒体中携带信息的极其重要的媒体。如上所述，人们获取信息 70% 来自视觉系统，实际就是图像和电视。但是，人们面临着一个很棘手的问题，那就是图像数字化之后的数据量非常大，无论是进入计算机，还是保存其数据都是困难的。特别是图像的传输，首先碰到的困难是图像数字信号占频带太宽，通常称为"信息容量"问题，在互联网上传输时很费时间，在盘上存储时很占"地盘"，因此，就必须要对图像数据进行压缩。压缩的目的就是要满足存储容量和传输带宽的要求，而付出的代价是大量的计算。几十年来。人们一直在孜孜不倦地寻找更有效的方法，力求用尽可能少的数据量来表达原始的图像。

图像数据压缩主要是根据下面两个基本事实来实现的。一个事实是图像数据中有许多重复的数据，用数学手段来表示这些重复数据就可以减少数据量；另一个事实是人的眼睛对图像细节和颜色的辨认有一个极限，把超过极限的部分去掉，这也就达到了压缩数据的目的，利用前一个事实的压缩技术称为无损数据压缩技术，利用后一个事实的压缩技术称为有损数据压缩技术。

实际的图像数据压缩是综合使用各种有损和无损数据压缩技术来实现的，本章将讨论数字图像的基本特性和类型以及图像压缩的主要实现方式，进而介绍使用得相当广泛的二值图像压缩编码——数字传真和 JPEC-压缩标准和图像文件的存储格式。并简单介绍新一代静态图像压缩标准——JPEG2000。

Chapter 5　Image Compression Coding – Still Image Compression

There is a saying both in Chinese and in English. That is "a picture is worth a thousand words". This is kindly gives us an idea that how important the image is for us to get the information we need in our life.

In our beautiful world, as we know the way we accept the information around us via our five sensitive organs. Among those five, our eyes are getting 70% of the information we need. The sound we get occupies 20% of the information we need. Other touching and smelling will help us to get the left 10% of the information we need.

Images are very important information carrier media. As indicated, 70% of the information we got comes from our visual system, which are images and TV. But we'll have a tough problem. That is the data quantity will be huge after the images being digitalized. Whether get into the computer or save the data will be a big issue. Especially for the image transmission, the digital images will occupy a much wider bandwidth, which is called "information capacity" problem. It takes a long time to transmit them on the internet and takes a big disk space to store them. That's why we need to do the data compression to meet both the transmission bandwidth and storage capacity requirements. The cost is the lots of mathematical calculations. It has been for years for people to seek the more efficient algorithms to use as less as possible data to represent the original image.

The images' two major characters are the bases of the image data compression. One of them is that there are many repeated data in the digital images. Using mathematical way to represent the repeated data will reduce the data quantity. The other one is that there are certain extreme limits for our eyes to identify some image's details and colors. Getting off the data representing the part outside those limits will reduce the data quantity. The data compression technologies based on the first character are lossless compression ones and on the second are the lossy compression ones.

The practical data compression consists of both lossless and lossy ones and usually put several of each together. In this chapter we're going to discuss the basic features and types of digital images and the main approaches to image compression. This is followed by a description of the widely common used bi-level image compression encoding, digital facsimile, and JPEG-compression standard and its image storage format. Finally, we'll give some briefly description of the newer compression standard, JPEG-2000.

5.1 图像的基本特性和类型

5.1.1 像素

人们周围到处都是图像，通常所见到的一般是具有高分辨率的彩色图像，图像中的很多东西（尤其是艺术加工过的东西）看上去非常的平滑，没有过渡的边角和颗粒。而计算机上的图像则不同，这些图像是由许多点（像素）组成的数字图像。数字图像实际上是一个矩形的点或说图像元素阵组成的，有 *m* 行 *n* 列。表达式 *m×n* 称为图像的分辨率。点称为像素（Pixels）。在传真图像和录像压缩中点称为 pels。分辨率有时也指图像中的单位长度上像素的个数，这时，dpi 指每英寸里共有多少个像素。

当第一次听到计算机图像时，可能会忽略整个画面，会自然而然地想象由点（像素）组成的图像看上去总会是点状的，不平滑的，粗糙的。不会像眼睛看到的那么好。但实际上随着岁月的推移和技术的发展，尽管计算机产生的图像是离散不连续的，由像素（点）组成的，也已经很难把计算机产生的图像和它对应的实际图像区分开来了。

大部分的工程师、程序员和用户把每个像素看成是一个个小方块。当像素在计算机的监视器上显示时，这是对的。当像素在其他的数字输出设备（显示器或打印机）上时可能是矩形的或圆的。但从原理上来说，像素应该是一个没有尺寸的数学点。直观来想，从一个离散的像素点组成的矩阵去恢复一个连续的图像几乎是不可能的事情。但前边讲到的 Nyquist 采样定理已经告诉人们这是可能的，并讨论过了 Nyquist 采样定理对数字声音信号的应用，即一维数字信号的应用。这里讨论 Nyquist 采样定理对二维数字语音信号的应用。

在第 4 章讲到了 Nyquist 采样定理，声音信号是理解 Nyquist 采样定理的一个很好的切入点。当声音送入到话筒的时候，它是被转换成了随时间而变化的模拟电压波的。那么波是有频率的，一个变化的波会有多种不同的频率。把电压波中最高的频率标记为 B（每秒的周期数或 Hz）。Nyquist 采样定理告诉我们只要采样频率大于或等于电压波中最高的频率标记为 B 的两倍，就可以不损失地重建原始信号。

图像就是一个由采样点（像素）组成的矩阵。采样定理保证了只要采样时在单位长度内的采样频率大于 $2B$，就可以恢复原始的图像（即算出图像中每一个数学点的颜色）。这是一种理想的状态，在现实中，像素们的值和它们的频率取决于抽取它们的设备的精确度。一个理想的设备应该能够取一个像素点的颜色到一定的精度。但实际设备中（如照相机和扫描仪）的图像传感器（CCD 们和 CMOS）可取到的是远离这些理想值的。这中间的主要原因是由于一些不可达到的客观条件的限制，生产过程中出现的瑕疵，摄取足够的光源，等等。这使得图像传感器常常测试的是一个小面积的平均颜色（密度），而不是一个具体点的颜色。

5.1   The Image's Basic Attributes and Types

5.1.1   Pixels

Images are all around us. We usually see them in color and in high resolution. Many objects (especially artificial objects) seem perfectly smooth, with no jagged edges and no graininess. Computer graphics, on the other hand, deals with digital images that consist of small dots, pixels. A digital image is actually a rectangular array of dots, or picture elements, arranged in m rows and n columns. The expression $m \times n$ is called the *resolution* of the image, and the dots are called *pixels* (except in the cases of fax images and video compression, where they are referred to as *pels*). The term "resolution" is sometimes also used to indicate the number of pixels per unit length of the image. Thus, dpi stands for dots per inch.

When we first hear of this feature of computer graphics, we tend to dismiss the entire field as trivial. It seems intuitively obvious that an image that consists of dots would always look pixelated, grainy, rough, and inferior to what we see with our eyes. Yet state-of-the-art computer-generated images are often difficult or impossible to distinguish from their real counterparts, even though they are discrete, made of pixels, and not continuous.

Most engineers, programmers, and users think of pixels as small squares, and this is generally true for pixels on computer monitors. Pixels in other digital output devices (displays or printers) may be rectangular or circular. However, in principle, a pixel should be considered a mathematical, dimensionless, point. It seems impossible to reconstruct a continuous image from an array of discrete pixels, but this is precisely what the surprising Shannon-Nyquist sampling theorem tells us (in fact, what it guarantees). We discussed the application of this theorem to digitized audio (a one-dimensional signal), while here we apply it to two-dimensional images.

We talked the Shannon-Nyquist sampling theorem in Chapter 4. Audio is a good starting point to understand the sampling theorem. Sound fed into a microphone is converted to an analog electrical voltage that varies with time; it becomes a wave. A wave has a frequency, and a wave that varies all the time consists of many frequencies. We denote the maximum frequency contained in a wave by B (cycles per second, or Hertz). The sampling theorem says that it is possible to reconstruct the original wave if it is sampled at a rate greater than $2B$ samples per second.

As mentioned, an image is a rectilinear array of point samples (pixels). The sampling theorem guarantees that we'll be able to reconstruct the image (i.e., to compute the color of every mathematical point in the image) if we sample the image at a rate greater than $2B$ pixels per unit length, where B is the maximum pixel frequency in the image. But in practice, pixels, their values, and their frequencies depend on the accuracy of the capturing device. An ideal device should measure the color of an image at certain points, but image sensors (CCDs and CMOS) used in real devices (cameras and scanners) are often far from ideal. Because of physical limitations, manufacturing defects, and the need to capture enough light, an image sensor often measures the average color (or intensity) of a small area of the image, instead of the color at a point.

5.1.2　图像种类

为了更好地理解图像压缩，先来区分一下以下几种图像的类型。

1．二值（单色）图像

在这种图像中，每个像素的取值只有两个值中的一个。通常称为要么黑要么白。这种图像中的像素只需要二进制数中的一位来表示，这使得它成为最简单的一种图像类型。

2．灰度级图像

在这种图像中，每个像素可以有 n 个值中的一个，从 $0\sim n-1$。意味着 $2n$ 次的灰度级（或 $2n$ 次的其他颜色级）。n 的值通常可和一个二进制数的字节匹配，如 4,8,12,16,24。或者是其他方便的 4 或 8 的乘积。在所有像素中，有一组最重要的比特位称为比特板。这样的话，一个灰度级图像有 n 个比特板。

3．连续色调图像。

这种类型的图像可以有很多种相似的颜色（或灰度级）。当相邻的像素仅差一个单位时，眼睛很难或者说不可能区分出它们的颜色来。这样的结果是这种图像中的可能会包含这样的一些区域，当眼睛扫过时会看到似乎是非常连续的颜色。这种图像中的像素要么用一个单一的大数字表示（在有许多灰度级的情况下），要么用三个分量表示（在彩色图像的情况下）。连续色彩图像通常是自然图像（相对于人工图像而言）。它们可以通过数字相机的拍摄得到，也可以通过对照片或绘画的扫描得到。

4．离散色调图像

离散色调图像也称为图形图像或合成图像，它们通常是人工图像。它可能有几种或多种颜色但没有像自然图像所有的那种噪声或模糊存在。这方面的例子有人工画的物体或机器、一张文稿、一张图、一张卡通或是一个计算机屏幕的内容（这里有一点需要说明，那就是不是所有的人工图像都是离散色彩图像）。典型的例子是计算机产生的锁定自然的图像，尽管它是人工产生的，但它不是离散色彩图像而是连续色彩图像。人工图像、文字以及线条图具有很明显和完美的边角，这使得它们与图的其他部分（背景）区分得很明显。在离散色彩图像中，相邻的像素要么完全一样，要么在像素值上差得很远。这类图像不适于用有损压缩方法处理，因为即使丢掉几个像素都有可能导致看不清楚或是把一个很熟悉的模式变成一个不认识的模式。在连续色彩图像上使用的压缩方法往往不能很好地处理明显的棱角。所以需要用特殊的压缩方法来处理这类离散色彩图像。有一点值得提到的是由于同一个字符或同一个图案可能会在一个离散色彩图像中出现多次，这使得一个离散色彩图像可能会有许多数据冗余。

5．卡通图

这是一种包含一些颜色均匀分布区域的彩色图像。在每个独立区域中颜色均匀分布，而在相邻区域中，颜色相互相差又很大。卡通图的这种特性导致了它会有非常好的压缩机会。

5.1.2 Image Types

For the purpose of image compression it is useful to distinguish the following types of images.

1. Bi-level (or monochromatic) image

This is an image where the pixels can have one of two values, normally referred to as black and white. Each pixel in such an image is represented by one bit, making this the simplest type of image.

2. Grayscale image

A pixel in such an image can have one of the n values $0 \sim n - 1$, indicating one of 2n shades of gray (or shades of some other color). The value of n is normally compatible with a byte size; i.e., it is 4, 8, 12, 16, 24, or some other convenient multiple of 4 or of 8. The set of the most-significant bits of all the pixels is the most-significant bitplane. Thus, a grayscale image has n bitplanes.

3. Continuous-tone image

This type of image can have many similar colors (or grayscales). When adjacent pixels differ by just one unit, it is hard or even impossible for the eye to distinguish their colors. As a result, such an image may contain areas with colors that seem to vary continuously as the eye moves along the area. A pixel in such an image is represented by either a single large number (in the case of many grayscales) or three components (in the case of a color image). A continuous-tone image is normally a natural image (natural as opposed to artificial) and is obtained by taking a photograph with a digital camera, or by scanning a photograph or a painting.

4. Discrete-tone image

Discrete-tone image also called a graphical image or a synthetic image. This is normally an artificial image. It may have a few colors or many colors, but it does not have the noise and blurring of a natural image. Examples are an artificial object or machine, a page of text, a chart, a cartoon, or the contents of a computer screen. (Not every artificial image is discrete-tone. A computer-generated image that's meant to look natural is a continuous-tone image in spite of its being artificially generated.) Artificial objects, text, and line drawings have sharp, well-defined edges, and are therefore highly contrasted from the rest of the image (the background). Adjacent pixels in a discretetone image often are either identical or vary significantly in value. Such an image does not compress well with lossy methods, because the loss of just a few pixels may render a letter illegible, or change a familiar pattern to an unrecognizable one. Compression methods for continuous-tone images often do not handle sharp edges very well, so special methods are needed for efficient compression of these images. Notice that a discrete-tone image may be highly redundant, since the same character or pattern may appear many times in the image. Figure 7.59 is a typical example of a discrete-tone image.

5. Cartoon-like image. This is a color image that consists of uniform areas. Each area has a uniform color but adjacent areas may have very different colors. This feature may be exploited to obtain excellent compression.

5.1.3 图像数据压缩方法

通过上面的讨论可以很容易地看到每一种图像都可能有数据冗余，而各自冗余的方式又不尽相同。这就是为什么一种单一的数据压缩方法不可能适用于所有图像，也是为什么不同图像类型的图像需要不同数据压缩方法的原因。现有的数据压缩方法中有一类是专门某一类图像的，如针对二值图像的，连续色彩图像的和离散色彩图像的；还有一类是把一副图像分割成连续色彩图像的和离散色彩图像的不同的部分，然后对各个部分进行压缩。

从最广义的角度来看，图像就是照片，是以一种视觉的方式来记录和显示信息的。照片之所以对人们很重要，是因为它们是一种可以以特别有效地方式存储和传输信息的媒体。在人们日常生活中，使用照片来永久性地记录所看到的东西，同时也使用照片去和别人分享所看到的东西。通过给别人看一张照片，可以省去对所看到的东西进行很多冗长、乏味甚至是模糊的口头描述。这进一步强调了人类主要是视觉生物的要点。人们依赖眼睛获取所需要的周围的大部分信息，大脑也特别适应于对视觉信息的处理，这就是在本章开头提到过那个说法："一幅图胜过千言万语"的科学依据。

摄影是最熟悉的图像获取办法了。原因很简单，就是因为用这种手段记录的信息和眼睛看到的相似。眼睛的视觉和摄影都需要光源照亮要看或摄的实景。光和实景中的实体相互作用，其中的一部分会到达观察者由人的眼睛或照相机捕捉到。实景中的实体的信息就以这部分光的密度和颜色的变化的形式被记录下来了。这里很关键的一点是：尽管实景一般情况下都是三维的，但同一实景的图像却总是两维的。

5.1.3 Image Data Compression Methods

From the discussions above, we could see that it is intuitively clear that each type of image may feature redundancy, but they are redundant in different ways. This is why any given compression method may not perform well for all images, and why different methods are needed to compress the different image types. There are compression methods for bi-level images, for continuous-tone images, and for discrete-tone images. There are also methods that try to break an image up into continuous-tone and discrete-tone parts, and compress each separately.

In the broadest possible sense, images are pictures: a way of recording and presenting information visually. Pictures are important to us because they can be an extraordinarily effective medium for the storage and communication of information. We use photography in everyday life to create a permanent record of our visual experiences, and to help us share those experiences with others. In showing someone a photograph, we avoid the need for a lengthy, tedious and, in all likelihood, ambiguous verbal description of what was seen. This emphasises the point that humans are primarily visual creatures. We rely on our eyes for most of the information we receive concerning our surroundings, and our brains are particularly adept at visual data processing. There is thus a scientific basis for the well-known saying we mentioned at the beginning of this chapter that "a picture is worth a thousand words".

Photography is the imaging technique with which we are most familiar, simply because the information it records is similar to that which we receive using our eyes. Both human vision and photography require a light source to illuminate a scene. The light interacts with the objects in the scene and some of it reaches the observer, whereupon it is detected by the eyes or by a camera. Information about the objects in the scene is recorded as variations in the intensity and color of the detected light. A key point is that, although a scene is (typically) three-dimensional, the image of that scene is always two-dimensional.

5.2　二值图像压缩编码——数字传真

前面已经谈到过图像的种类，静止图像（Still Images）基本上包括两类：

（1）黑白（二值）静止图像；

（2）连续色调（彩色或灰度）静止图像。

也可把图像分为彩色图像和灰度图像两大类，前面已经提到二值图像就是只有黑白两种灰度级的特殊灰度图像，典型的例子有文件、气象图、工程图、指纹卡片、手写文字、地图、报纸，等等。此外，为了报纸的印刷，即使原来为灰度的图像，也要做成网纹的二值图像。传真只能一点一点地传送二值数据，也要把灰度的图像转化为二值的图像。二值图像信源编码的目的和灰度图像的编码一样，也是为了去掉数据冗余，减少表示图像所需的比特数。

二值图像压缩方法主要用于不包含任何连续色调图像的文档。例如，办公/商业文档，手写文本、线条图形、工程图等。

除了只有黑白两个灰度外，二值图像还有一些如下的其他特征：

首先，在统计特性上，由于只有两种灰度，即只有两种信源符号，所以只对应两种信源概率 P_0 和 P_1，且满足 $P_1 = 1 - P_0$，也就是说信源符号的概率可以只用一种概率来表示。

其次，图像数据量较小，单个像素既可以用其灰度值（例如 0 和 255）来表示，也可以用二进制值（0 和 1）来表示，显然后一种表示方法在存储和对图像进行数据处理时会比较简便。

此外，二值图像的结构也往往比较简单，黑、白像素区域多为连续分布，划分明显。

这些特征对于二值图像的压缩编码都具有重要的意义，大部分编码方法都是直接利用这些特征或者建立在这些特征的基础上的。

5.2　Bi-Level Image Compression Coding – Digital Facsimile

We have mentioned the image types. The still images basically could be classified into two types:

(1) Black-White (Bi-Level) still image;

(2) Continuous tone (color or grayscale) still image.

We could also classify the imaged into two major types, color and grayscale. The mentioned bi-level image is a special kind of grayscale image with only black and white two grayscales. The typical examples of this kind include text document, meteorological graphic, engineering drawing, fingerprint card, handwriting text, map, newspaper, and etc. Besides above, to print the newspaper, the original grayscale images will be converted to net-like bi-level images first. Since the facsimile can only transmit the bi-level data point by point, we have to convert the grayscale image to bi-level image as well before faxing them. The purpose of bi-level image encoding is the same as the grayscale's to reduce the redundancy and use as less as possible bits to represent the image.

Bi-level image compression method majorly applies on the documents which doesn't have any continuous tone images, such as official/commercial documents, handwriting text, line graphic, engineering drawing, and etc.

Besides only has two gray scales, the bi-level image has the following characters as well.

First of all, since it only has two gray scales that is to say it only has two source symbols, statistically, this is only correspondent to two possible probabilities P_0 and P_1. Furthermore, P_1 and P_0 has the relationship of $P_1 = 1 - P_0$ which means that the probabilities of the source symbols could be represented by only one probability.

Secondly, the image's data quantity is relatively small. The single pixel could be either represented by its gray scale such as 0 and 255 or by binary values 0 and 1. Obviously, the latter one is more convenient for image storage and image data processing.

Finally, the bi-level image's structure is simple. The either black or white areas are easy to be distinguished and continuously distributed.

Those characters are very helpful in bi-level image's compression coding. Most compression coding methods directly utilize those characters or created based on those characters.

前面提到，由于灰度级别只有两种，所以用于表示二值图像的数据量本身就远小于同等尺寸的灰度图像和彩色图像。但是，这并不意味着对它们就不必再进行压缩处理了。二值图像同一般的图像一样，也有着很大的压缩空间。如果每一像素用一位二进制码 0 或 1（白像素为 1，黑像素为 0）表示，则称为直接编码。一位二进制码为 1 比特，因而直接编码时表示一帧图像的比特数就等于该图像的像素数。直接编码对数据量是没有压缩效果的，因而通常把直接编码得到数据比特数作为该二值图像的原始数据大小，例如，二值图像以 Windows 操作系统中的标准图像文件 BMP 格式存储就是这样一种情况。如前面提到的，和其他图像形式类似，二值图像在结构和统计上也都有冗余特性，直接编码所形成的符号数据中必然包含了所需要的信息之外的冗余成分，所以，经过各种编码处理，去掉这些冗余成分，能够使表示二值图像的比特数小于该图像的像素数（即小于图像原始大小），达到压缩的目的。

二值图像在日常生活与科学研究中都大量存在，其传输和存储都占用着相当多的资源，因此，不断研究二值图像压缩和编码技术，想方设法提高现有算法的压缩性能，创新编码方法，都有着极其重要的理论和现实意义。众所周知，传真是一种静止图像通信方式，除照片传真外，一般的文件传真都是二值图像。为了缩短传输每帧传真图片所需的时间，就应通过有效编码减少表示每帧图片所需的比特数。例如，本来每帧图片需用 2Mb 表示，若用 2400bit/s 的数据调解器在电话线上传输，则需约 15min。如果能够经过编码，在一分钟内把它传完，则压缩比应为 15。另外，在运动跟踪、目标检测、视频压缩等课题的研究过程中，也会产生大量的二值图像，存于数据库或者用于分析研究，把这些图像经过压缩后再存储则可以节省很多空间。

As noticed, the data quantity would be much smaller for a bi-level image compared to either grayscale or continuous tone one with the same size. This doesn't mean we don't need to do the image compression on bi-level images anymore. Similar to other types of images, the bi-level image also has redundancy and therefore has room for the compression. If each pixel is represented by one digit of binary code 0 or 1, where white pixel is represented by 1 and the black pixel is represented by 0, this kind of coding is called directly encoding. Since one digit of binary code occupies one bit, the total number of the bits will be the same as the total number of the pixels in the image under this kind of directly encoding. As we can see that there is no data compression in the directly encoding. That's why we always take the number of the bits of the image as this kind of bi-level image's original data size. One example is the bi-level image saved under the Windows system's standard image file format bmp. As we mentioned, similar to other types of images, the bi-level image has redundancy both structurally and statistically as well. The directly encoded bi-level image data will definitely contain redundancy besides the information we need. We'd certainly like to apply kinds of encoding method on the data to get rid of the redundancy and left the only information we need. As a result, the number of total bits will be smaller than the number of total pixels of the image. This way, the data has been compressed.

The bi-level images are all around us. They occupy quite a bit resource in storage and transmission. That's why it is significant both theoretically and practically to continuously research the bi-level image's compression and encoding methods, try different ways to improve the efficiency of the existing compression and encoding methods, and find the new ones at the same time. As we know, facsimile is one of the still image communication formats. Except the photograph, the entire general documents facsimile are bi-level images. To cut the time of transmitting each frame of image, we need to reduce the number of bits needed to represent each frame of image via increasing the efficiency of the compression encoding method. For example, originally a frame of image has to be represented by 2Mb data. If we use 2400bp/s data modem to transmit it via a telephone line, it will take 15min. If we could finish the transmission within 1min after the compressing encoding, the compression rate will be 15. Besides the above, quantity of bi-level images will also be created in the process of some researches such as movement tracing, object detection, and video compression. Those bi-level images will stored in database for later research and analyzing. Compress those bi-level images before the storage will save quite bits spaces.

5.3 主要编码方法

实际当中要处理的二值图像可以分为多种，例如文本类、表格类、工程图类等。如同前面已经提到过的，不同的二值图像往往需要有不同的编码方法才能达到最好的压缩效果。目前，常用的二值图像编码方法主要有游程长度编码、跳白块编码、方块编码、识别编码、边界编码等等。这里主要介绍游程长度编码及糅合哈夫曼编码的 CCITT（Group 3）3 类数字传真标准及 CCITT（Group 4）4 类数字传真标准。

5.3 Major Encoding Methods

In practice, the bi-level images could be classified into many types, such as text files, tables, engineering drawings, and etc. As we mentioned, different type of bi-level images need different encoding methods to get the best compression effects. Currently, the most common bi-level image encoding methods include run-length encoding, jump white block encoding, square encoding, boundary encoding, and etc. Here we're going to mainly discuss the run-length encoding. The combination of run-length and huffman encoding, which are CCITT (Group 3) type 3 facsimile standard and CCITT (Group 4) type 4 facsimile standard.

5.3.1 游程长度编码

游程长度编码的基本思想是将具有相同数值、连续出现的信源符号构成的符号串用它们共同的数值及串的长度表示。当这种方法应用在图像上时，如果有连续的 L 个像素具有相同的灰度值 G，则对其作游程编码后，不论存储或传输，只需要一个数组（G, L）就可代替这一串像素的灰度值。这些连续的相同像素的数目称为游程。

很明显，游程长度越长，游程编码效率越高，因而这种编码方法特别适用于灰度等级少，灰度值变化小的二值图像。在实际应用当中，游程编码往往与其他编码方法结合使用，即把所有游程对应的数组（G, L）作为信源符号再进行编码，每一个（G, L）分配一个码字，例如，被 CCITT 选作文件传真三类机及四类机（T3 及 T4）一维标准码的修正哈夫曼编码（MH），就对不同长度的黑游程和白游程采用了最佳哈夫曼编码。

Run Length Encoding（RLE）天生是压缩压缩图像数据的备胎。前面已经谈到过，数据图像是由称为像素的一个个小点组成的，每个像素点可以像是在二值图像中似的由一位比特位表示，要么是黑要么是白，或者可以像在灰度级图像或彩色图像中似的由几位比特位表示几种灰度中的一种或几种彩色中的一种。假设像素以一维数组一位图的形式存储在内存中（后边我们将进一步讨论这方面的内容），如此一来，位图就成了图像压缩的输入流。在位图里，通常我们把位图里的像素按照扫描行来排列，所以位图里的第一个像素是图像中最左上角的一个像素，而位图里的最后一个像素是图像中最右下角的一个像素。

前面已提到，利用 RLE 来压缩图像数据是基于一个基本事实的，那就是随机地在图像中取一个像素，那么这个像素和它周围的像素的颜色一样的概率很大。这样的话，压缩处理器就可以一行一行地扫描位图，寻找连续的具有同样颜色的像素。下面来看一个例子，如果位图开始有 12 个白色像素，跟着有一个黑色像素，又跟着 61 个白色像素等等。这时候如前面所说的，只需要在输出编码序列中记录下 12,1,61 即可。

压缩处理器的具体工作过程为：它假设位图是以白色像素开始的，如果这个假设与实际情况不符，那么说明位图是以 0 个白色像素起始的，这时在输出序列中相应的以 0 为起始码。位图的分辨率也应该存储在输出序列的起始。

输出编码序列的尺寸大小取决于图像的复杂程度。对图像的细节要求的越多，压缩会越差。图 5-1 给出了扫描线扫过图像均匀区域的情况。一条扫描线从均匀区域周长的某一点进入该区域，从另外一点出去，而这两点是这条扫描线独用的，任何其他的扫描线都不会使用这同样的两个点。可以清楚地看到：穿过这个均匀区域的扫描线的总数大约等于它周长长度的一半（以像素来度量的话）。由于所扫描的区域是均匀的，所以每一条扫描线在输出序列中给出的是一个数字。所以一个均匀区域的数据压缩率大约是：

$$\frac{半周长的长度}{整个均匀区域总的像素数}$$

也可以用游程长度编码来压缩灰度级图像。如前面所提到的，具有同一个灰度密度（灰度级）的一个游程长度的像素编码为一对数据（游程长度 L，像素值 C）。游程长度 L 通常占一个字节，可表示的最大游程长度为 255 个像素。像素值占几个比特，具体几位取决于表示灰度值的总数所需要的位数（典型的值是 4～8 个比特）。

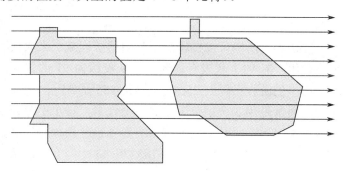

图 5-1　图像均匀区域及其扫描线

下面再来看一个例子。

假设有一个深度为 8 位的灰度级位图如下：

　　15, 15, 15, 15, 15, 15, 15, 15, 33, 88, 109, 56, 91, 91, 91, 3, 3, 3, 3, 3, 3, 2,…

它被压缩编码为

　　　　　　[8], 15,33,88,109,56, [3], 91, [6], 3, 2, …

中括号里的数字表示连续同样像素值的数目。面临的问题是怎样来区分具体的像素值，比如说 15 和表示连续同样像素值的数目值，比如说 8.下面列出了一些可行的解决方案。当然这些方案不是仅有的解决方案。

（1）如果图像的灰度级限制到 128 个，那么可以用每个字节中的一位作为标志位来区分这个字节是具体的像素值还是连续同样像素值的数目值。

（2）如果图像的灰度级是 256 个，可以把它减少到 255 个，而拿出 256 值中的一个作为标识值放在表示连续同样像素值的数目值字节的前面以告知后面紧跟着的字节是一个计数的数目值而不是具体的像素值。假如用一个字节所表示的 256 个数中的一个，255 来作为标识值，那么上面的序列就可以表示为

255, 8, 15, 33, 88, 109, 56, 255, 3, 91, 255, 6, 3, 2, …

（3）类似于 1 的情况，仍用一个比特来区分是具体的像素值还是连续同样像素值的数目值，但这次把这些额外的比特位每 8 个分成一组，在输出序列中，把每一组这样的 8 位写在它们所"属于"的像素之前或之后。

这时上边的例子中序列：[8],15,33,88,109,56, [3] ,91, [6] ,3,2, …就变成了：

[10000010] ,8,15,35,88,109,56,3,91, [100…] ,6,3,2, …

不难看到，因为每一个比特占了输出码流每个字节中的一位，这些额外比特数所形成的字节的总数是输出码流的 1/8，这时它们增加了 12.5%输出码流的尺寸大小。

（4）这种方法中我们把 m 个不同的像素之前加上一个字节，这个字节的值是–m。上面的序列就编码成了：

8, 15, –4, 33, 88, 109, 56, 3, 91, 6, 3, ?, 1 …

其中问号的位置是不定的，它的正负取决于 1 后面跟的是什么像素值。最坏的情况是像素串（p_1, p_2, p_3）在整个位图中重复 n 次。根据规则，它就将会被编码成（–1, p_1,2, p_2），原来的 3 个数字变成了 4 个，数据不但没有变少反而变多了！如果每个具体的像素是用一个字节来表示的，那么原始的三个字节就扩张成了四个字节。如果每个具体的像素是用三个字节来表示的，那么表示原来三个像素的九个字节就被压缩成了 1+3+1+3=8 个字节，只比原来的数据少了一个字节。

另外有三点是值得提到的。

（1）因为游程不可能是零，所以可以在输出序列中用游程减一来取代实际游程的值显得更为合理。这样一来，数组（3, 91）代表有连续 4 个灰度级为 91 的像素，而不是 3 个。这样的话，最长可以表示出来的游程可以到 256 而不再是 255 了。

（2）在彩色图像中，通常是用 3 个字节来存储一个像素，每个字节分别代表像素的红、绿、蓝、三种颜色。这种情况下，每种颜色的游程应当各自分别编码。这时，像素

（183, 76, 45），（184, 76, 46），（185, 76, 43），（181, 77, 41）

应当分解成三个序列：

（183, 184, 185, 181, …），（76, 76, 76, 77, …）和（45, 46, 43, 41, …）

然后对每一个序列分别进行游程长度编码。这就意味着任何可以用于灰度图像编码的方法也可以用于彩色图像编码。

（3）如果可能，则倾向于对位图的每一行进行独立编码。这样一来，如果在一行结束时有 5 个灰度级为 91 的像素，紧跟着的一行的起始有 8 个同样灰度级的像素，那么最好在输出序列中写出…5，91，8，91…，而不是…13，91…。更好的办法是在输出序列中写出…5，91，eol，8，91…，其中的 eol 是一个特殊的行标志结束符。这样做的原因是有些时候读者粗略地看一下大致而不需要了解图像的细节的时候就决定把它放弃了。如果每一行是独立编码的，那么解码算法就可以从第 1,7,13，…行开始解码并显示。接下来解码并显示行 2，8，14，…，这样持续下去。图像的独立行们之间是相互糅合关联的，图像在屏幕上是经过几个步骤后逐渐建立的。这种情形下就有可能在图像建立的初级阶段大致看到它是一个什么样的图像。图 5-2（c）给出了一个这样扫描的例子。

每行独立编码的另一个优点是可以有选择地获取一副图像的一部分，如从 m 行到 n 行的图像。除此之外，还可以在正常的解压之前就把两幅图像糅合在一起。

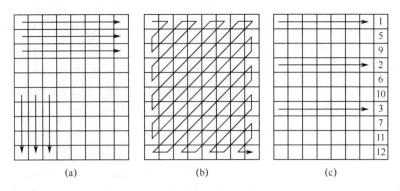

图 5-2　游程长度编码（RLE）扫描

如果接受这种每行独立编码的想法，那么在输出码流里就需要给出图像中的每一行是从哪里开始的信息？可以用如下的方法来实现，在输出码流的起始为位图中的每一行给出一组有 4 个字节即 32 位的码头。第 x 个这样的组包含着从输出码流的起始到第 y 行内容的开始所相差的字节数。很容易看到这样做的结果会增加压缩后码流的长度，但它仍然可能提供了一种在空间（压缩后的字节数）和时间（决定是否接受所处理的图像）之间很好的交换。

任何事情都有两面性，图像游程长度编码也有其不利之处。

（1）当图像被修改后，游程长度通常需要全部重做。

（2）对于复杂图像，游程长度编码的输出有时会比没有压缩以前的按一个个像素存储的尺寸还大。可以想象一个具有很多纵向线的图像，当对它进行横向扫描时，我们会得到一个很糟糕的压缩结果，甚至是数据的膨胀。

一个好的和实用的游程长度编码图像压缩器应该能够对位图进行逐行、逐列或 Zigzag 式的扫描，如图 5-2（a）和（b）所示。它甚至可能能够自动地对所压缩的图像进行三种扫描方式的尝试，以取得最佳的压缩结果。

有损图像压缩是一个比较大的话题，它涉及到许多种算法，这里暂时只谈和游程长度编码有关的内容。

可以直观地想象到如果可以丢掉一些短长度游程的像素，压缩率就会更高。这种方法是会在压缩中丢掉一些信息的，但这种丢失对用户来说有时是可接受的。当然在一些特定的情况下是不可接受的，这些特定的情况包括 X 光图像，由大型望远镜获取的天文图像等，每一幅这样图像的造价也是天文数字。

有损压缩图像游程长度编码算法在起始时就要问用户可忽略掉的最大游程是多少。比如如果用户说是 3，那么程序应该把所有游程长度为 1,2 或 3 的同样像素都和它们相邻的像素合并。这种情况下，压缩后的数据 9,10,1,2,5,3,12,2 将被存储为 9,10,8,17，其中 8 是 1+2+5=8 的和，而 17 是 3+12+2=17 的和。这种方法对大的高分辨率的图像是和适合的，因为人眼觉察不到这种丢失，而数据量却可大大减少。

5.3.1　Run Length Encoding

The basic idea of the run length encoding is to represent the source symbols which have the same value and consecutively appeared with their common value and the length of the symbol string. When this applies to the image, if there are L pixels which have the same grayscale C, after the run length encoding, either for storage or transmission, we could use one array (C, L) to represent those L pixels with the same value C. The number of the total pixels is call the run length.

Obviously, the longer the run length, the more efficient the encoding is. This makes the method is really suitable for the bi-level image which has few grayscales with little value changing. In practice, the run-length encoding is applied to the image with others. That is to encode the run-length encoding codes once more. Each pair of (C, L) will be encoded by another encoding method and assigned with another code. The typical example is the one dimensional standard codes' modified Huffman encoding accepted by CCITT as the document facsimile type 3 and type 4 codes. What it did is to encode the variable length of the black or white run-length via optimized Huffman encoding.

RLE is a natural candidate for compressing graphical data. As we mentioned, a digital image consists of small dots called pixels. Each pixel can be either one bit, indicating a black or a white dot as in bi-level image, or several bits, indicating one of several colors or shades of gray as in grayscale image. We assume that the pixels are stored in an array called a bitmap in memory (We'll talk about this further later.), so the bitmap is the input stream for the image compression. Pixels are normally arranged in the bitmap in scan lines, so the first bitmap pixel is the dot at the top left corner of the image, and the last pixel is the one at the bottom right corner.

As we indicated, compressing an image using RLE is based on the observation that if we select a pixel in the image at random, there is a good chance that its neighbours will have the same color. The compressing processor thus scans the bitmap row by row, looking for runs of pixels of the same color. If the bitmap starts, e.g., with 12 white pixels, followed by 1 black one, followed by 61 white ones, etc., then only the numbers 12, 1, 61, $\cdots$ need to be written on the output encoded codes stream as we indicated.

The compressing processor is working on the way as follows. It assumes that the bitmap starts with white pixels. If this is not true, then the bitmap starts with zero white pixels, and the output stream should start with 0. The resolution of the bitmap should also be saved at the start of the output stream.

The size of the compressed output stream depends on the complexity of the image. The more detail, the worse the compression. However, Fig. 5-1 shows how scan lines go through a uniform area. A line enters through one point on the perimeter of the area and exits through another point, and these two points are not "used" by any other scan lines. It is now clear that the number of scan lines traversing a uniform area is roughly equal to half the length (measured in pixels) of its perimeter. Since the area is uniform, each scan line contributes one number to the output stream. The compression ratio of a uniform area thus roughly equals the ratio

$$\frac{\text{half the length of the perimeter}}{\text{total number of pixels in the area}}$$

RLE can also be used to compress grayscale images. As we mentioned above, each run of pixels of the same intensity (gray level) is encoded as a pair (run length L, pixel value C). The run length usually occupies one byte, allowing for runs of up to 255 pixels. The pixel value occupies several bits, depending on the number of gray levels (typically between 4 and 8 bits).

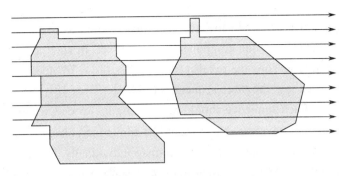

Fig.5-1   Uniform Area and Scan Lines

Example: A 8-bit deep grayscale bitmap that starts with:

15, 15, 15, 15, 15, 15, 15, 15, 33, 88, 109, 56, 91, 91, 91, 3, 3, 3, 3, 3, 3, 2,···

is compressing encoded into [8], 15,33,88,109,56, [3] ,91, [6] ,3,2, ···, where the boxed numbers indicate counts. The problem is to distinguish between a byte containing a grayscale value (such as 15)and one containing a count (such as [8]). The following are some solutions (although not the only possible ones):

(1) If the image is limited to just 128 grayscales, we can devote one bit in each byte to indicate whether the byte contains a grayscale value or a count.

(2) If the number of grayscales is 256, it can be reduced to 255 with one value reserved as a flag to precede every byte with a count. If the flag is, say, 255, then the sequence above becomes

255, 8, 15, 33, 88, 109, 56, 255, 3, 91, 255, 6, 3, 2, ···

(3) Again, one bit is devoted to each byte to indicate whether the byte contains a grayscale value or a count. This time, however, these extra bits are accumulated in groups of 8, and each group is written on the output stream preceding (or following) the 8 bytes it "belongs to."

The example: the sequence [8],15,33,88,109,56, [3] ,91, [6] ,3,2, ··· becomes:

[10000010] ,8,15,35,88,109,56,3,91, [100 ···] ,6,3,2, ···

The total size of the extra bytes is, of course, 1/8 the size of the output stream (they contain one bit for each byte of the output stream), so they increase the size of that stream by 12.5%.

(4) A group of m pixels that are all different is preceded by a byte with the negative value $-m$. The sequence above is encoded by

8, 15, −4, 33, 88, 109, 56, 3, 91, 6, 3, ? , 1, ···

The value of the byte with ? is positive or negative depending on what follows the pixel of 1. The worst case is a sequence of pixels (p_1, p_2, p_2) repeated n times throughout the bitmap. By the rule it will be encoded as $(-1, p_1, 2, p_2)$, four numbers instead of the original three. The number becomes more instead of getting less! If each pixel requires one byte, then the original three bytes will be expanded into four bytes. If each pixel requires three bytes, then the original three pixels (comprising 9 bytes) are compressed into $1 + 3 + 1 + 3 = 8$ bytes which are only one byte less than the original one's.

Three more points need to be mentioned:

(1) Since the run length cannot be 0, it makes sense to write the "run length minus one" on the output stream. Thus the pair (3, 91) means a run of four pixels with intensity 91 instead of three. This way, a run can be up to 256 pixels long.

233

(2) In color images it is common to have each pixel stored as three bytes, representing the intensities of the red, green, and blue components of the pixel. In such a case, runs of each color should be encoded separately. Thus the pixels (183, 76, 45), (184, 76, 46), (185, 76, 43), and (181, 77, 41) should be separated into the three sequences (183, 184, 185, 181, ···), (76, 76, 76, 77, ···), and (45, 46, 43, 41, ···). Each sequence should be run-length encoded separately. This means that any method for compressing grayscale images could be applied to color images as well.

(3)It is preferable to encode each row of the bitmap individually. Thus if a row ends with five pixels of intensity 91 and the following row starts with 8 such pixels, it is better to write ···, 5, 91, 8, 91, ···. on the output stream rather than ···, 13, 91, ··· It is even better to write the sequence ···, 5, 91, eol, 8, 91, ···, where "eol" is a special end-of-line code. The reason is that sometimes the user may decide to accept or reject an image just by examining its general shape, without any details. If each line is encoded individually, the decoding algorithm can start by decoding and displaying lines 1, 7, 13, ···, continue with lines 2, 8, 14, ···, etc. The individual rows of the image are interlaced, and the image is built on the screen gradually, in several steps. This way, it is possible to get an idea of what is in the image at an early stage. Fig. 5-2(c)shows an example of such a scan.

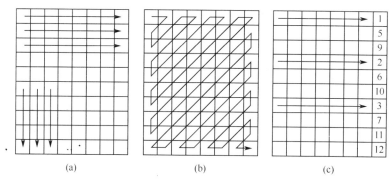

Fig.5-2 RLE Scanning

Another advantage of individual encoding of rows is to make it possible to extract just part of an encoded image (such as rows *m* through *n*). Furthermore, another application is to merge two compressed images without having to decompress them first.

If this idea that encodes each bitmap row individually is adopted, then the output compressed stream of the encoder shall contain information on where each bitmap row starts in the stream. This can be done by writing a header at the start of the stream that contains a group of 4 bytes (32 bits) for each bitmap row. The *x* th such group contains the offset (in bytes) from the start of the stream to the start of the content for row y. It's easy to see that this increases the size of the compressed stream but may still offer a good trade-off between space (size of compressed stream) and time (time to decide whether to accept or reject the image).

Anything has two sides. The image RLE has some disadvantages as well.

(1) When the image is modified, the run lengths normally have to be completely redone.

(2) The RLE output can sometimes be bigger than pixel-by-pixel storage (i.e., an uncompressed image, a raw dump of the bitmap) for complex pictures. Imagine a picture with many vertical lines.

When it is scanned horizontally, it produces very short runs, resulting in very bad compression, or even in expansion.

A good, practical RLE image compressing processor should be able to scan the bitmap by rows, columns, or in zigzag (Fig. 5-1 a,b) and it may automatically try all three ways on every image bitmap compressed to achieve the best compression.

The Lossy Image Compression is a big topic. There are many algorithms included. Here we only talk about the RLE related.

Directly thinking, it is possible to get even better compression ratios if short runs are ignored. Such a method loses information when compressing an image, but sometimes this is acceptable to the user. (Notable examples where no loss is acceptable are X-ray images and astronomical images taken by large telescopes, where the price of an image is astronomical.)

A lossy run length encoding algorithm should start by asking the user for the longest run that should still be ignored. If the user specifies, for example, 3, then the program merges all runs of 1, 2, or 3 identical pixels with their neighbours. The compressed data "9,10,1,2,5,3,12,2" would be saved, in this case, as "9,10,8,17" where 8 is the sum 1+2+5 and 17 is the sum 3 + 12 + 2. This makes sense for large high-resolution images where the loss of some detail may be invisible to the eye, but may significantly reduce the size of the output stream.

5.3.2 CCITT Group 3 和 Group 4 传真编码

通过传真机传输的文件是位图，这也是为什么它需要特殊编码方式的原因。随着传真机的普及应用，这种特殊的编码方式显得极为需要。ITU-T 开发和提出了几种实用的方法。

ITU-T 是隶属于 ITU（International Telecommunication Union）的四个长期机构之一。总部在瑞士日内瓦。它经常性的发布有关调制解调器、包交换接口、V.24 接口件和其他一些类似器件/设备标准的建议。虽然 ITU-T 本身没有强制执行的权利，但它所提出的有关标准性的建议都会在这个领域中采用。ITU-T 直到 1993 年 3 月更名为 CCITT（Consultative Committee for International Telephone and Telegraph）。

ITU-T 开发并提出的第一个相应数据压缩标准是 T2，也即 Group 1 和 T3，也即 Group 2。这两个标准目前已不再使用了，取而代之的是 T4，也即 Group 3 和 T6，也即 Group 4。Group 3 标准是目前所有使用 PSTN（Public Switched Telephone Network）网络传真机所使用的标准。而这种传真机正是人们每天所使用的传真机。在写这个标准的时候，传真机的最高传输速率为 9600baud。Group 4 标准应用于为数字网络设计的传真机，例如 ISDN 网络。它们的典型速率为 64kbaud。这两种方法都可以达到大约 10 倍的数据压缩率。对每页纸的传真时间，Group 3 约可减少一分钟的时间，Group 4 约可减少几秒钟的时间。

5.3.2.1 Group 3 & 4 传真编码方式——扫描、尺寸和传输

传真机逐行扫描传真文件，把每一行转换成称之为 pels 的黑白点，其扫描的格式如下。

光栅扫描，文档位置垂直于取像器的平面，扫描过程从左至右逐行进行，从页的顶端不间断地扫到底部。

水平方向分辨率总是 8.05pels/mm（约 205pels/inch），因此对于标准的 8.5 英寸宽（约 218mm）的扫描行来说，每行有 1728 个 pels（元素，白或黑）。但 T4，Group 3 标准建议每

行只扫描 8.2 英寸的宽度，这样每行就减少到 1664 个 pels（元素，白或黑）。

标准扫描方式，垂直方向 3.85 行/mm；可选精扫模式，垂直方向为 7.7 行/mm 。如今的很多传真机还可选高度精扫模式，垂直方向为 15.4 行/mm。

表 5-1 中的数据假设被传真的纸是 10 英寸（254mm）高，给出的是每页纸的总 pels 数，如前面所提到的，在这里像素称为 pels，表中给出了在没有数据压缩时典型的三种传真方式所需要的传输时间。可以看出所用时间很长，这也很生动地说明数据压缩在传真传输中的重要性。在实际生活中，没人有耐心为一张纸的传真等 3～11min。

表 5-1　传真传输时间

扫描行数	每行的 pels 数	每页的 pels 数	时间/s	时间/min
978	1664	1.670M	170	2.82
1956	1664	3.255M	339	5.65
3912	1664	6.510M	678	11.3
10 英寸等于 254mm。pels 数的单位是百万。传输时间在 9600baud 的信道下未压缩的数据约需 3～11min，不同的模式所需时间不同。 如果页面比 10 英寸短，或者其大部分地方是白的，那么我们可取得 10 或更好的压缩率，传输时间可缩短到 17～68s。				

为了开发 Group 3 码，ITU-T（原来的 CCITT）统计了他们认为能够代表那些在传真中典型使用的文字和图像的所有白、黑行程，并把它们放在了一套 8 个"训练"文件里，用哈夫曼算法为每一个不同的行程赋予了一个不同长度的编码。他们发现最常用到的行程是 2、3、4 个黑色 pels 的行程，所以给它们赋予了最短的码字（见表 5-2）。紧跟着最常用到的行程是 2 到 7 个白色 pels 的行程，所以给这些行程赋予了比上述码字略长一点的码字。大部分的行程很少出现，所以赋予了 12 个比特的长码字。这样，Group 3 使用了行程长度编码和哈夫曼编码的组合。

由于行程可能会很长，所以哈夫曼编码需要修改。1 到 63 个 pels 的黑白行程被赋予终端码，如表 5-2（a）所示，64 个 pels 的整数倍的黑白行程被赋予补充码，如表 5-2（b）所示。这样一来，Group 3 就形成了修正的哈夫曼编码，也称为 MH。任意一个行程的编码要么是一个单一的终端码（当这个行程的长度比 64 短的时候），要么是一个或多个补充码再加上一个终端码（当这个行程的长度等于或大于 64 的时候）。

表 5-2（a）　Group 3 和 Group 4 传真码字　终端码

行程	白码字	黑码字	行程	白码字	黑码字
0	00110101	0000110111	11	01000	0000101
1	000111	010	12	001000	0000111
2	0111	11	13	000011	00000100
3	1000	10	14	110100	00000111
4	1011	011	15	110101	000001100
5	1100	0011	16	101010	0000010111
6	1110	0010	17	101011	0000011000
7	1111	00011	18	0100111	0000001000
8	10011	000101	19	0001100	00001100111
9	10100	000100	20	0001000	00001101000
10	00111	0000100	21	0010111	00001101100

行程	白码字	黑码字	行程	白码字	黑码字
22	0000011	00000110111	43	00101100	000011011011
23	0000100	00000101000	44	00101101	000000101010
24	0101000	00000010111	45	00000100	000001010101
25	0101011	00000011000	46	00000101	000001010110
26	0010011	000011001010	47	00001010	000001010111
27	0100100	000011001011	48	00001011	000001100100
28	0011000	000011001100	49	01010010	000001100101
29	00000010	000011001101	50	01010011	000001010010
30	00000011	000001101000	51	01010100	000001010011
31	00011010	000001101001	52	01010101	000000100100
32	00011011	000001101010	53	00100100	000000110111
33	00010010	000001101011	54	00100101	000000111000
34	00010011	000011010010	55	01011000	000000100111
35	00010100	000011010011	56	01011001	000000101000
36	00010101	000011010100	57	01011010	000001011000
37	00010110	000011010101	58	01011011	000001011001
38	00010111	000011010110	59	01001010	000000101011
39	00101000	000011010111	60	01001011	000000101100
40	00101001	000001101100	61	00110010	000001011010
41	00101010	000001101101	62	00110011	000001100110
42	00101011	000011011010	63	00110100	000001100111

小结一下采用一维行程长度的编码程序。

（1）"一维"：每一连续的水平行都独立于其他行编码；

（2）"行程"：每行都可以分成交替的同色像素点段——白行程、黑行程、白行程、黑行程等。

CCITT (ITU-T) Group 3 和 Group 4 利用哈夫曼编码为给定的位流产生了一套补充码和一套终端码，其码字如表 5-2（a）和（b）所列。

表 5-2（b） Group 3 和 Group 4 传真码字补充码

行程	白码字	黑码字	行程	白码字	黑码字
64	11011	0000001111	640	01100111	0000001001010
128	10010	000011001000	704	011001100	0000001001011
192	010111	000011001001	768	011001101	0000001001100
256	0110111	000000101011	832	011010010	0000001001101
320	00110110	000000110011	896	011010011	0000001110010
384	00110111	000000110100	960	011010100	0000001110011
448	01100100	000000110101	1024	011010101	0000001110100
512	01100101	0000001101100	1088	011010110	0000001110101
576	01101000	0000001101101	1152	011010111	0000001110110

行程	白码字	黑码字	行程	白码字	黑码字
1216	011011000	0000001110111	1920	00000001101	
1280	011011001	0000001010010	1984	000000010010	
1344	011011010	0000001010011	2048	000000010011	
1408	011011011	0000001010100	2112	000000010100	
1472	010011000	0000001010101	2176	000000010101	
1536	010011001	0000001011010	2240	000000010110	
1600	010011010	0000001011011	2304	000000010111	
1664	011000	0000001100100	2368	000000011100	
1728	010011011	0000001100101	2432	000000011101	
1792	00000001000	从这里开始和白码字相同。	2496	000000011110	
1856	00000001100		2560	000000011111	

（1）补充码用于表示大于 64 个像素的行程；

（2）终端码用于表示小于 64 个像素的行程；

（3）黑像素的行程编码与白像素的行程编码不同。

例 5-1 传真扫描得到的结果如下：

白（10）黑（2）白（13）黑（3）白（14）黑（4）白(96)。

求其 Group 3 & 4 编码。

解：由于白（10）黑（2）白（13）黑（3）白（14）黑（4）的行程均小于 64，所以可由表 5-1（a）直接得到它们的码字为

$$00111, 11, 000011, 10, 110100, 011$$

白（96）的行程大于 64，可分为 64+32，由表 5-2（b）可得到白（64）的补充码为

$$11011$$

由表 5-2（a）可得到白（32）的终端码为

$$00011011$$

因此，白（96）的码字为

$$11011/00011011$$

整个传真扫描的 Group 3 & 4 编码为

$$00111, 11, 000011, 10, 110100, 011，11011 00011011$$

总码长为 37bit。如果采用直接编码，传真扫描得到 142 个 pels 则需要 142 bit，Group3 & 4 编码实现了数据压缩。在这一组短数据中，压缩比约为 3.84。

下面来看几个相关的小问题。

a. 问：3 类传真机只能用 A4 幅面的图幅吗？

答：否。可以用 A4 大小或更小一点的图幅，如果需要页面更宽，水平扫描的密度就会相应降低。

b. 问：有无时间的限制？

答：有。最小行传输时间为 20ms，是指对每行扫描编码后形成数据位、填充位和行结束符号的时间总和；

最大行传输时间为 5s，因为要保持同步，同时为了防止断线检测电路误动作而断线。

5.3.2　CCITT Group 3 and Group 4 Facsimile Encoding

Documents transferred between fax machines are sent as bitmaps, so a standard data compression method was needed when those machines became popular. Several methods were developed and proposed by the ITU-T.

The ITU-T is one of four permanent parts of the International Telecommunications Union (ITU), based in Geneva, Switzerland (http://www.itu.ch/). It issues recommendations for standards applying to modems, packet switched interfaces, V.24 connectors, and similar devices. Although it has no power of enforcement, the standards it recommends are generally accepted and adopted by industry. Until March 1993, the ITU-T was known as the Consultative Committee for International Telephone and Telegraph (CCITT).

The first data compression standards developed by the ITU-T were T2 (also known as Group 1) and T3 (Group 2). These are now obsolete and have been replaced by T4 (Group 3) and T6 (Group 4). Group 3 is currently used by all fax machines designed to operate with the Public Switched Telephone Network (PSTN). These are the machines we have at home, and at the time of writing, they operate at maximum speeds of 9,600 baud. Group 4 is used by fax machines designed to operate on a digital network, such as ISDN. They have typical speeds of 64k baud. Both methods can produce compression factors of 10 or better, reducing the transmission time of a typical page to about a minute with Group 3, and a few seconds with Group 4.

5.3.2.1　Group 3 & 4 Facsimile Encoding Format – Scan, Size, and Transmission

A fax machine scans a document line by line, converting each line to small black and white dots called pels (from Picture ELement). Its scan formats are as follows.

Grid scan, in this style, the document is located vertical to the machine's scan capture plane. The scanning is from left to right line by line, consecutively from the top of the page to the bottom of the page.

The horizontal resolution is always 8.05 pels per millimeter (about 205 pels per inch). A standard 8.5-inch-wide(about 218mm) scan line is therefore converted to 1728 pels. The T4 group 3 standard, though, recommends to scan only about 8.2 inches wide, thereby reducing the pels to 1664. (either white or black) per scan line (these numbers, as well as those in the next paragraph, are all to within ±1% accuracy).

In standard scan mode, the vertical resolution is 3.85 scan lines per millimetre. In fine scan mode, the vertical resolution is 7.7 scan lines/mm. Many fax machines today also have a very-fine mode, where they vertically scan 15.4 lines/mm.

Table 5-1 assumes the faxed page is a 10-inch-high page (254 mm), and shows the total number of pels per page, as we indicated; here the pixels are called pels. It also shows the typical transmission times for the three fax modes without compression. The times are long, illustrating how important data compression is in fax transmissions. In our real life, nobody has the patience to wait 3 to 11 minutes for one page of fax.

Table 5-1 Fax Transmission Times

Scan lines	Pels per line	Pels per page	Time (sec.)	Time (min.)
978	1664	1.670M	170	2.82
1956	1664	3.255M	339	5.65
3912	1664	6.510M	678	11.3

Ten inches equal 254 mm. The number of pels is in the millions, and the transmission times, at 9600 baud without compression, are between 3 and 11 minutes, depending on the mode. How-ever, if the page is shorter than 10 inches, or if most of it is white, the compression factor can be 10 or better, resulting in transmission times of between 17~68 seconds.

To develop the Group 3 code, the ITU-T (formerly CCITT) counted all the run lengths of white and black pels in a set of eight "training" documents that they felt could represent typical text and images sent by fax, and used the Huffman algorithm to assign a variable-length code to each run length. The most common run lengths were found to be 2, 3, and 4 black pixels, so they were assigned the shortest codes as shown in Table 5-2. The next most come run lengths are 2–7 white pixels, which were assigned slightly longer codes. Most run lengths were rare and were assigned long 12-bit codes. This way, Group 3 uses a combination of RLE and Huffman coding.

Since run lengths could be very long, the Huffman algorithm needs to be modified. The termination codes were assigned to black or white run lengths of 1 to 63 pels as shown in Table 5-2(a) and the make-up codes assigned to black or white run lengths that are multiples of 64 pels as shown in Table 5-2(b). Group 3 is therefore a modified Huffman code which is also called MH. The encoded code of a run length is either a single termination code when the run length is shorter than 64 or one or more make-up codes, followed by one termination code when it is longer than 64 including 64.

Table 5-2（a） Group 3& 4 Fax Codes Termination Codes

Run length	White code-word	Black code-word	Run length	White code-word	Black code-word
0	00110101	0000110111	19	0001100	00001100111
1	000111	010	20	0001000	00001101000
2	0111	11	21	0010111	00001101100
3	1000	10	22	0000011	00000110111
4	1011	011	23	0000100	00000101000
5	1100	0011	24	0101000	00000010111
6	1110	0010	25	0101011	00000011000
7	1111	00011	26	0010011	000011001010
8	10011	000101	27	0100100	000011001011
9	10100	000100	28	0011000	000011001100
10	00111	0000100	29	00000010	000011001101
11	01000	0000101	30	00000011	000001101000
12	001000	0000111	31	00011010	000001101001
13	000011	00000100	32	00011011	000001101010
14	110100	00000111	33	00010010	000001101011
15	110101	000001100	34	00010011	000011010010
16	101010	0000010111	35	00010100	000011010011
17	101011	0000011000	36	00010101	000011010100
18	0100111	0000001000	37	00010110	000011010101

Run length	White code-word	Black code-word	Run length	White code-word	Black code-word
38	00010111	000011010110	51	01010100	000001010011
39	00101000	000011010111	52	01010101	000000100100
40	00101001	000001101100	53	00100100	000000110111
41	00101010	000001101101	54	00100101	000000111000
42	00101011	000011011010	55	01011000	000000100111
43	00101100	000011011011	56	01011001	000000101000
44	00101101	000001010100	57	01011010	000001011000
45	00000100	000001010101	58	01011011	000001011001
46	00000101	000001010110	59	01001010	000000101011
47	00001010	000001010111	60	01001011	000000101100
48	00001011	000001100100	61	00110010	000001011010
49	01010010	000001100101	62	00110011	000001100110
50	01010011	000001010010	63	00110100	000001100111

Let's put together what we talked about the one-dimensional run length encoding procedure so far.

(1) One-dimensional means that each horizontal line is encoded independently.

(2) Run-length means that each scan line could be divided into one or the other either white or black same color pixel sections, such as white run-length, black run-length, white run-length, and etc.

As we indicated, CCITT Group 3 and Group 4 proposed one set of Make-up codes and one set of Termination codes for the input codes string based on Huffman encoding algorithm.

The encoded codes are shown in Table 5-2(a) & (b).

Table 5-2（b） Group 3& 4 Fax Codes　Make-up Codes

Run length	White code-word	Black code-word	Run length	White code-word	Black code-word
64	11011	0000001111	1344	011011010	0000001010011
128	10010	000011001000	1408	011011011	0000001010100
192	010111	000011001001	1472	010011000	0000001010101
256	0110111	000001011011	1536	010011001	0000001011010
320	00110110	000000110011	1600	010011010	0000001011011
384	00110111	000000110100	1664	011000	0000001100100
448	01100100	000000110101	1728	010011011	0000001100101
512	01100101	0000001101100	1792	00000001000	Same aswhite from this point
576	01101000	0000001101101	1856	00000001100	
640	01100111	0000001001010	1920	00000001101	
704	011001100	0000001001011	1984	000000010010	
768	011001101	0000001001100	2048	000000010011	
832	011010010	0000001001101	2112	000000010100	
896	011010011	0000001110010	2176	000000010101	
960	011010100	0000001110011	2240	000000010110	
1024	011010101	0000001110100	2304	000000010111	
1088	011010110	0000001110101	2368	000000011100	
1152	011010111	0000001110110	2432	000000011101	
1216	011011000	0000001110111	2496	000000011110	
1280	011011001	0000001010010	2560	000000011111	

(1) The Make-up codes are used to represent the run-lengths which are longer than 64 pixels.

(2) The Terminate codes are used to represent the run-lengths which are shorter than 64 pixels.

(3) The run-length codes of the while pixels and the black pixels are different from each other.

Example 5-1. One of the scans result is as follows:

white (10), black (2), white (13), black (3), white (14), black (4), white (96).

Find out its Group 3 & 4 codes.

Solution: Since each of the run-length of white (10), black (2), white (13), black (3), white (14), black (4) is shorter than 64, we could directly get their termination codes from Table 5-2 (a) and they are as follows.

00111, 11, 000011, 10, 110100, 011

The run-length of white (96) is longer than 64; we could decompose it into 64 + 32. From Table 5-2(b), we could get the white 64's make-up code as 11011.

From Table 5-2(a), we could get the white 32's termination code as 00011011.

So, the white (96)'s code shall be 11011 00011011.

The whole scan's Group 3 & 4 codes will be as follows.

00111, 11, 000011, 10, 110100, 011, 11011 00011011

The total length of the codes is 37bits. If we use the binary codes directly encode the same scan, we'll get total 142bits for the 142 pels. From this we can see that the Group 3 & 4 encoding did compress the data from 142bits to 37bits, the compression ratio here is roughly 3.84.

Let's have a look at several relation questions below.

a. Asking: Is the type 3 fax machine only can use the image size of A4 paper?

Answer: No, it could use either A4 size or smaller than A4 size. If one need wider than A4 size, the horizontal scan density will be correspondently lower.

b. Asking: Is there time limit?

Answer: There is. The smallest line transmission time is 20ms. It means the total time of the line scanning, encoding, and forming the data bits, compensation bits, and the end of line bits.

The maximum transmission time is 5 seconds to keep the synchronization and to prevent the connection cut off detection circuit misacting leading the connection getting lost.

5.3.2.2 一维编码

基于上述讨论，进一步给出几个更为典型的例子。

从表 5-2 （a）和（b）可以很容易得到下述行程长度的编码码字。

（1）15 个白 pels 的行程的编码为 110101，这里只用到了终端码。

（2）79 个白 pels 的行程的编码为（79=64+15）11011 110101，这里同时用到了补充码和终端码。

（3）143 个白 pels 的行程的编码为（143=128+15）10010 110101，这里同时用到了补充码和终端码。

（4）64 个黑 pels 的行程的编码为（64=64+0）0000001111 0000110111，这里同时用到了补充码和终端码。还有一点值得一提的是这里用到了零个黑 pels 的行程编码的终端码。

（5）2566 个黑 pels 的行程的编码为（2566=2560+6）000000011111　0010，这里同时用到了补充码和终端码。

已经知道每一个扫描行的编码是独立完成的。在实际应用中，每一行的编码由一个特殊12 位的 EOL（我们提到过的行结束符）码 000000000001 来结束。除此之外，在扫描时每一行的最左边加了一位额外的白 pel，这样做的目的是去掉接收端解码时可能存在的行的模糊性。当读完了上一行的 EOL 结束符后，接收器就假设新的一行以一串白 pels 开始了，并且忽略其中的第一个。下面是两个这样的例子。

a. 13 个 pels 的扫描行 ■■■□□■■□□□□□□ 编码为行程 1w、3b、 2w、2b、6w、EOL 的码字：000111　10　0111　11　1110　000000000001。译码器在译码时会忽略第一个白 pel。

b. 12 个 pels 的扫描行 □□□■■■■■□□■■ 编码为行程 3w、5b、 3w、2b、EOL 的码字：1000　0011　1000　11　000000000001。

Group 3 编码没有特殊的纠错码，但许多错误是可以被检测到的。由于哈夫曼编码的特性，在传输过程中，即使有一位坏码出现也会致使接收机失去同步，产生出一串错误的 pels，这也是为什么每一行要独立扫描的原因。如果接收机检测到了错误，它会跳过错误位而寻找行结束符 EOL。这样的话，一个错误最多能引起一行扫描行接收错误。如果接收机在经过一定量的行后仍然找不到行结束符 EOL，它将假设有较大的错误率发生了，它会放弃接收过程并通知发送端。从表 5-2（a）和（b）中可以看到，每一个码字的码长都在 2 到 12 位之间，所以如果接收机在接收到 12 位数据之后，仍然解码不出一个有效的码字，那么它知道出错了。

每一页编好码的文件之前有一个行结束符 EOL，之后有六个行结束符 EOL。如前面所提到的，由于每一行都是独立编码的，所以这种方法称为一维编码方式。压缩率取决于图像本身，具有大片连续黑或白区域的图像（这里指文字或黑白图像）可以达到很高的压缩率。反之，具有很多短行程的图像可能会产生负值的压缩率。特别是对于具有灰色阴影的图像（比如扫描后的照片），这种效应更为明显。这种阴影是由半调产生的，所谓半调就是说有很多一黑一白 pels 的黑白 pels 交替出现的区域。

T4 标准还允许在数据和行结束符 EOL 之间插入附加位。这样做的原因是有时需要暂停或者是要传输的总位数要求是 8 的倍数。附加位是一个或多个 '0'。下面是一个这样的例子。

二进制码串

000111　10　0111　11　1111　000000000001

加了附加位后变为

000111　10　0111　11　1111　00　0000000001

在加了两位附加位之后，总码长加长成了 32 个比特（=8×4）。当接收器看到了两位附加码的 0，和紧跟着的 EOL 的 11 个 0，然后是一个单一的 1，它知道它是收到了一组附加码和一个行结束符 EOL。

5.3.2.2　One–Dimensional Coding

Based on the discussion above, we could give some more typical examples as follows.

From the shown Table 5–2 (a) & (b), we could easily get the following run-lengths encoded codes.

(1) A run length of 15 white pels is coded as 110101 (Termination code only).

(2) A run length of 79 white pels (79 = 64 + 15) is coded as 11011　110101 (Make-up code

plus the termination code).

(3) A run length of 143 white pels (143 = 128 + 15) is coded as 10010 110101 (Make-up code plus the termination code).

(4) A run length of 64 black pels (64 = 64 + 0) is coded as 0000001111 0000110111 (Make-up code plus the termination code). Notice that the black zero's termination code applied here.

(5) A run length of 2566 black pels (2566 = 2560 + 6) is coded as 000000011111 0010 (Make-up code plus the termination code).

As we have known, each scan line is coded independently. In reality, each line's code is terminated by the special 12-bit EOL(end of line, we mentioned at an earlier time) code 000000000001. Besides this, each line also gets one white pel appended to it on the left when it is scanned. This is done to remove any ambiguity when the line is decoded on the receiving end. After reading the EOL for the previous line, the receiver assumes that the new line starts with a run of white pels, and it ignores the first of them. Here are some examples:

a. The 13-pel line  is encoded as the run lengths 1w 3b 2w 2b 6w EOL, which are the codes 000111 10 0111 11 1110 000000000001. The decoder ignores the single white pel at the start.

b. The 12-pel line  is encoded as the run lengths 3w 5b 3w 2b EOL, which are the binary string codes 1000 0011 1000 11 000000000001.

The Group 3 code has no particular error correction, but many errors could be detected. Because of the nature of the Huffman code, even one bad bit in the transmission can cause the receiver to get out of synchronization, and to produce a string of wrong pels. This is why each scan line is encoded independently. If the receiver detects an error, it skips bits, looking for an EOL. This way, one error can cause at most one scan line to be received incorrectly. If the receiver does not see an EOL after a certain number of lines, it assumes a high error rate, and it aborts the process, notifying the transmitter. From the Table 5-2(a) & (b), we could see that the codes are between 2 and 12 bits long, so, the receiver detects an error if it cannot decode a valid code after reading 12 bits.

Each page of the coded document is preceded by one EOL and is followed by six EOL codes (as we mentioned). As indicated, because each line is coded independently, this method is so called a one-dimensional coding scheme. The compression ratio depends on the image. Images with large contiguous black or white areas (text or black and white images) can be highly compressed. Images with many short runs can sometimes produce negative compression. This is especially true in the case of images with shades of gray (such as scanned photographs). Such shades are produced by halftoning, which covers areas with many alternating black and white pels (runs of length one).

The T4 standard also allows for fill bits to be inserted between the data bits and the EOL. This is done in cases where a pause is necessary, or where the total number of bits transmitted for a scan line must be a multiple of 8. The fill bits are zeros.

Here is an example: The binary string

000111 10 0111 11 1111 000000000001

becomes

000111   10   0111   11   1111   00   0000000001

after two zeros are added as fill bits, bringing the total length of the string to 32 bits (= 8×4). The decoder sees the two zeros of the fill, followed by the 11 zeros of the EOL, followed by the single 1, so it knows that it has encountered a fill followed by an EOL.

5.4 连续色调（彩色或灰度）图像压缩编码

5.4.1 视觉系统对颜色的感知

颜色不是自然存在的东西，它是视觉系统对可见光的感知结果。可见光是波长在 380～780nm 之间的电磁波，人们看到的大多数光不是一种波长的光，而是由许多不同波长的光组合成的。人的视网膜含有对红、绿、蓝颜色敏感程度不同的三种锥体细胞，这也是为什么这三种颜色很自然地被选成主要颜色。更为准确地说，三种锥体细胞对于波长的最为敏感度的约为 420nm（蓝色），534nm（偏蓝的绿色）和 564nm（偏黄的绿色）。另外还有一种在光功率极端低的条件下才起作用的杆状体细胞，因此颜色只存在于眼睛和大脑，而不是大自然存在的东西。这个听起来可能有些奇怪，但想一想，当向别人描述我们所看到的颜色时，是无法像别的物体特性一样给出准确的描述的。想象一下，当讲述看到的东西是红色的，一个盲人就得不到一个准确的概念，这个红色是什么样子的？而当说一个东西是方的，一个盲人就可以得到一个准确的概念，因为这个方的是可以摸到的，想一想这些，就不会感到奇怪了。目前在计算机图像处理中，杆状细胞还没有扮演什么重要角色。

人的视觉系统对颜色的感知可归纳出如下几个特性。

（1）眼睛本质上是一个照相机。人的视网膜通过神经元来感知外部世界的颜色，每个神经元或者是一个对颜色敏感的锥体，或者是一个对颜色不敏感的杆状体。

（2）红、绿和蓝三种锥体细胞对不同频率的光的感知程度不同，对不同亮度的感知程度也不同，这就意味着，人们可以使用数字图像处理技术来降低表示图像的数据量而不使人感到图像质量明显下降。

（3）颜色只存在于人们心中，三种对颜色敏感的锥体特殊对红、绿、蓝三种颜色敏感，这也是为什么人们总是很自然地把这三种颜色选为主要色，自然界中的任何一种颜色都可以由 RGB 这三种颜色值之和来确定，它们构成一个三维的 RGB 矢量空间。这就是说，RGB 的数值不同混合得到的颜色就不同，也就是光波的波长不同。

5.4.2 图像的三个基本属性

描述一幅图像需要使用图像的属性。图像的属性包含分辨率、像素深度、真、伪彩色、图像的表示法和种类等，本节只介绍前三个属性。

5.4.2.1 分辨率

我们经常遇到的分辨率有两种：显示分辨率和图像分辨率。让我们来分别看一看这两种分辨率。

1. 显示分辨率

显示分辨率是指显示屏上能够显示出的像素数目。例如，显示分辨率为 640×480 表示显示屏分成 480 行，每行显示 640 个像素，整个显示屏就含有 307 200 个显像点。屏幕能够显示的像素越多，说明显示设备的分辨率越高，显示的图像质量也就越高。早期除像笔记本计算机用液晶显示（Liquid Crystal Display，LCD）外，一般都采用 CRT 显示，它类似于彩色电视机中的 CRT。显示屏上的每个彩色像点由代表 R G B 三种模拟信号的相对强度决定，这些彩色像点就构成一幅彩色图像。

早期计算机用的 CRT 和家用电视机用的 CRT 之间的主要差别是显像管玻璃面上的孔眼掩膜和所涂的荧光物不同。孔眼之间的距离称为点距（Dot Pitch）。因此常用点距来衡量一个显示屏的分辨率。电视机用的 CRT 的平均分辨率为 0.76mm，而标准 SVGA 显示器的分辨率为 0.28mm。孔眼越小，分辨率就越高，这就需要更小更精细的荧光点。这也就是为什么同样尺寸的计算机显示器比电视机的价格贵得多的原因。

更早期用的计算机显示器的分辨率是 0.41mm。随着技术的进步，分辨率由 0.41→0.38→0.35→0.31→0.28→直到 0.26mm 以下。这种显示器的价格主要集中体现在分辨率上，因此在购买这种显示器时应在价格和性能上综合考虑。

如今的计算机大多都使用 LED 显示器了。

2．图像分辨率

图像分辨率是指组成一幅图像的像素密度的度量方法。对同样大小的一幅图，如果组成该图的图像像素数目越多，则说明图像的分辨率越高，看起来就越逼真。相反，图像显得越粗糙。

在用扫描仪扫描彩色图像时，通常要指定图像的分辨率，用每英寸多少点（Dots Per Inch，DPI）表示。如果用 300DPI 来扫描一幅 8 "×10" 的彩色图像，就得到一幅 2400×3000 个像素的图像。分辨率越高，像素就越多。

图像分辨率与显示分辨率是两个不同的概念。图像分辨率是确定组成一幅图像的像素数目，而显示分辨率是确定显示图像的区域大小。如果显示屏的分辨率为 640×480，那么，一幅 320×240 的图像只占显示屏的 1/4；相反，2400×3000 的图像在这个显示屏上就不能显示一个完整的画面。

这里顺便说一下，在显示一幅图像时，有可能会出现图像的宽高比（Aspect Ratio）与显示屏上显示出的图像的宽高比不一致这种现象。这是由于显示设备中定义的宽高比与图像的宽高比不一致造成的。例如，一幅 200×200 像素的方形图，有可能在显示设备上显示的不再是方形图，而变成了矩形图。这种现象在 20 世纪 80 年代的显示设备上经常遇到。

5.4.2.2　像素深度

像素深度是指存储每个像素所用的位数，它也是用来度量图像的分辨率。像素深度决定彩色图像的每个像素可能有的颜色数，或者确定灰度图像的每个像素可能有的灰度级数，例如，一幅彩色图像的每个像素用 RGB 三个分量表示．若每个分量用 8 位，那么一个像素共用 24 位表示，就说像素的深度为 24，每个像素的颜色可以是 2 的 24 次方＝16777216 种颜色中的一种。在这个意义上，往往把像素深度说成是图像深度。表示一个像素的位数越多，它能表达的颜色数目就越多，而它的深度就越深。

虽然像素深度或图像深度可以很深，但各种 VGA 的颜色深度却受到限制。例如，标准 VGA 支持 4 位 16 种颜色的彩色图像，多媒体应用中推荐至少用 8 位 256 种颜色。由于设备的限制，加上人眼分辨率的限制，一般情况下，不一定要追求特别深的像素深度。此外，像素深度越深，所占用的存储空间越大。相反，如果像素深度太浅，那也影响图像的质量，图像看起来让人觉得很粗糙和很不自然。

在用二进制数表示彩色图像的像素时，除 RGB 分量用固定位数表示外，往往还增加 1 位或几位作为属性位。例如，RGB 5∶5∶5 表示一个像素时，用两个字节共 16 位表示，其中 RGB 各占 5 位，剩下一位作为属性位。在这种情况下，像素深度为 16 位，而图像深度为 15 位。

属性位用来指定该像素应具有的性质。例如，在 CD-I 系统中，用 RGB 5∶5∶5 表示的像素共 16 位，其最高位（b15）用作属性位。并把它称为透明位，记为 T。T 的含义可以这样来理解：假如显示屏上已经有一幅图存在，当前这幅图或者当前这幅图的一部分要重叠在上面时，T 位就用来控制原图是否能看得见。例如，定义 $T=1$，原图完全看不见；$T=0$，原图能完全看见。

在用 32 位表示一个像素时，若 RGB 分别用 8 位表示，剩下的 8 位常称为 α 通道（Alpha Channel）位，或称为覆盖（Overlay）位、中断位或属性位。它的用法可用一个预乘 α 通道（Premultiplied Alpha）的例子说明。假如一个像素（A，R，G，B）的四个分量都用规一化的数值表示，当（A，R，G，B）为（1，1，0，0）时显示红色。当像素为（0.5，1，0，0）时，预乘的结果就变成（0.5，0.5，0．0），这表示原来该像素显示的红色的强度为 1，而现在显示的红色的强度降了一半。

用这种办法定义一个像素的属性在实际中很有用。例如，在一幅彩色图像上叠加文字说明，而又不想让文字把图覆盖掉，就可以用这种办法来定义像素，而该像素显示的颜色又称为混合色（Key Color）。

在图像产品生产中，也往往把数字电视图像和计算机生产的图像混合在一起，这种技术称为视图混合（Video Keying）技术，它也采用 α 通道。

5.4.2.3 真彩色、伪彩色与直接色

搞清真彩色和伪彩色与直接色的含义，对于编写图像显示程序和理解图像文件的存储格式有直接的指导意义，也不会对出现诸如这样的现象感到困惑：本来是用真彩色表示的图像，但在 VGA 显示器上显示的图像颜色却不是原来图像的颜色。

1. 真彩色（True Color）

真彩色是指在组成一幅彩色图像的每个像素值中，有 RGB 三个基色分量，每个基色分量直接决定显示设备的基色强度，这样产生的彩色称为真彩色。例如，用 RGB 5∶5∶5 表示的彩色图像，RGB 各用 5 位，用 RGB 分量大小的值直接确定三个基色的强度，这样得到的彩色是真实的原图彩色。

如果用 RGB8∶8∶8 方式表示一幅彩色图像，就是 RGB 都用 8 位来表示，每个基色分量占一个字节，共 3 个字节，每个像素的颜色就是由这 3 个字节中的数值直接决定，则可生成的颜色数就是 2 的 24 次方=16777216 种。用 3 个字节表示的真彩色图像所需要的存储空间很大，而人的眼睛是很难分辨出这么多种颜色的。因此在许多场合往往用 RGB 5∶5∶5 来表示，每个彩色分量占 5 个位，再加 1 位显示属性控制位共 2 个字节，生成的真颜色数目为 2 的 15 次方=32k（32768）。

在许多场合，真彩色图通常是指 RGB 8∶8∶8，即图像的颜色数等于 2 的 24 次方，也常称为全彩色（Full Color）图像。但在显示器上显示的颜色就不一定是真彩色，要得到真彩色图像需要有真彩色显示适配器．以前在 PC 上用的 VGA 适配器是很难得到真彩色图像的。而如今的 LED 显示器上是可以实现的。

2．伪彩色（Pseudo Color）

伪彩色图像是每个像素的颜色不是由每个基色分量的数值直接决定，而是把像素值当作彩色查找表（Color Look Up Table，CLUT）的表项人口地址，去查找一个显示图像时使用的RGB值，用查找出的RGB值产生的彩色称为伪彩色。

彩色查找表是一个事先做好的表，表项人口地址也称为索引号。例如，256种颜色的查找表，0号索引对应黑色，…，255号索引对应白色。彩色图像本身的像素数值和彩色查找表的索引号有一个变换关系，这个关系可以使用Windows定义的变换关系，也可以使用你自己定义的变换关系。使用查找得到的数值显示的彩色是真的，但不是图像本身真正的颜色，它没有完全反映原图的彩色。

3．直接色（Direct Color）

每个像素值分成RGB分量，每个分量作为单独的索引值对它做变换。也就是通过相应的彩色变换表找出基色强度，用变换后得到的RGB强度值产生的彩色称为直接色。它的特点是对每个基色进行变换。

用这种系统产生颜色与真彩色系统相比，相同之处是都采用RGB分量决定基色强度，不同之处是前者的基色强度直接用RGB决定，而后者的基色强度由RGB经变换后决定。因而这两种系统产生的颜色就有差别。试验结果表明，使用直接色在显示器上显示的彩色图像看起来真实、很自然。

这种系统与伪彩色系统相比，相同之处是都采用查找表，不同之处是前者对RGB分量分别进行变换，后者是把整个像素当作查找表的索引值进行彩色变换。

5.4　Continuous Tone (Color or Gray Level) Image Compression Encoding

5.4.1　The Color Sensitivity of the Vision System

Colors do not exist in nature. What does exist is light of different wavelengths. When such light enters our eye, the light-sensitive cells in the retina send signals that the brain interprets as color. Thus, colors exist only in our minds. The visible magnetic wave's wavelength is between 380nm and 780nm. The most light we see is not a single wavelength light and is a composed light with many different wavelengths. There are three kinds of cones inside the retina, which have different sensitivities to the color red, green, and blue. This is also the reason why these three colors are a natural choice for primary colors. More accurately, the three types of cones feature maximum sensitivity at wavelengths of about 420 nm (blue), 534 nm (Bluish-Green), and 564 nm (Yellowish-Green). There is another kind of light-sensitive (photoreceptor) cell, rods inside the retina, which are most sensitive to variations of light and dark, shape and movement. They contain one type of light-sensitive pigment and respond to even a few photons. So, colors exist only in our eyes and brains and it doesn't exist in nature. This may sound strange, but when we think it over, we can find that it is difficult to describe a color to another person. All we can say is something like "this object is red," but a color blind person has no idea of redness and is left uncomprehending. The reason we cannot describe colors is that they do not exist in nature. We cannot compare a color to any known attribute of the objects around us. When we say one object is square, even a blind person could get an exact idea what it is right away sine he/she could physically feels it. After we think about those, we won't be surprised anymore. Currently, in computer image processing, the rods haven't played a crucial role yet.

Several characteristics of human being's vision color sensitivity could be concluded as follows.

(1) Our eyes are actually a camera. The retina is the back part of our eye. It contains the light-sensitive (photoreceptor) cells which enable us to sense light and color. There are two types of photoreceptors, rods and cones. The cones are sensitive to the color and the rods are not.

(2) There are three types of cones. They have different sensitivities to the lights with different frequencies (wavelengths). They also have different sensitivities to the lights with different lightness. This gives people a hint that they can reduce the data quantity representing the image without noticeable image quality decreasing via digital image processing technology.

(3) As we talked, the colors exist only in our mind. The three types of cones are sensitive to red, green, and blue, which is why these colors are a natural choice for primary colors. Any one of other colors could be gotten by adding those three primary colors together. They form a three dimension RGB space. This means that with different quantity of the three primary colors mixed together, we could get different colors. In another words, we could get the lights with different wavelengths.

5.4.2 The Three Characteristics of the Images

We need to use the image's characteristics to describe an image. The image's characteristics include image resolution, pixel depth, true and pseudo colors, the representation way and classification of the images, and etc. Here we only talk about the first three of them.

5.4.2.1 Resolution

There are two kinds of resolutions we have met. One of them is the displaying resolution and the other is the image resolution. Let's have a look at both of them separately.

1. Displaying Resolution

The displaying resolution means the pixels could be displayed in the displaying screen. For example, the displaying resolution of 640×480 means that the displaying screen is divided into 480 lines vertically and each line could display 640 pixels. This way the whole display screen contains total 307200 pixels. The more pixels the screen could display, the higher displaying resolution the equipment has and the higher quality images the equipment could show. At the earlier stage, except the laptop using liquid crystal display, LED, all the computers were using CRT display screen which is similar to the CRT used in color TV. Each color pixel on the screen is formed based on the relative strength of three analogue signals representing R, G, and B. All those pixels construct a color image.

The major differences between the CRT used in the earlier stage computer and the CRT used in home usage color TV are the film used to cover the dots on the glass surface and the lustrous surface substance applied. The distance between the dots is called dot pitch witch is often used to measure the display screen's resolution. The average resolution of the CRT used in the home color TV is 0.76mm. The resolution of the standard SVGA display screen is 0.28mm. The smaller the dot is, the higher the resolution is, where needs the smaller and more precisely dots. This is why the same size computer CRT monitor costs a lot more than the same size home color TV.

The resolution of the even earlier stage computer CRT monitor is 0.41mm. Along with the development of the technologies, the resolution increases from 0.41mm to 0.38mm, to 0.35mm, to 0.31mm, to 0.28mm, and to under 0.26mm. The major cost on this kind of monitor is its resolution. When buying this kind of monitor, you should balance the price and functions at the same time.

The most computers today are using LED monitors.

2. Image Resolution

The image resolution means the measurement method of pixel density constructing the image. For the same size image, the bigger the quantity of the pixels consisting of the image, the higher resolution the image has and the closer to the real the image looks. On the other hand, the image looks rougher.

When we use a scanner to scan a color image, we usually set up the image resolution represented in dots per inch, DPI in advance. If we use 300DPI setting to scan an 8"×10" color image, we could get an image with 2400×3000 pixels. The higher the resolution, the more pixels will be.

The image resolution and the display resolution are two different ideas. The image resolution is the quantity of the pixels consisting of the image. The display resolution on the other hand represents the size of the image displaying area. If the screen's display resolution is 640×480, the image with 320×240 pixels resolution will occupy a quarter of the screen. Under the same condition, the screen can't even display the whole image that has 2400×3000 pixels resolution.

One thing worthies to be mentioned is that when display an image, we may have the situation that the image's aspect ratio doesn't match with the display screen's aspect ratio. This is because that the aspect ratio defined in the displaying equipment is different from the one defined in the image. For example, a 200×200 pixels square image may appear on the displaying equipment as a rectangle image instead of the original square image. You could see this often on the displaying equipment in 1990's, the 20th century.

5.4.2.2　Pixel Depth

The pixel depth means the bits used for every single pixel, which is also used to measure the image's resolution. The pixel depth could determine how many different colors the color image's each pixel may have or how many grey levels the grey image's each pixel could have. Here is an example. Each pixel of a color image is represented by R G B three components. If each component needs 8 bits, every pixel will need total 24 bits. We'll say this image's pixel depth is 24. Each pixel's color could be one of the values of the power 24 of 2 = 16777216. For this reason, we often call the pixel depth as image depth. The more bits used to represent a pixel, the more color could be represented and the deeper the depth.

Although the pixel depth or say image depth could be very deep, the VGA's color depth is limited. For example, the standard VGA supports the color image with 4 bits, 16 different colors.

In multimedia applications, at least the 8 bits, 256 different colors is recommended. Based on the equipment's limit and the human being's eyes' limit, general speaking, the very deep pixel depth is not recommended. Besides this, it is a fact that the deeper the pixel depth, the more space will be needed to storage the data. On the other hand, if the pixel depth is too shallow, the image's quality will be affected. The image will look rough and unreal.

When we use binary numbers to represent pixels of the color images, except the RGB components need certain bits to represent, we often need to add one or several extra attribute bits. Let's have a look at an example. RGB has 5 : 5 : 5 bits to be represented. This will occupy two bytes which has 16 bits. Within those 16 bits, the R G B occupies total 15 bits and the extra one bit will be the attribute bit. In this situation, the pixel depth is 16 bits and the image depth is 15.

The attribute bit/s is/are used to indicate the characteristics of the pixels. Example: in CD-I system, the pixels represented by RGB5 : 5 : 5 have 16bits. Among those 16bits, the top bit, b15 is the attribute bit which is called as transparency bit and noted as T. We could understand T as follows. Suppose there is an image on a display screen already. When current image or part of the current image will overlap on that existing one, the T is used to control whether the existing one will be seen or not. Let's say when $T=1$, we can't see the existing image at all and when $T=0$, we can completely see it.

When we use 32 bits to represent a pixel, if RGB each occupies 8 bits, the rest 8 bits are called α channel bits, or overplay bits, or interrupt bits, and or attribute bits. The following pre-multiplying alpha example explains how to use it. Suppose a pixel's four components, (A, R, G, B) are represented by a standardization values, when (A, R, G, B) takes (1, 1, 0, 0), it shows red. When (A, R, G, B) takes (0.5, 1, 0, 0), after the pre-multiplying, it becomes (0.5, 0.5, 0, 0). This means that the pixel's showing strength of the original redness is 1 and now reduced to half.

This pixel attribute definition is useful in the real world. Let's have a look at the following example. Suppose we need to add texts on an color image and we don't want the texts completely cover the image so that we can't see the image anymore, we could use this way to define the pixels. This kind of pixels is sometimes called key color.

During the image products manufacture, people usually mix the TV color images with the computer produced color images together. This kind of technology is known as video keying technology which applies α channel as well.

5.4.2.3   True Color, Pseudo Color, and Direct Color

Precisely knowing the meaning of true color, pseudo color, and direct color is really helpful to programming the image display program and understanding the image files storage formats. We won't be confused by the things such as true color represented image when displayed on the VGA monitor, we won't get the original image's color.

1. True Color

The true color means that each and all the pixels which consist of the color image have RGB three components. Every component directly determines the base color strength of the display equipment. The color produced this way is called true color. Example: The color image has RGB values as 5 : 5 : 5. The RGB use 5 bits respectively. The values of RGB directly determine the strength of three base colors. The produced colors this way will be the real original image colors.

If take RGB as 5 : 5 : 8 to represent a color image, that is RGB occupy 8 bits respectively, each component occupies one byte and total three bytes are occupied, each pixel's color will be decided directly by the values of those three bytes. The total quantities of the colors could get is the power 24 of 2, 16777216. The storage space for three bytes represented true color image is big. Our eyes actually rare can recognize that many different colors. That's why we use RGB as 5 : 5 : 5 more often. Each component occupies 5 bits respectively. Plus one display attribute control bit and make the total 2 bytes data. The true colors produced total will be the power 15 of 2, 32k(32768).

Under many circumstances, the true color usually means the RGB as 8 : 8 : 8 and the total quantities of the colors are the power 24 of 2, 16777216, which is usually referred to full color image. But the colors displayed on the monitor may not be the true colors. To get the true color images on the monitor needs the true color display adapter. It's very hard to get the true color images on the VGA monitors used in the computers. It's practical now though to get the true color images on LED monitors.

2. Pseudo Color

The colors of the pixels in pseudo color image are not directly determined by the base color component's values. It takes the values of the pixels as the entry addresses of a color look up table, CLUT. Based on those addresses, look for the RGB values in the table used to display the image. The colors produced by those checking out RGB values are known as pseudo colors.

The color look up table, CLUT is created in advance. The table's entry addresses are also called indexes. For a 256 different colors CLUT, 0 index is correspondent to black color, ... , and index 255 is correspondent to white color. There is a transform relationship between the pixels of the color image its own and the indexes of the color look up table, CLUT. You could take the transform relationship defined by the Windows. Or you can define the particular transform relationship yourself. The color of the images displayed based on the indexed data is true. But it's not the image's real color. It doesn't reflect the original color the image had.

3. Direct Color

Every pixel contains RGB, three components. Each component as an independent index is transformed. The correspondent base color strength is found via the related color transform table. The created colors through those transformed base RGB strengths are called direct colors. Its characteristic is to do the color transform to each base color respectively.

Comparing the colors created with this direct color system and the ones created with true color system, they all use RGB components to control the strengths of the base colors; but the true color system directly uses the RGB components to determine the strengths of the base colors and the direct color system uses the transformed RGB instead of the direct ones to determine to strengths of the base colors. For this reason, the colors produced from those two different systems are different. The results from the experiments show that the colors produced by the direct color system look more real and natural on the monitors.

Comparing this direct color system with the pseudo color system, they are in common to check the table for the necessary base colors. But the direct color system indexing the table and transforming the base colors based on individual RGB components and the pseudo color system indexing the table and transforming the base colors based on every integrated pixel as a whole unit.

5.4.3 图像的种类

5.4.3.1 矢量图与点位图

在计算机中，表达图像和计算机生成的图形图像有两种常用的方法：一种称为矢量图（Vector Based Image）法，另一种叫点位图（Bit Mapped Image）法。虽然这两种生成图的方法不同，但在显示器上显示的结果几乎没有什么差别。

矢量图是用一系列计算机指令来表示一幅图，如画点、画线、画曲线、画圆、画矩形等、这种方法实际上是用数学方法来描述一幅图，把一幅图变成许许多多的数学表达式，再编程，用计算机语言来表达。在计算显示图时。也往往能看到画图的过程。绘制和显示这种图的软件通常称为绘图程序（Draw Programs）。

矢量图有许多优点。例如，当需要管理每一小块图像时，矢量图法非常有效；目标图像的移动、缩小放大、旋转、复制、属性的改变（如线条变宽变细、颜色的改变）也很容易做到；相同的或类似的图可以把它们当作图的构造块，并把它们存到图库中，这样不仅可以加速画的生成，而且可以减小矢量图文件的大小。

然而，当矢量图变得很复杂时，计算机就要花费很长的时间去执行绘图指令。此外，对于一幅复杂的彩色照片，如一幅真实世界的彩照，恐怕就很难用数学方法来描述，因而就不用矢量法表示，而是采用点位图法表示。

点位图法与矢量图法很不相同。其实，点位图已经在前面的章节中作了详细介绍，它是把一幅彩色图分成许许多多的像素，每个像素用若干个二进制位来指定该像素的颜色、亮度和属性。因此一幅图由许许多多描述每个像素的数据组成，这些数据通常称为图像数据。而这些数据作为一个文件来存储，这种文件又称为图像文件。如要画点位图，或者编辑点位图，则用类似于绘制矢量图的软件工具，这种软件称为画图程序（Paint Programs）。

点位图的获取通常用扫描仪，以及摄像机、录像机、激光视盘与视频信号数字化卡一类设备，通过这些设备把模拟的图像信号变成数字图像数据。

点位图文件占据的存储器空间比较大。影响点位图文件大小的因素主要有两个：即前面介绍的图像分辨率和像素深度。分辨率越高，就是组成一幅图的像素越多，则图像文件越大；像素深度越深，就是表达单个像素的颜色和亮度的位数越多，图像文件也就越大，而矢量图文件的大小则主要取决于图的复杂程度。

矢量图与点位图相比，显示点位图文件比显示矢量图文件要快；矢量图侧重于"绘制"、去创造，而点位图偏重于"获取"、去"复制"；矢量图和点位图之间可以用软件进行转换，由矢量图转换成点位图采用光栅化（Rasterizing）技术，这种转换也相对容易；由点位图转换成矢量图用跟踪（Tracing）技术，这种技术在理论上说容易，但在实际中很难实现，对复杂的彩色图像尤其如此。

5.4.3.2 灰度图与彩色图

灰度图（Gray-scale Image）按照灰度等级的数目来划分。只有黑白两种颜色的图像称为单色图像（Monochrome Image），如图 5-3 所示的标准图像。图中的每个像素值用 1 位二进制数存储，它的值只有 "0" 或者 "1"。一幅 640×480 的单色图像需要占据 37.5KB 的存储空间。

图 5-4 是一幅标准灰度图像。如果每个像素的像素值用一个字节表示，灰度值级数就等于 256 级，每个像素可以是 0～255 之间的任何一个值，存储一幅 640×480 的灰度图像就需要占据 300KB 的存储空间。

图 5-3　标准单色图像

图 5-4　标准灰度图像

彩色图像可按照颜色的数目来划分。例如，256 色图像和真彩色（2 的 24 次方＝16777216 种颜色）等。图 5-5 是一幅用 256 色标准图像转换成的 256 级灰度图像，彩色图像的每个像素的 RGB 值用二进制数的一个字节来表示，则存储一幅 640×480 的 8 位彩色图像需要 300KB 的存储空间；图 5-6 是一幅真彩色图像转换成的 256 级灰度图像，一幅真彩色图像的每个像素的 RGB 分量分别用二进制数的一个字节表示，则一幅 640×480 的真彩色图像需要 900KB 的存储空间。

许多 24 位彩色图像是用 32 位存储的，这个附加的 8 位称为 α 通道，它的值称为 α 值，它用来表示该像素如何产生特技效果。

使用真彩色表示的图像需要很大的存储空间，在网络传输也很费时间。由于人的视角系统的颜色分辨率不高，因此在没有必要使用真彩色的情况下就尽可能不用。

图 5-5　256 色标准图像转换成的灰度图像

图 5-6　24 位标准图像转换成的灰度图像

5.4.3　The Images Classifications

5.4.3.1　Vector Based Image and Bit Mapped Image

There are two most common methods to represent and produce images in computer. One of them is known as vector based image and the other is known as bit mapped image. Although the

254

images produced differently, you won't see much difference on the monitors.

The vector based image represents an image using a series computer commons, such as drawing a dot, drawing a line, drawing a curve, drawing circle, drawing a rectangle, and etc. This method is actually using a mathematical method to describe an image. It converts an image into a lot of mathematical formulas, programs it, and describes the image in computer programming languages. While calculating and displaying an image, you could see the process drawing the image as well. The software drawing and displaying this kind of images is usually called draw programs.

The vector based image has many advantages. It is very efficient to manage a small area of a given image. With the vector based image, it's also easy to move the image, compress and expand the image, turn the image, copy the image, and change the attributes such as line width and color variation of the image. The same or similar images could be saved into the image database as composition images. This way not only can speed up the image produce process, but also can reduce the vector based image file size.

The computer has to spend a long time to finish the drawing instructions when the image is getting complicated. Besides this, it'll be very hard for the vector based image system to mathematically describe a complicated color photo such as a real world picture. That's why we need bit mapped image system instead of the vector based image system here.

The bit mapped image is totally different from the vector based image. We actually mentioned this kind of image at the earlier chapters. What it does is that it set up the image with a lot of pixels. For ach pixel, some binary bits are used to indicate the pixel's color, brightness, and attribute. So, each image is formed by a lot of data describing the pixels of the image, which is usually called image data. The data will be saved as a file which is known as image file. If we want to draw or edit a bit mapped image, we could use the software tool similar to the vector based image's one, which is called paint programs.

To get the bit mapped images, usually use the equipments like scanner, video camera, VCR, laser video disk, video signal digitalizing card, and etc. Those equipments could convert the analogue image signals into digital image data.

The bit mapped image occupies bigger storage space. There are two major factors we have mentioned affecting the bit mapped image data file size. One of them is the image resolution and the other is the pixel depth. The higher the image resolution, the more pixels the image has and the bigger the image file is. The deeper the pixel depth is, the more the bits representing the single pixel's color and brightness are and the bigger the image file is. On the other hand, the size of the vector based image file majorly depends on the complexity of the image.

Comparing the vector based image and the bit mapped image, it's faster to display the bit mapped image than to display the vector based image. The vector based image emphasizes the "making", to create. The bit mapped image emphasizes the "getting", to copy. The vector based image and the bit mapped image are convertible via certain drawing software. The rasterizing technology is used to convert the vector based image to the bit mapped image. This directional converting is relatively easy. The tracing technology is used to convert the bit mapped image to the vector based image. This technology is theoretically easy. But it's hard to put it into practice especially for the complicated color images.

5.4.3.2 Gray Scale Image and Color Image

The gray scale images are classified by the numbers of the grey scales the images have. The images that only have black and white two colors are called monochrome image such as the standard monochrome image shown in Fig. 5-3. Each pixel in the image only needs one bit binary number to save, whose value is either 0 or 1. Saving one 640×480 monochrome image will need 37.5KB storage space.

Fig. 5-4 is a standard gray scale image. If each pixel uses one byte to represent, the gray scale image's scale values are 256, the power 8 of 2. Each pixel could be any one of the values of 0 to 255. Saving one 640×480 gray scale image will need 300KB storage space.

Fig. 5-3   Standard Monochrome Image   Fig. 5-4   Standard Gray Scale Image

The color image could be classified by the quantities of the colors the image has. The typical examples are the 256 different color image, the power 24 of 2, 16777216 different colors, known as real color image, and etc. The Fig. 5-5 a 256 levels gray scale image transferred from a 256 colors standard color image. The RGB, values of the color image's every pixel is represented by one binary byte. Then to save a 640×480 resolution, 8 bits pixel depth color image will need 300kB storage space. The Fig. 5-6 shows a 256 levels gray scale image converted from a real color image. In a real color image, the RGB three components of each pixel are represented by one byte binary number respectively. This way, saving one 640×480 resolution real color image will need 900KB storage space.

Many 24 bits color images are saved in 32 bits format. The extra one byte, 8 bits are called α(alpha) channel. Its values are called α(alpha) values which are used to show how this pixel create special technique effects.

The real color images need very big storage spaces. It takes a long time for the network transmission as well. Since the human being's vision system is not that sensitive to the color, we won't use the real colors unless we really have to.

Fig. 5-5　The Gray scale image Converted From 256 colors Standard Color Image

Fig. 5-6　The Gray scale image Coverted From 24 bits Standard Color Image

5.5　JPEG 压缩编码

5.5.1　JPEG 压缩编码算法概要

JPEG 标准：全称"连续色调静止图像的数字压缩和编码（Digital Compression and Coding of Continuous-tone Still Images）. JPEG 代表由 ISO 和 IEC 两个组织机构联合组成的一个专家组（Joint Photographic Experts Group，JPEG），这个专家组负责制定静态的数字图像数据压缩编码标准，这个专家组开发的算法即称为 JPEG 算法，并且成为国际上通用的标准. 因此又称为 JPEG 标准。

JPEG 花费了大量的时间，致力于图像的压缩和实现。他们在思维上创新并且拥有精湛的技术，终于使 JPEG 静止图片压缩技术成为一种最广泛认可的标准。JPEG 的基本压缩方式已成为一种通用的技术，很多应用程序都采用了与之相配套的软硬件。

JPEG 是一个适用范围很广的静态图像数据压缩标准，既可用于灰度图像又可用于彩色图像。

和相同图像质量的其他常用文件格式（如 GIF，TIFF，PNG）相比，JPEG 是目前静态图像中压缩比最高的。我们给出具体的数据来对比一下。例如一张大小为 1024×768，24 位色的原图，用 Microsoft 图画工具将其分别转成 24 位色 BMP、24 位色 JPEG、24 位色 TIF、24 位色 PNG 压缩格式。得到的文件大小（以 KB 为单位）分别为：2593，214，2398，1751。可见 JPEG 比其他几种压缩比要高得多，而图像质量都差不多。

可以从图 5-7～图 5-10 中看到这种效果。

图 5-7　24 位色 BMP 格式　2593KB

图 5-8　24 位色 JPEG 格式　214KB

JPEG 专家组开发了两种基本的压缩算法，一种是采用以离散余弦变换（Discrete Cosine Transform，DCT）为基础的有损压缩算法，另一种是采用以预测技术为基础的无损压缩算法。使用有损压缩算法时，在压缩比为 25∶1 的情况下，压缩后还原得到的图像与原始图像相比较，非图像专家难于找出它们之间的区别，因此得到了广泛的应用。其中的例子之一是在 VCD 和 DVD-Video 电视图像压缩技术中，就使用 JPEG 的有损压缩算法来取消空间方向上的冗余数据。为了在保证图像质量的前提下进一步提高压缩比，近年来 JPEG 专家组正在制定 JPEG 2000（简称 JP 2000）标准，这个标准中将采用小波变换（Wavelet）算法。

图 5-9　24 位色 PNG 格式 1751KB

图 5-10　24 位色 TIF 格式 2398KB

JPEG 压缩是有损压缩，它利用了人的视角系统的特性，使用量化和无损压缩编码相结合来去掉视角的冗余信息和数据本身的冗余信息。JPEG 算法框图如图 5-11 所示。压缩编码大致分成三个步骤。

（1）使用正向离散余弦变换（Forward Discrete Cosine Transform，FDCT）把空间域表示的图变换成频率域表示的图。

（2）使用加权函数对 DCT 系数进行量化，这个加权函数对于人的视觉系统是最佳的。

（3）使用哈夫曼可变字长编码器对量化系数进行编码。译码或者称为解压缩的过程与压缩编码过程正好相反。

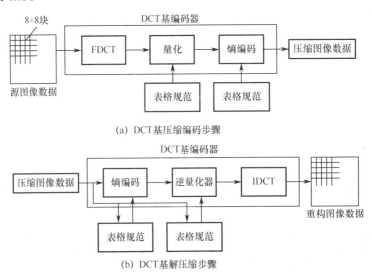

图 5-11　JPEG 压缩编码-解压缩算法框图

JPEG 算法与彩色空间无关，因此"RGB 到 YUV 变换"和"YUV 到 RGB 变换"不包含在 JPEG 算法中。JPEG 算法处理的彩色图像是单独的彩色分量图像，因此，它可以压缩来自不同彩色空间的数据，如 RGB，YCbCr 和 CMYK。

5.5.2 JPEC 算法的主要计算步骤

JPEG 压缩编码算法的主要计算步骤如下。

（1）把每一个颜色分量的像素按 8×8 的序列排好，每一个 8×8 的像素块称为一个数据单元。每一个数据单元将被分别压缩。如果图像的行或列不是 8 的整数倍，那么，图像的底行或最右边的行可以按照需要多次重复使用。

（2）对每一个数据单元实行正向离散余弦变换（FDCT），产生一个 8×8 的相应的频率分量。

（3）量化。

（4）Z 字形编码（Zigzag Scan）。

（5）使用差分脉冲编码调制（Deferential Pulse Code Modulation，DPCM）对直流系数（DC）进行编码。

（6）使用行程长度编码（Run Length Encoding，RLE）对交流系数（AC）进行编码。

（7）熵编码（Entropy Coding）。

JPEG 的编码和解码过程示于图 5-11 中。

下面对正向离散余弦变换（FDCT）作几点进一步的说明。

对每一个单独的颜色分量，把整个颜色分量的像素分成多个 8×8 的数据块，如图 5-12 所示。把分好后的 8×8 数据块作为二维离散余弦变换，DCT 的输入。通过 DCT 变换，或说 FDCT 变换，把能量集中在有限的几个系数上。

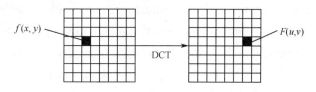

图 5-12 离散余弦变换

5.5 JPEG Compression Encoding

5.5.1 JPEG Compression Encoding Algorithm Outline

JPEG is the first international image compression standard for continuous tone still images-both grayscale and color images. Its formal name is digital compression and coding for continuous tone still images. JPEG stands for Joint Photographic Experts Group (JPEG).

The two standardization groups involved, the CCITT and the ISO, worked actively to get input from both industry and academic groups concerned with image compression, and they seem to have avoided the potentially negative consequences of their actions.

The standards group created by these two organizations is the Joint Photographic Experts Group (JPEG). The JPEG standard was developed over the curse of several years, and is now firmly entrenched as the leading format for lossy graphics compression.

JPEG is a sophisticated lossy/lossless compression method for color or grayscale still images (not videos). It does not handle bi-level (black and white) images very well.

It also works best on continuous-tone images, where adjacent pixels have similar colors. An important feature of JPEG is its use of many parameters, allowing the user to adjust the amount of the data lost (and thus also the compression ratio) over a very wide range. Often, the eye cannot see any image degradation even at compression factors of 10 or 20. There are two operating modes, lossy (also called baseline) and lossless (which typically produces compression ratios of around 0.5). Most implementations support just the lossy mode. This mode includes progressive and hierarchical coding.

Comparing with other similar quality file formats such as GIF, TIFF, and PNG, JPEG has the highest compression ratio to deal with the still images. Let's have a look at a couple examples and get some detailed idea. For an original color image whose resolution is 1024×768 and pixel depth is 24, we use Microsoft drawing tool to convert it to 24 bits BMP, 24 bits JPEG, 24 bits TIF, 24 bits PNG compression formats respectively. We take the kB as the file size unit and get the files as BMP - 2593, JPEG - 214, TIF - 2398, and PNG - 1751. From the data we got, we could see that the JPEG has much less data than other formats'. This is to say that the JPEG has much higher data compression ratio than others' and the images' qualities got are very similar.

Fig. 5-7 to Fig. 5-10 shows the effects of the above several compression formats.

The JPEG experts group developed two different basic compression algorithms. One of them is the lossy compression algorithm based on the discrete cosine transform, DCT and the other is the lossless algorithm based on the prediction technology. When using the loosy compression algorithm, with the compression ratio 25 : 1, comparing the recovered image from the compressed data with the original image, the normal people will feel hard to tell the difference. That's why it is widely accepted. One of the examples is the JPEG application in VCD and DVD-Video TV image compression technology. The JPEG lossy compression algorithm is used to cancel the spacious redundancy data. To increase the data compression ratio with the same image quality, JPEG experts group is working on the JPEG 2000 (simplified as JP 2000) standards. In this standard, the wavelet algorithm will be used.

Fig. 5-7　24bits BMP Format　2593KB　　　　Fig. 5-8　24bits JPEG Format　214KB

Fig. 5-9   24bits PNG Format   1751KB   Fig. 5-10   24bits TIF Format   2398KB

The JPEG compression is a lossy compression method. It takes the advantage of the human being's vision angle's characteristic and combines the quantifying and lossless compression encoding to take off the vision angle redundancy and data's own redundancy. The JPEG algorithm calculation frame diagram is shown in Fig. 5-11. The compression algorithm roughly has three steps.

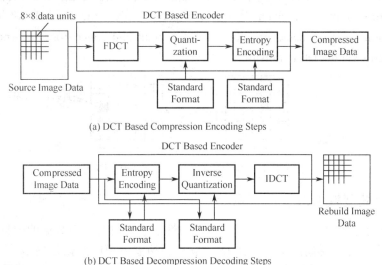

Fig.5-11   JPEG Compression Encoding-Decompression Algorithm Frame Diagram

(1) Transform the image in space area to the image in frequency area via the forward discrete cosine transform, FDCT.

(2) Use the quantization coefficients to quantify the DCT frequency components. The quantization coefficients used are the optimum ones to human being vision system.

(3) Apply the Huffman variable length encoding to the quantified DCT frequency components and get the compressed codes. The process of the decoding or decompression is reversed from the process of encoding or compression.

The JPEG algorithm has nothing to do with the color spaces. That's why the RGB to YUV transform and YUV to RGB transform are not included in the JPEG algorithm. The color images the JPEG processes is the sole color component image. For this reason, it can compress the data from different color spaces such as RGB, YCbCr, and CMYK.

5.5.2 The Major Calculation Steps of JPEG Algorithm

The major calculation steps of the JPEG compression encoding algorithm are as follows.

(1) The pixels of each color component are organized in groups of 8×8 pixels called **data units**, and each data unit is compressed separately. If the number of image rows or columns is not a multiple of 8, the bottom row and the rightmost column are duplicated as many times as necessary.

(2) The forward discrete cosine transform, FDCT, is then applied to each data unit to create an 8×8 map of frequency components.

(3) Quantization.

(4) Zigzag scan encoding.

(5) Apply the deferential pulse code modulation, DPCM to the DC coefficients to encode.

(6) Apply the run length encoding, RLE to the AC coefficients to encode.

(7) Entropy encoding.

The process of the encoding and decoding of the JPEG are illustrated in Fig. 5-11.

Let make a little bit more explanations about the forward discrete cosine transform, FDCT.

As we mentioned, to every separate color component, divide the whole color component image pixels into 8x8 image blocks as shown in Fig. 5-12. Take the divided 8x8 image blocks as the two-dimension discrete cosine transform, DCT's input and concentrate the energy on limited several coefficients via the DCT transform.

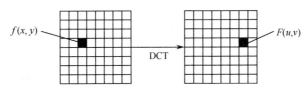

Fig. 5-12   Discrete Cosine Transform, DCT

5.5.3 离散余弦变换 DCT

JPEG 委员会选择使用 DCT 是因为它良好的性能，它对要变换数据的数据结构没有任何特定的假设要求，比如说离散余弦变换 DFT 就假设所要变换的数据是周期性的。

JPEG 标准不是一下子对整个图像进行离散余弦变换，而是对 8×8 个像素的数据块进行离散余弦变换，这样做有如下几个原因。

（1）对大块的数据进行离散余弦变换时包含了很多数学运算，因而使得整个过程很慢。而对小块的数据进行离散余弦变换时会快很多。

（2）经验证明，对于连续色调的图像，像素间的相关性是小范围内的。在这个范围内的像素的值（它的颜色值或灰度级值）与它邻近的像素的值是很接近的，而与远离它的像素的值是毫不相干的。因而 JPEG 的 DCT 由下面的方程式来完成，这里 n 的个数为 8。

$$F(u,v) = \frac{1}{4}C(u)C(v)\sum_{x=0}^{7}\sum_{y=0}^{7}f(x,y)\cos\left(\frac{(2x+1)u\pi}{16}\right)\cos\left(\frac{(2y+1)v\pi}{16}\right)$$

其中
$$C(u),\ C(v) = \begin{cases} \dfrac{1}{\sqrt{2}}, & u=0, v=0 \\ 1, & u>0, v>0 \end{cases} \text{且} 0 \leqslant u,v \leqslant 7$$

DCT 是 JPEG 有损压缩的关键步骤。通过量化 64 个 DCT 系数，不重要的图像信息被减少或去掉了，尤其是数据块右下角的部分。如果图像的像素是相互关联的，量化就不会明显降低图像的质量。为了得到最好的结果，64 个 DCT 系数的每一个都用一个不同的量化系数（QC）去除。如我们所提到的，从原理上来说，所有的 64 个量化系数（QC）都是可以由用户来控制的参数。

JPEG 解码器通过实施逆 DCT 来工作。DCT 逆变换的计算公式如下，其中 n 取值为 8。

$$f(x,y) = \frac{1}{4} C(u)C(v) \sum_{u=0}^{7} \sum_{v=0}^{7} F(u,v) \cos\left(\frac{(2x+1)u\pi}{16}\right) \cos\left(\frac{(2y+1)v\pi}{16}\right)$$

其中

$$C(u),\ C(v) \begin{cases} \dfrac{1}{\sqrt{2}}, & u=0,\ v=0 \\ 1, & u>0,\ v>0 \end{cases}$$

$f(x,y)$ 经 DCT 变换之后，$F(0,0)$ 是直流系数，其他为交流系数。

在计算两维的 DCT 变换时，可使用下面的计算式把两维的 DCT 变换变成一维的 DCT 变换，如图 5-13 所示。

$$F(u,v) = \frac{1}{2} C(u) \left[\sum_{i=0}^{7} G(i,v) \cos\frac{(2i+1)u\pi}{16} \right]$$

$$G(i,v) = \frac{1}{2} C(v) \left[\sum_{j=0}^{7} f(i,j) \cos\frac{(2j+1)v\pi}{16} \right]$$

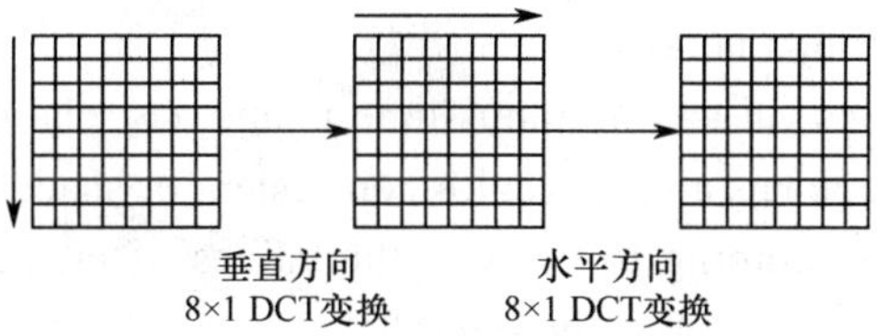

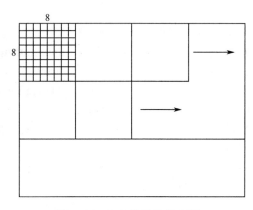

垂直方向　　　水平方向
8×1 DCT变换　　8×1 DCT变换

图 5-13　两维 DCT 变换方法

具体操作时，将图像分为 8×8 的像素块，按照从左到右，从上到下的光栅扫描方式进行排序，如图 5-14 所示。

图 5-14　8×8 像素块的分割

对每个单独的 8×8 的图像块，作两维离散余弦变换 DCT，变换公式如上。

通过 FDCT 变换，把能量集中在少数几个系数上，如图 5-15 所示。

U 代表水平像素号，V 代表垂直像素号。如当 $U=0$，$V=0$ 时，$F(0,0)$ 是原 64 个样值的平均，相当于直流分量，随着 u、v 值增加，相应系数分别代表逐步增加的水平空间频率分量和垂直空间频率分量的大小。

79	75	79	82	82	86	94	94
76	78	76	82	83	86	85	94
72	75	67	78	80	78	74	82
74	76	75	75	86	80	81	79
73	70	75	67	78	78	79	85
69	63	68	69	75	78	82	80
76	76	71	71	67	79	80	83
72	77	78	69	75	75	78	78

(a)

619	−29	8	2	1	−3	0	1
22	−6	−6	0	7	0	−2	−3
11	0	5	−4	−3	4	0	−3
2	−10	5	0	0	7	3	2
6	2	−1	−1	−3	0	0	8
1	2	1	2	0	2	−2	−2
−8	−2	−4	1	2	1	−1	1
−3	1	5	−2	1	−1	1	−3

(b)

图 5-15　源像素矩阵 $f(x, y)$ 和 DCT 变换后系数矩阵 $F(u, v)$

5.5.3　Discrete Cosine Transform, DCT

The JPEG committee elected to use the DCT because of its good performance, because it does not assume anything about the structure of the data, the DFT, for example, assumes that the data to be transformed is periodic.

The JPEG standard calls for applying the DCT not to the entire image but to data units of 8×8 pixels, the reasons for this are.

(1) Applying DCT to large blocks involves many arithmetic operations and is therefore slow. Applying DCT to small data units is faster.

(2) Experience shows that, in a continuous-tone image, correlations between pixels are short range. A pixel in such an image has a value (color component or shade of gray) that's close to those of its near neighbors, but has nothing to do with the values of far neighbors. The JPEG DCT is therefore executed by the equation below, here $n = 8$.

$$F(u,v) = \frac{1}{4}C(u)C(v)\sum_{x=0}^{7}\sum_{y=0}^{7}f(x,y)\cos\left(\frac{(2x+1)u\pi}{16}\right)\cos\left(\frac{(2y+1)v\pi}{16}\right)$$

$$\text{where } C(u)C(v) = \begin{cases} \dfrac{1}{\sqrt{2}}, & u = 0, v = 0 \\ 1, & u > 0, v > 0 \end{cases} \text{ and } 0 \leqslant u, v \leqslant 7$$

The DCT is JPEG's key to lossy compression. The unimportant image information is reduced or removed by quantifying the 64 DCT coefficients, especially the ones located toward the lower-right. If the pixels of the image are correlated, quantification won't degrade the image quality much. For best results, each of the 64 DCT coefficients is quantified by dividing it by a different quantification coefficient (QC). As mentioned, all 64 QCs are parameters that can be controlled, in principle, by the user.

The JPEG decoder works the opposite way by computing the inverse DCT (IDCT), the inverse DCT equation is shown below, where the $n = 8$.

$$f(x,y) = \frac{1}{4}C(u)C(v)\sum_{u=0}^{7}\sum_{v=0}^{7}F(u,v)\cos\left(\frac{(2x+1)u\pi}{16}\right)\cos\left(\frac{(2y+1)v\pi}{16}\right)$$

$$\text{where } C(u), C(v) = \begin{cases} \dfrac{1}{\sqrt{2}}, & u = 0, v = 0 \\ 1, & u > 0, v > 0 \end{cases}$$

Within the data unit, the $f(x, y)$ after DCT, $F(0, 0)$ is the DC coefficient and the others are AC coefficients.

When process the two dimensional DCT calculation, we could use the calculating formulas below to convert the two dimensional DCT to one dimensional DCT as shown in Fig. 5-13.

$$F(u,v)=\frac{1}{2}C(u)\left[\sum_{i=0}^{7}G(i,v)\cos\frac{(2i+1)u\pi}{16}\right]$$

$$G(i,v)=\frac{1}{2}C(v)\left[\sum_{j=0}^{7}f(i,j)\cos\frac{(2j+1)v\pi}{16}\right]$$

Vertical   Horizontal
8×1 DCT   8×1 DCT

Fig. 5-13   Two Dimension DCT Method

In real operation, divides the image into 8×8 data units arranging from left to right and from top to bottom scan as shown in Fig. 5-14.

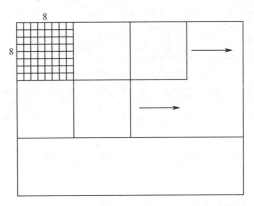

Fig. 5-14   The Dividing of 8x8 Data Units

To each 8x8 data unit, process the two dimensional DCT using the formulas above.

After the DCT, concentrate the energy to several limited coefficients as shown in Fig. 5-15.

79	75	79	82	82	86	94	94
76	78	76	82	83	86	85	94
72	75	67	78	80	78	74	82
74	76	75	75	86	80	81	79
73	70	75	67	78	78	79	85
69	63	68	69	75	78	82	80
76	76	71	71	67	79	80	83
72	77	78	69	75	75	78	78

(a)

619	−29	8	2	1	−3	0	1
22	−6	−6	0	7	0	−2	−3
11	0	5	−4	−3	4	0	−3
2	−10	5	0	0	7	3	2
6	2	−1	−1	−3	0	0	8
1	2	1	2	0	2	−2	−2
−8	−2	−4	1	2	1	−1	1
−3	1	5	−2	1	−1	1	−3

(b)

Fig. 5-15   Source Pixels Matrix $f(x, y)$ and After DCT Coefficients Matrix

U represents the horizontal pixel's numbers and V represents the vertical pixel's numbers. When $U=0$ and $V=0$, $F(0, 0)$ is the average value of the 64 pixels and equivalent to the DC component. Along with the increasing of the *u, v* values, the correspondent coefficients represent sizes of the horizontal space frequency component and vertical space frequency component respectively.

5.5.4 JPEG 算法的具体步骤

（1）如前所述，对每个单独的彩色图像分量，把整个分量图像像素分成 8×8 的图像数据块，如图 5-14 所示，并作为二维离散余弦变换 DCT 的输入分别进行压缩。如果图像的行数和列数不是 8 的整数倍，最下边的行和最右边的行可以根据需要重复多次使用。

（2）对每一个数据单元实行正向离散余弦变换（FDCT）。

他们代表了平均的像素值和在这组数据里相继的较高频率的变化。通过 DCT 变换，把能量集中在少数几个系数上，如图 5-16 所示。

这里 *U* 代表水平像素号，*V* 代表垂直像素号。如当 *U*=0，*V*=0 时，*F*(0, 0)是原 64 个样值的平均，相当于直流分量，随着 *U*、*V* 值增加，相应系数分别代表逐步增加的水平空间频率分量和垂直空间频率分量的大小。

79	75	79	82	82	86	94	94
76	78	76	82	83	86	85	94
72	75	67	78	80	78	74	82
74	76	75	75	86	80	81	79
73	70	75	67	78	78	79	85
69	63	68	69	75	78	82	80
76	76	71	71	67	79	80	83
72	77	78	69	75	75	78	78

(a)

619	−29	8	2	1	−3	0	1
22	−6	−6	0	7	0	−2	−3
11	0	5	−4	−3	4	0	−3
2	−10	5	0	0	7	3	2
6	2	−1	−1	−3	0	0	8
1	2	1	2	0	2	−2	−2
−8	−2	−4	1	2	1	−1	1
−3	1	5	−2	1	−1	1	−3

(b)

图 5-16　源像素矩阵 $f(i, j)$ 和 DCT 系数 $F(u, v)$

（3）量化。

量化是对经过 FDCT 变换后的频率系数进行量化。

使用量化矩阵对 FDCT 变换后的频率系数进行量化。在 JPEG 标准中，有对亮度和色度推荐使用的量化矩阵。用户也可以根据自己的需要而给出自己独有的量化矩阵。

$$K(U,V) = \left[\frac{F(U,V)}{Q(U,V)} \right]$$

这样一来，数据块里的每一个频率系数都分别被一个在量化系数矩阵 *Q*(*U, V*)里的称为量化系数的数字来除，把结果再四舍五入，就如上边的式子所示即如图 5-17 所示。正是在这里，部分信息不可逆地丢失了。大的量化系数引起较大的丢失，所以高频率系数通常用较大的量化系数去除。每一个量化系数都是 JPEG 参数，从原理上来说，可以由用户自己根据需要去选择。但在实际当中，大部分的 JPEG 实施中，都选用 JPEG 标准中推荐的亮度分量和色度分量的量化系数。

量化的目的是减小非"0"系数的幅度以及增加"0"值系数的数目。量化是图像质量下降的最主要原因。

表 5-3 和表 5-4 中的两个矩阵是在 JPEG 标准中，对亮度和色度推荐使用的量化矩阵。

16	11	10	16	24	40	51	61
12	12	14	19	26	58	60	55
14	13	16	24	40	57	69	56
14	17	22	29	51	87	80	62
18	22	37	56	68	109	103	77
24	35	55	64	81	104	113	92
49	64	78	87	103	121	120	101
72	92	95	98	112	100	103	99

39	-3	1	0	0	0	0	0
2	-1	0	0	0	0	0	0
1	0	0	0	0	0	0	0
0	-1	0	0	0	0	0	0
0	0	0	0	0	0	0	0
0	0	0	0	0	0	0	0
0	0	0	0	0	0	0	0
0	0	0	0	0	0	0	0

(a)　　　　　　　　　　　(b)

图 5-17　量化矩阵 $Q(u, v)$ 和量化后 DCT 矩阵 $K(u, v)$

表 5-3　色度量化矩阵

17	18	24	47	99	99	99	99
18	21	26	66	99	99	99	99
24	26	56	99	99	99	99	99
47	66	99	99	99	99	99	99
99	99	99	99	99	99	99	99
99	99	99	99	99	99	99	99
99	99	99	99	99	99	99	99
99	99	99	99	99	99	99	99

表 5-4　亮度量化矩阵

16	11	10	16	24	40	51	61
12	12	14	19	26	58	60	55
14	13	16	24	40	57	69	56
14	17	22	29	54	87	80	62
18	22	37	56	68	109	103	77
24	35	55	64	81	104	113	92
49	64	78	87	103	121	120	101
72	92	95	98	112	100	103	99

（4）Z 字形扫描。

量化后的系数要重新编排，目的是为了增加连续的"0"系数的个数，就是"0"的游程长度，方法是按照 Z 字形的式样编排。这样就把一个 8×8 的矩阵变成一个 1×64 的矢量，频率较低的系数放在矢量的顶部。图 5-18 示意出了这样的一个扫描过程。

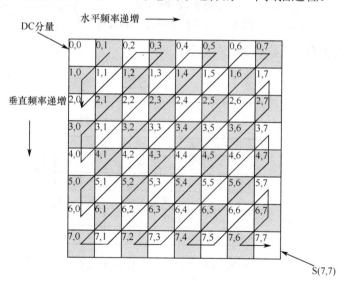

图 5-18　Zigzag 扫描

图 5-19 给出了 JPEG 运算中各个矩阵之间的关系。

（5）使用差分脉冲编码调制对直流系数（DC）进行编码。

8×8 图像块经过 DCT 变换之后得到的 DC 直流系数有两个特点，一是系数的数值比较大，二是相邻 8×8 图像块的 DC 系数值变化不大。根据这个特点，JPEG 算法使用了差分脉冲调制编码（DPCM）技术，对相邻图像块之间量化 DC 系数的差值进行编码，格式为（符号 1）（符号2），符号1为尺寸；符号2为差值的幅度值。那么符号1的尺寸也就是用二进制数表示差值的幅度值的位数是如何得来的呢?换句话说,怎么决定用几位二进制数来表示某一个差值的幅度值呢？将在5.5.5.节详细介绍编码过程。

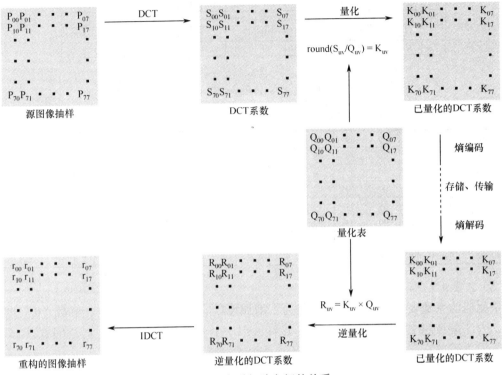

图 5-19　各种矩阵之间的关系

图 5-20 显示出了 DC 系数和 AC 系数的位置。

DC系数				AC系数			
39	−3	1	0	0	0	0	0
2	−1	0	0	0	0	0	0
1	0	0	0	0	0	0	0
0	−1	0	0	0	0	0	0
0	0	0	0	0	0	0	0
0	0	0	0	0	0	0	0
0	0	0	0	0	0	0	0
0	0	0	0	0	0	0	0

图 5-20　DC 系数和 AC 系数的位置

　　如前所述，每一个 8×8 的图像块都有一个 DC 系数和 63 个 AC 系数。DC 系数位于数据块的最左上角 K（0,0）的位置，其余的 63 个元素均为 AC 系数。

　　举例：假设前一个 8×8 子块 DC 系数的量化值为 36，则本块 DC 系数与它的差值为 3，

　　根据"符号码表"，其中间格式为 DC（2）（3）。符号 1 为尺寸；符号 2 为幅度值（"符号码表"将在 5.5.5 节中讨论）。符号 1 尺寸选为 2 的原因如前文所指，也将可在 5.5.5 里详细的编码过程中找到解释。

5.5.4 The Detailed JPEG compression steps

(1) As indicated above, the pixels of each color component are divided in groups of 8×8 pixels called data units, as shown in Fig. 5-14, and each data unit as the input of the two dimension discrete cosine transform, DCT is compressed separately. If the number of image rows or columns is not a multiple of 8, the bottom row and the rightmost column are duplicated as many times as necessary.

(2) Apply the discrete cosine transform, DCT or precisely speaking the FDCT to each data unit to create an 8×8 map of frequency coefficients.

They represent the average pixel value and successive higher-frequency changes within the group. After the DCT transform, concentrate the energy to the limited several coefficients. As shown in Fig.5-16.

79	75	79	82	82	86	94	94
76	78	76	82	83	86	85	94
72	75	67	78	80	78	74	82
74	76	75	75	86	80	81	79
73	70	75	67	78	78	79	85
69	63	68	69	75	78	82	80
76	76	71	71	67	79	80	83
72	77	78	69	75	75	78	78

(a)

619	−29	8	2	1	−3	0	1
22	−6	−6	0	7	0	−2	−3
11	0	5	−4	−3	4	0	−3
2	−10	5	0	0	7	3	2
6	2	−1	−1	−3	0	0	8
1	2	1	2	0	2	−2	−2
−8	−2	−4	1	2	1	−1	1
−3	1	5	−2	1	−1	1	−3

(b)

Fig.5-16　Source pixels vector $f(i, j)$ and DCT coefficients $F(u, v)$

Here the U represents the horizontal pixel numbers and the v represents the vertical pixel numbers. When $U=0$ and $V=0$, $F(0, 0)$ is the average number of the 64 pixel values that is equivalent to the DC component. Along with the increasing of U and V, the correspondent coefficients represent amount of the horizontal frequency component and vertical frequency component respectively.

(3) Quantification.

The quantification is to quantify the frequencies got from the FDCT.

The quantification coefficients matrix is used to quantify the frequencies got from the FDCT. The JPEG standard gives out the recommended luminance and chrominance quantification coefficients matrixes. But the user could have their own quantification coefficients matrixes base on their particular requirements.

$$K(U,V) = \left[\frac{F(U,V)}{Q(U,V)} \right]$$

This way, each of the 64 frequency coefficients in a data unit are divided by a separate number called its quantification coefficient (QC) in the quantification coefficients matrix $Q(U, V)$, and then rounded to an integer as shown above. The results are clsoin Fig.5-17.

lost. Large QCs cause more loss, so the high-frequency coefficients typically have larger QCs. As mentioned, each of the 64 QCs is a JPEG parameter and can, in principle, be specified by the user. But in practice, most JPEG implementations use the QC recommended by the JPEG standard for the luminance and chrominance image components.

16	11	10	16	24	40	51	61
12	12	14	19	26	58	60	55
14	13	16	24	40	57	69	56
14	17	22	29	51	87	80	62
18	22	37	56	68	109	103	77
24	35	55	64	81	104	113	92
49	64	78	87	103	121	120	101
72	92	95	98	112	100	103	99

39	-3	1	0	0	0	0	0
2	-1	0	0	0	0	0	0
1	0	0	0	0	0	0	0
0	-1	0	0	0	0	0	0
0	0	0	0	0	0	0	0
0	0	0	0	0	0	0	0
0	0	0	0	0	0	0	0
0	0	0	0	0	0	0	0

(a)   (b)

Fig. 5-17　Quantification Coefficients Matrix $Q(U, V)$ and Quantified DCT Matrix $K(U, V)$

The purpose of the quantification is to reduce the amplitudes of the non-zero coefficients and to increase the number of the "o" value coefficients. As indicated, the quantification is the major reason the image quality decreasing.

The two matrixes shown in Table 5-3 and Table 5-4 are the two JPEG standard recommended luminance and chrominance quantification coefficients matrixes.

Table5-3　Chrominance

17	18	24	47	99	99	99	99
18	21	26	66	99	99	99	99
24	26	56	99	99	99	99	99
47	66	99	99	99	99	99	99
99	99	99	99	99	99	99	99
99	99	99	99	99	99	99	99
99	99	99	99	99	99	99	99
99	99	99	99	99	99	99	99

Table5-4　Luminance

16	11	10	16	24	40	51	61
12	12	14	19	26	58	60	55
14	13	16	24	40	57	69	56
14	17	22	29	54	87	80	62
18	22	37	56	68	109	103	77
24	35	55	64	81	104	113	92
49	64	78	87	103	121	120	101
72	92	95	98	112	100	103	99

(4) Zigzag Scan.

The quantified coefficients will be rearranged. The purpose is to increase the numbers of the consecutive "0"s which is the run length of "0". This will be done by zigzag scan. This will transform a 8x8 matrix into a 1x64 vector. The lower frequency coefficients set on the top part of the vector. The Fig. 5-18 shows such a scan process.

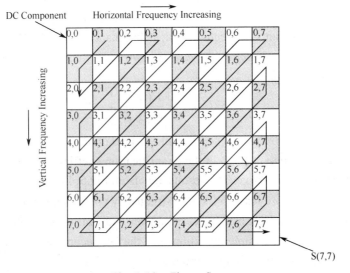

Fig. 5-18　Zigzag Scan

270

Fig. 5-19 shows out the relationships among the matrixes used in JPEG calculation process.

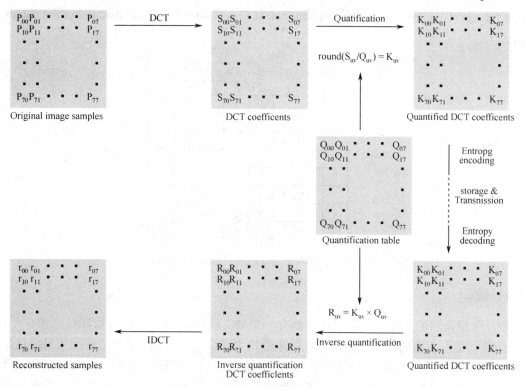

Fig. 5-19   The Relationships Among The Different Matrixes

(5) Encoding the DC component coefficient using the differential pulse code modulation, DPCM.

The DC component coefficient of the 8×8 data unit after the DCT transform has two characteristics. One of them is its value relatively bigger. The other one is that the changing between the DCs between the neighboring data units is not big. Based on those characteristics, JPEG algorithm applies the differential pulse code modulation, DPCM to encode the different values, delta of the neighboring data units. The format is (sign 1) and (sign 2). The sign 1 is the code size and the sigh 2 is the value of delta. If we think further, we could find out that we didn't give the reason that how the sign 1, the code size which is the binary digits used to represent the value of the DC difference is determined. In another words, how could decide to use how many bits to represent a DC difference value? We will answer this question next in section 5.5.5 while we talking about the encoding process in detail.

The Fig. 5-20 shows the locations of DC and AC coefficients within the data unit.

Each data unit has one DC coefficient and 63 AC coefficients. The DC coefficient is located at the top left corner of the data unit $K(0, 0)$ and the others are AC coefficients.

Example: Suppose one data unit's quantified DC coefficient is 36, the difference of this data unit's DC coefficient with it will be 3. Based on the "symbol codes table", its code intermediate format will be DC (2) (3) where the sign 1, 2 is the size and the sign 2, 3 is the actual value. We're going to talk about the "symbol codes table" further in section 5.5.5. As we indicated above, the reason for sign 1 taking as 2 will be explained next in section 5.5.5. while discussing the encoding process.

DC Coefficient				AC Coefficients			
39	–3	1	0	0	0	0	0
2	–1	0	0	0	0	0	0
1	0	0	0	0	0	0	0
0	–1	0	0	0	0	0	0
0	0	0	0	0	0	0	0
0	0	0	0	0	0	0	0
0	0	0	0	0	0	0	0
0	0	0	0	0	0	0	0

Fig. 5-20   The Locations of DC and AC Coefficients Within a Data Unit

5.5.5   编码

每一个量化后的 8×8DCT 系数矩阵都有一个 DC 系数和 63 个 AC 系数。DC 系数位于（0，0），矩阵的最左上角的位置，其余位置的系数均为 AC 系数。DC 系数是原始的形成数据块的 64 个像素的平均值的度量。经验证明，对一个连续色调的图像，相邻数据块的像素以某种方式相关联着，这种关联方式下，相邻数据块的平均值很接近。我们已经知道，数据块的直流系数 DC 是组成数据块的 64 个像素平均值的整数倍，这暗示着相邻数据块的直流系数 DC 相差不多。JPEG 输出第一个数据块的直流系数 DC 编码，接下来输出接续的数据块的直流系数 DC 的差值编码。

下面列举一个例子。假设一副图像的前三个数据单元量化后的直流系数 DC 为 1118、1115、1121，那么 JPEG 算法对于第一个数据单元的输出是如下所示的 1118 的哈夫曼编码，跟在后面的是这个数据单元的其他 63 个交流系数 AC 的编码。对第二个数据单元的输出是 1115 和 1118 的差值 –3（1115–1118= –3）的哈夫曼编码，紧跟着的是这个单元的 63 个交流系数 AC 的编码。对第三个数据单元的输出是 1121 和 1115 的差值 6（1121–1115=6）的哈夫曼编码，紧跟着的是这个单元的 63 个交流系数 AC 的编码。以这种方式处理直流系数 DC 所带来的一些额外的麻烦是值得的，因为这些差值都大大小于那些直流系数 DC 本身的实际值。

编码直流系数 DC 本身及它们的差值是通过表 5-5 来实现的。首先仔细地看一下表 5-5。表中，每一行有一个行号在表的左边，一个每一行对应的唯一的一个一元码在表的右边，中间有几个数字列。每一行都包含着比前一行更大更多的数字，但同时又不包含着前一行已有的数字。第 j 行所包含的数字的范围是整数 $[-(2^j-1),+(2^j-1)]$ 但没有包含中间部分 $[-(2^{j-1}-1)$，$+(2^{j-1}-1)]$ 的内容。这样的话，行会变得很长，这意味着简单的二维矩阵不是一个理想的数据结构用来存储这样的表。事实上，没有必要来存储这些整数，因为在实际的运算中，计算机程序可以通过分析给定整数 i 的比特位数从而找出 i 在表中应处的位置。

现在可以回到我们的例子上来，看看如何通过表 5-5 来对各个数据块的直流系数 DC 及其他们的差值进行编码。

表 5-5　编码直流系数 DC 及它们的差值

	0	1	2	3	4	5	6	7	...	...	...
0:	0										0
1:	−1	1									10
2:	−3	−2	2	3							110
3:	−7	−6	−5	−4	4	5	6	7			1110
4:	−15	−14	...	−9	−8	8	9	10	...	15	11110
5:	−31	−30	−29	...	−17	−16	16	17	...	31	111110
6:	−63	−62	−61	...	−33	−32	32	33	...	63	1111110
7:	−127	−126	−125	...	−65	−64	64	65	...	127	11111110
⋮				⋮							
14:	−16383	−16382	−16381	...	−8193	−8192	8192	8193	...	16383	111111111111110
15:	−32767	−32766	−32765	...	−16385	−16384	16384	16385	...	32767	1111111111111110
16:	32768										1111111111111111

正如前面所分析的，第一个需要在例子中编码的直流系数 DC 是 1118。它位于表 5-5 的第 11 行第 930 列。基于上面所讨论的，它的编码是 111111111110 01110100010。码字包含两个部分，前面一部分是第 11 行的独有的一元码，跟着的是 11 位（因为 1118 位于表 5-5 的第 11 行）的 930（因为 1118 位于表 5-5 的第 930 列）的二进制码值。这里同时回答了上节提出的问题，DC（sign 1）（sign 2)的模式中，尺寸 sign 1 是怎样决定的。它是直流系数 DC 或直流系数 DC 的差值在表 5-5 中所在位置的行号。

第二个需要编码的是直流系数 DC1115 和 1118 的差值−3(1115−1118=−3)。它位于表 5-5 的第 2 行第 0 列，编码为 110 00。和上边类似，码字包含两个部分，前面是第 2 行独有的一元码，跟着的是 2 位（因为−3 位于表 5-5 的第 2 行）的 0（因为−3 位于表 5-5 的第 0 列）的二进制码值。

第三个需要编码的是直流系数 DC1121 和 1115 的差值 6(1121−1115=6)。它位于表 5-5 的第 3 行第 6 列，编码为 1110 110。同样原因，码字包含两个部分，前面是第 3 行独有的一元码，跟着的是 3 位（因为 6 位于表 5-5 的第 3 行）的 6（因为 6 位于表 5-5 的第 6 列）的二进制码值。

在讨论过了直流系数 DC 和它们的差值编码之后，来看看交流系数 AC 的编码。对每一个数据块的交流系数 AC，这种编码方式是一样的。

对每一个数据单元的 63 个交流系数 AC，精确的编码和压缩方式是混用 RLE 和要么哈夫曼编码要么算术编码。之所以这样说，因为交流系数 AC 通常只包含有有限个非零的数字，在这些有限的非零数字之间，是一连串的零，在这之外是更长串的轨迹零。对每一个非零的数字 m，编码器做下面的几件事情。

（1）找出 m 前面 0 的个数 Z；

（2）找出 m 在表 5-5 中的位置，准备好相应的行号和列号 R 和 C；

（3）一对数字（R，Z）（注意是（R，Z）而不是（R，C））将被用作查找表 5-6 时所用的行号和列号；

（4）在表 5-6 中，（R，Z）位置找到的哈夫曼编码和 C 的编码组合在一起形成 JPEG 编码器对于交流系数 m 及其他前边直到前一个非零数之间的所有的 0 的最终输出编码，其中，

C 的编码即是把 C（m 在表 5-5 中的列数）写成 R（m 在表 5-5 中的行数）位的二进制数。

表 5-6　编码交流系数 AC

R Z:	0	1	…	15
0:	1010			11111111001（ZRL）
1:	00	1100	…	11111111111110101
2:	01	11011	…	11111111111110110
3:	100	1111001	…	11111111111110111
4:	1011	111110110	…	11111111111111000
5:	11010	11111110110	…	11111111111111001
⋮	⋮			

表 5-6 中的哈夫曼编码实际上不是 JPEG 标准所推荐的。JPEG 标准推荐使用另外两张表，在这里因为它们有些复杂而省略掉了。这样看来，表 5-6 中的码字取得就有些随意。但这不影响理解 JPEG 标准的编码方法。读者应该注意到了位于（0,0)EOB 码字和位于（0,15）的 ZRL 码字。前者指出数据块的结束位置，而后者 JPEG 编码器当连续 0 的个数超过 15 个时发出 15 个连续 0 的码字。

列举下面一个例子，来考虑下面的序列。

$$1118, 3, 0, -2, \underbrace{0, \cdots, 0}_{10}, -1, 0, \cdots$$

把此序列编码如下。

在直流系数 DC1118 之后，第一个非零交流系数 AC 是 3，它前边 0 的个数是 0，所以 $Z=0$。在表 5-5 中，3 位于第二行第三列，所以 $R=2$，$C=3$。在表 5-6 中，位于位置（R，Z）=（2,0）的哈夫曼码字是 01，所以编码器为 3 编码输出的最终码字是 01 11，其中 11 是 2（第二行，$R=2$）位 3（第三列，$C=3$）的二进制数。

本例中，直流系数 DC 后的第二个非零交流系数 AC 是–2，它前边有一个 0，所以 $Z=1$。在表 5-5 中，–2 位于第二行第一列，可得 $R=2$，$C=1$。在表 5-6 中，位于位置（R，Z）=（2,1）的哈夫曼码字是 11011，所以编码器为 3 及它前边的一个 0 编码输出的最终码字是 1101101，其中 01 是 2（第二行，$R=2$）位 1（第一列，$C=1$）的二进制数。

本例中，直流系数 DC 后的第三个非零交流系数 AC 是–1，它前边有十个 0，所以 $Z=10$。在表 5-5 中，–1 位于第一行第零列，可得 $R=1$，$C=0$。在表 5-6 中，假设位于位置（R，Z）=（1,10）的哈夫曼码字是 1001101（为简化起见，我们在这里没有给出完整的表 5-6），所以编码器为–1 及它前边的十个 0 编码输出的最终码字是 1001101 0，其中 0 是 1（第一行，$R=1$）位 0（第零列，$C=0$）的二进制数。

最后，余下的 0 的序列编码为 1010（数据块结束码）。最后得到 JPEG 编码器对上述交流系数 AC 序列的输出码字为

011111011011001101101010

在前边已得到了此例子中直流系数 DC 编码为

111111111110 01110100010

把它们放在一起，就得到了一个 47 位的数字，此即 JPEG 编码器对这整个 64 个像素数据块的最后输出。

11111111111001110100010011111011011001101101010

这 47 位编码了 64 个像素数据块的一个颜色分量。可以合理地假设其他的另外两个颜色分量也可以编码为另外两个各 47 位的二进制数。如果每个像素原始包含 24 位（如前面量化部分所讨论到的），那么这就隐含着我们获得了 64×24/（47×3）≈10.89 的数据压缩率，非常可观。

从这个例子中可以看到，数据块的前三个直流系数 DC 中的第一个 1118，这样一个直流系数 DC 就占据了最终 47 位编码中的 23 位。另外两个跟着的数据块编码它们直流系数 DC 的差值，而不是直流系数 DC 本身，结果可能会使编码位数少于 10，更不会是 23，因此会得到高得多的数据压缩率。

当然，在编码端使用的同样的码表（表 5-5 和表 5-6）也要用在解码端的解码器中。

这种表可以是事先定义好后作为默认表而为 JPEG 编码器使用，也可以是在一个特定的过程当中为一个给定的图像在压缩之前特殊计算出来的。JPEG 标准不特殊指定任何一个码表，所以任何 JPEG 编码器都要使用自己的码表。

有的读者可能会感到这种编码框架有些复杂，但当你看到用在 H.264 标准中复杂得多的 CAVLC 编码方法时，你就不会感到奇怪了。这种 CAVLC 编码方法也是编码类似的 8x8DCT 变换系数的。

5.5.5 Coding

As discussed, each 8×8 matrix of quantified DCT coefficients contains one DC coefficient and 63 AC coefficients. The DC coefficient is at the position of (0, 0), the top left corner of the matrix and all the others are AC coefficients. The DC coefficient is a measure of the average value of the 64 original pixels, constituting the data unit. As mentioned, experience shows that in a continuous-tone image, adjacent data units of pixels are normally correlated in the sense that the average values of the pixels in adjacent data units are close. We already know that the DC coefficient of a data unit is a multiple of the average of the 64 pixels constituting the unit. This implies that the DC coefficients of adjacent data units don't differ much. JPEG outputs the DC coefficient of the first data unit, encoded, followed by differences, also encoded of the DC coefficients of consecutive data units.

Here is an example: Suppose the first three 8×8 data units of an image have quantified DC coefficients of 1118, 1115, and 1121, then the JPEG algorithm output for the first data unit is 1118, Huffman encoded as shown below followed by the 63 encoded AC coefficients of that data unit. The output for the second data unit will be 1115 − 1118 = −3, also Huffman encoded, followed by the 63 encoded AC coefficients of that data unit, and the output for the third data unit will be 1121 − 1115 = 6, also Huffman encoded, again followed by the 63 encoded AC coefficients of that data unit. This way of handling the DC coefficients is worth the extra trouble, because the differences are much smaller than the DC coefficient itself.

Encoding the DC coefficients and their differences is done with Table 5-5. Let's have a close look at the table first. Inside the table, each row has a row number on the left, the unary code for the row on the right, and several columns in between. Each row contains greater numbers and also more numbers than its predecessor row but not the numbers contained in previous rows. The row j

contains the range of integers $[-(2^j-1), +(2^j-1)]$ but is missing the middle range $[-(2^{j-1}-1), +(2^{j-1}-1)]$. This way, the rows could get very long, which means that a simple two-dimensional array is not a good data structure for this kind of table. In fact, it's not necessary to store these integers in a data structure, since during the calculation, the computer program can figure out where in the table any given integer i is supposed to reside by analyzing the bits of i.

Table 5-5　Encoding the DC Coefficients and Their Differences

	0	1	2	3	4	5	6	7	...	...	...	...	...
0:	0												0
1:	-1	1											10
2:	-3	-2	2	3									110
3:	-7	-6	-5	-4	4	5	6	7					1110
4:	-15	-14	...	-9	-8	8	9	10	...			15	11110
5:	-31	-30	-29	...	-17	-16	16	17	...			31	111110
6:	-63	-62	-61	...	-33	-32	32	33	...			63	1111110
7:	-127	-126	-125	...	-65	-64	64	65	...			127	11111110
⋮			⋮										
14:	-16383	-16382	-16381	...	-8193	-8192	8192	8193	...			16383	111111111111110
15:	-32767	-32766	-32765	...	-16385	-16384	16384	16385	...			32767	1111111111111110
16:	32768												1111111111111111

We now come back to the example and see how to encode the DC coefficients and their differences via Table 5-5.

As we have analyzed above, the first DC coefficient to be encoded in our example is 1118. It resides in row 11 column 930 of the table, based on what we talked, it is encoded as 111111111110 01110100010. There are two parts in the codes, the first part is the unary code for row 11, followed by the 11-bits (row 11) binary value of 930 (column 930). Here we answer the question proposed in last section. In the format DC(sign 1)(sign 2), how the size of sign 1 is decided. It is the row number the DC coefficients or their differences located in the Table 5-5.

The second encoded will be the DC difference -3 (1115 - 1118 = -3). It resides in row 2 column 0 of Table 5-5, it is encoded as 110 00. Similar as above, there are two parts in the codes, the first part is the unary code for row 2, followed by the 2-bits (row 2) binary value of 0 (column 0).

The third encoded will be the DC difference 6 (1121 - 1115=6). It resides in row 3 column 6 of Table 5-5, it is encoded as 1110110. Same reason, there are two parts in the codes, the first part is the unary code for row 3, followed by the 3-bits (row 3) binary value of 6 (column 6).

After talking about the DC coefficients and their differences encoding, let's have a look at the AC coefficients encoding. It'll be the same to every data unit's AC coefficients encoding.

The precise way to encode and compress the 63 AC coefficients of a data unit is to use a combination of RLE and either Huffman or arithmetic coding. The idea is that the sequence of AC coefficients normally contains just a few nonzero numbers as we have seen, with runs of zeros between them, and with a long run of trailing zeros. For each nonzero number m, the encoder does several of the following things.

(1) finds the number Z of consecutive zeros preceding m;

(2) finds m in Table 5-5 and prepares its row and column numbers, R and C;

(3) the pair (R, Z) (be careful! that's (R, Z), not (R, C)) is used as row and column numbers for Table 5-6;

(4) the Huffman code found in position (R, Z) in the Table 5-6 is concatenated with C, where C is written as an R-bit binary number and the result is finally the code emitted by the JPEG encoder for the AC coefficient m and all the consecutive zeros preceding it.

Table 5-6　Encoding AC Coefficients

R Z:	0	1	...	15
0:	1010			11111111001 （ZRL）
1:	00	1100	...	1111111111110101
2:	01	11011	...	1111111111110110
3:	100	1111001	...	1111111111110111
4:	1011	111110110	...	1111111111111000
5:	11010	11111110110	...	1111111111111001
⋮	⋮			

The Huffman codes in Table 5-6 are not the ones actually recommended by the JPEG standard. The standard recommends the use of another two tables that are eliminated here since it's more complicated. The actual codes in Table 5-6 are thus arbitrary. But this won't affect us understanding the JPEG standard algorithm. The reader should notice the EOB code at position (0, 0) and the ZRL code at position (0, 15). The former indicates end-of-block, and the latter is the code emitted for 15 consecutive zeros when the number of consecutive zeros exceeds 15.

As an example let's consider the sequence

$$1118, 3, 0, -2, \underbrace{0, \cdots, 0}_{10}, -1, 0, \cdots$$

Based on what we talked, let's encode this given sequence as below.

After the DC coefficient 1118, the first non zero AC coefficient followed is 3 and has no zeros preceding it, then $Z = 0$. In Table 5-5, the 3 is located in row 2, column 3, so $R = 2$ and $C = 3$. The Huffman code in position $(R, Z) = (2, 0)$ of Table 5-6 is 01, so the final code emitted for 3 is 01 11, where 11 is the 2 (row 2, $R=2$) digits binary number of 3 (column 3, $C=3$).

The second nonzero AC coefficient in this example after the DC coefficient is -2 and has one zero preceding it, so $Z = 1$. In Table 5-5, the -2 is located in row 2, column 1, so $R = 2$ and $C = 1$. The Huffman code in position $(R, Z) = (2, 1)$ of Table 5-6 is 11011, so the final code JPEG emitted for -2 and its preceding one zero is 11011 01, where 01 is the 2 (row 2, $R=2$) digits binary number of 1 (column 1, $C=1$).

The third nonzero AC coefficient in this example after the DC coefficient is -1 and has ten zeroes preceding it, so $Z = 10$. In Table 5-5, the -1 is located in row 1, column 0, so $R = 1$ and $C = 0$. Suppose the Huffman code in position $(R, Z) = (1, 10)$ of Table 6-6 is 1001101 (For simplicity reason, we didn't give the whole Table 5-6 here.), so the final code JPEG emitted for -1 and its preceding ten zeroes is 1001101 0, where 0 is the 1 (row 1, $R=1$) digit binary number of 0 (column

0, $C=0$).

Finally, the sequence of trailing zeros is encoded as 1010 (EOB), so the JPEG output for the above sequence of AC coefficients is 011111011011001101101010. We saw earlier that the DC coefficient is encoded as 111111111110 01110100010, put them together, the final output for the entire 64-pixel data unit will be the 47-bit number

$$11111111111001110100010011111101101100110101010$$

These 47 bits encode one color component of the 64 pixels of a data unit. Let's reasonably assume that the other two color components are also encoded into 47-bit binary numbers. If each pixel originally consists of 24 bits (as we talked in the quantification section), then this corresponds to a compression ratio of $64 \times 24/(47 \times 3) \approx 10.89$; really impressive!

As we can see, in this example, the first of the three data units' DC coefficients, 1118 has contributed 23 of the total 47 bits of the final codes. The other two subsequent data units encode differences of their DC coefficients instead of the DC coefficients themselves, which may take fewer than 10 bits instead of 23. They may feature much higher compression ratios as a result.

The same tables (Tables 5-5 and 5-6) used by the encoder should, of course, be used by the decoder. The tables may be predefined and used by a JPEG codec as defaults, or they may be specifically calculated for a given image in a special pass preceding the actual compression. The JPEG standard does not specify any code tables, so any JPEG codec has to use its own.

Some readers may feel that this coding scheme is complicated. But when you see the much more complex CAVLC encoding method used in H.264 standard, you won't be surprised anymore. This CAVLC encoding method is also employed to encode the similar sequence of 8×8 DCT transform coefficients.

5.5.6　彩色空间

致力于光和颜色的主要国际组织是 International Committee on Illumination （Commission Internationale de l' Eclairage，CIE）。它的职责就是在这个领域中开发相关的标准和定义。它的高端成就之一就是其在 1931 年开发的彩色框图。在这个彩色框图中指出要定义一个颜色只需要三个参数。用一个三维的（x，y，z）来表达一个特定的颜色很类似于在一个三维空间里表达一个特定的点，由此称这个三维空间为颜色空间。最通用的彩色空间是 RGB，其中的三个参数是红、绿、蓝三种颜色的浓度。当在计算机上使用时，这些参数的范围一般在 0～255（8bits)之间。

CIE 把颜色定义为可见光频谱范围内光的接收结果，可见光的波长范围为 400～700nm。

CIE 把光亮度定义为视觉敏感特性，表现为一个区域释放光的多少。大脑接受光亮度的程度是无法定义的，所以 CIE 定义了一个更为实际的度量单位称为亮度（Luminance）。亮度定义为一种频谱敏感函数加权了的释放能量，其中的函数是视觉的一种特性。眼睛对绿色非常敏感，对红色略微逊色一些，对蓝色最不敏感。CIE 把标准观察者的亮度效率定义为波长的正函数，其中波长的最大值为 555nm。当频谱功率分布使用这个函数作为加权函数而综合起来时，其结果就是 CIE 亮度，用 Y 来表示。亮度在数字图像处理和压缩领域里是一个非常重要的量化值。

亮度和光源的功率成正比，它和浓度相似，但亮度的光谱成分是和人类视觉对光亮度的敏感度相关联的。

　　眼睛对亮度的微小变化都是很敏感的，这也是为什么把 *Y* 选为彩色空间中三个参数中的一个是很有用的原因。实现这种彩色空间最简单的方法是从 RGB 彩色空间的蓝和红两个分量中减去 *Y*，然后用 *Y*，*B-Y*，*R-Y* 作为三个彩色分量而形成一个新的彩色空间。其中后边的两个分量称为色度（Chrominance）。它们代表对于给定的亮度浓度在颜色中出现和缺少蓝（Cb）和红（Cr）的程度。

　　对于不同的应用，*B-Y* 和 *R-Y* 使用了不同的数字范围。YPbPr 的数字范围对于分量模拟录像来说是最佳的，而 YCbCr 的数字范围对于分量数字录像来说是最合适的，这些分量数字录像包括有录像制作室、JPEG、JPEG2000、MPEG。

　　YCbCr 彩色空间是作为 ITU-R BT.601（以前的 CCIR 601）提议的一部分而开发出来的，发生在开发世界上国际领域里数字分量录像标准期间。*Y* 的范围定义为 16 到 235，Cb 和 Cr 的范围定义为 16 到 240，其中的 128 等于零。现存在着几种 YCbCr 的采样模式，它们是 4：4：4、4：2：2、4：1：1 和 4：2：0，在提议中也描述了这些模式。

　　在 16～235 范围之内的 RGB 与 YCbCr 之间的转换是线性的因而也是简单的。从 RGB 到 YCbCr 的转换由下面的公式来完成（请留意蓝色的小的加权量）：

$$Y = (77/256)R + (150/256)G + (29/256)B$$
$$Cb = -(44/256)R - (87/256)G + (131/256)B + 128$$
$$Cr = (131/256)R - (110/256)G - (21/256)B + 128$$

反变换的公式为

$$R = Y + 1.371(Cr - 128)$$
$$G = Y - 0.698(Cr - 128) - 0.336(Cb - 128)$$
$$B = Y + 1.732(Cb - 128)$$

　　当执行从 YCbCr 到 RGB 的反变换时，结果中的 RGB 的取值通常是在 16～235 的范围内，但也不排除偶尔会出现 0～15 和 236～255 的值。

5.5.6　Color Space

The main international organization devoted to light and color is the International Committee on Illumination (Commission Internationale de l' Eclairage), abbreviated CIE. It is responsible for developing standards and definitions in this area. One of the earl achievements of the CIE was its chromaticity diagram, developed in 1931. It shows that no fewer than three parameters are required to define color. Expressing a certain color by the triplet (*x, y, z*) is similar to denoting a point in three-dimensional space, hence the term color space. The most common color space is RGB, where the three parameters are the intensities of red, green, and blue in a color. When used in computers, these parameters are normally in the range 0～255 (8 bits).

The CIE defines color as the perceptual result of light in the visible region of the spectrum, having wavelengths in the region of 400 nm to 700 nm.

The CIE defines brightness as the attribute of a visual sensation according to which an area appears to emit more or less light. The brain's perception of brightness is impossible to define, so the CIE defines a more practical quantity called luminance. It is defined as radiant power weighted by a spectral sensitivity function that is characteristic of vision, our eyes are very sensitive to green, slightly less sensitive to red, and much less sensitive to blue. The luminous efficiency of the Standard Observer is defined by the CIE as a positive function of the wavelength, which has a maximum at about 555 nm. When a spectral power distribution is integrated using this function as a weighting function, the result is CIE luminance, which is denoted by Y. Luminance is an important quantity in the fields of digital image processing and compression.

Luminance is proportional to the power of the light source. It is similar to intensity, but the spectral composition of luminance is related to the brightness sensitivity of human vision.

The eye is very sensitive to small changes in luminance, which is why it is useful to have color spaces that use Y as one of their three parameters. A simple way to do this is to subtract Y from the Blue and Red components of RGB, and use the three components Y, $B - Y$, and $R - Y$ as a new color space. The last two components are called chrominance. They represent color in terms of the presence or absence of blue (Cb) and red (Cr) for a given luminance intensity.

Various number ranges are used in $B - Y$ and $R - Y$ for different applications. The YPbPr ranges are optimized for component analog video. The YCbCr ranges are appropriate for component digital video such as studio video, JPEG, JPEG 2000, and MPEG.

The YCbCr color space was developed as part of Recommendation ITU-R BT.601 (formerly CCIR 601) during the development of a worldwide digital component video standard. Y is defined to have a range of 16 to 235; Cb and Cr are defined to have a range of 16 to 240, with 128 equal to zero. There are several YCbCr sampling formats, such as 4:4:4, 4:2:2, 4:1:1, and 4:2:0, which are also described in the recommendation.

Conversions between RGB with a $16\sim235$ range and YCbCr are linear and thus simple. Transforming RGB to YCbCr is done by (note the small weight of blue):

$$Y = (77/256)R + (150/256)G + (29/256)B$$
$$Cb = -(44/256)R - (87/256)G + (131/256)B + 128$$
$$Cr = (131/256)R - (110/256)G - (21/256)B + 128$$

while the opposite transformation is:

$$R = Y + 1.371(Cr - 128)$$
$$G = Y - 0.698(Cr - 128) - 0.336(Cb - 128)$$
$$B = Y + 1.732(Cb - 128)$$

When performing YCbCr to RGB conversion, the resulting RGB values have a nominal range of $16\sim235$, with possible occasional values in $0\sim15$ and $236\sim255$.

在图像处理及后来的多媒体技术领域发生了一件值得一提的有趣的事情，那就是下面的图 5-21 所示的 Lena 图像。

Lena 图像除了在图像压缩领域众所周知外，还广泛地应用于图像处理及后来的多媒体技术领域，由于对它的兴趣浓重，所以对于它的历史和出处也有了研究并很好地存了档。这张图像是花花公子（Playboy）杂志 1972 年 11 月版中间的插图的一部分。它描绘了瑞典陪伴女郎 Lena Soderberg（n'ee Sjooblom），在 1970 年早期，南加利福尼亚大学一位不知名的研究人员发现了这张图，把它剪切和扫描了下来，用在了他做图像压缩研究的测试图像。从那时起，这幅图就成了图像、电子通信以致后来的多媒体通信历史上众所周知，最重要、最通用的一幅图。结果是现在 Lena 已经被很多人当成为了网络上的第一夫人。花花公子杂志通常是要追究没有被授权就擅自使用其插图者的，而当他们发现了他们这幅有版权的图像这样被使用后，他们决定给这个特殊的'应用'以祝福，而不是去追究版权了。

图 5-21 Lena 图像及其细节

    Lena 本人如今住在瑞典。当她在 1988 年被告知她的'著名'时，她很奇怪同时又很高兴。她被邀请参加了 1977 年 5 月在波士顿举行的第 50 界 IS&T(The Society for Imaging Science and Technology)年会。在会上，她自己又照了一幅照片并发布了出来，并在会上讲了话（当然是有关于她自己，而不是有关于压缩了）。

    There is an interesting thing happened in the image processing and later multimedia technology area which is worth to mention. That is the Lena Image showing in Fig. 5-21 below.

Fig. 5-21 Lena and Detail

The Lena image is widely used by the image processing and later on the multimedia technology community, in addition to being popular in image compression. Because of the interest in it, its origin and history have been researched and are well documented. This image is part of the Playboy centerfold for November, 1972. It features the Swedish playmate Lena Soderberg (n′ee Sjooblom), and it was discovered, clipped, and scanned in the early 1970s by an unknown researcher at the University of Southern California for use as a test image for his image compression research. It has since become the most important, well-known, and commonly used image in the history of imaging, electronic communications, and later on the multimedia communications. As a result, Lena is currently considered by many the First Lady of the Internet. Playboy, which normally prosecutes unauthorized users of its images, has found out about the unusual use of one of its copyrighted images, but decided to give its blessing to this particular "application."

Lena herself currently lives in Sweden. She was told of her "fame" in 1988, was surprised and amused by it, and was invited to attend the 50th Anniversary IS&T (the society for Imaging Science and Technology) conference in Boston in May 1997. At the conference she autographed her picture, posed for new pictures, and gave a presentation (certainly about herself, not compression).

5.5.7　JPEG 2000

数据压缩领域是一个非常活跃的领域，一些新的逼近、新的想法、新的技术总是在不断地被开发出来并付之于应用。然而，在这个世界上没有一件事是完美无缺的，JPEG 算法也不例外，它是已被广泛地应用于图像压缩中了，但仍不是完美无缺的。它所使用的 8×8 个像素的数据块有时会导致在重构的图像中出现块的虚影，这在把 JPEG 的参数设置得丢失多些信息时显得更为明显。这也是为什么 JPEG 委员会早在 1995 就决定了开发一个新的，小波变换为基础的静止图像压缩标准，即 JPEG 2000。

在 JPEG 2000 的开发过程中，最重要的里程碑发生在 1999 年 12 月，当 JPEG 委员会在 Maui，夏威夷开会并通过了 JPEG 2000 第一部分的第一个委员会草案。在 2000 年 8 月 Rochester 的会议上，JPEG 委员会通过了这个国际标准的最后方案。2000 年 12 月 ISO 和 ITU-T 最后接受了这个方案并作为一个完整的国际标准。

被 JPEG 2000 引进的一个对于压缩很重要的一个逼近是"压缩一次，解压多样"的模式。这一点实际上也在下面期望改进的项目里列出来了。JPEG 2000 编码器选择一个最好的图像质量 Q 和一个最高的分辨率 R，然后用这些参数去压缩一幅图像。解码器可以以任意次于最高可达到 Q 的质量去解压这幅图像，所用的分辨率可以是任意一个小于或等于 R 的分辨率。假设一幅图像 I 被压缩成了 B 位，那么解码器可以从压缩的码字中抽出 A 位，这里 $A<B$，并由这 A 位数据可产生一副有损解压图像，这幅图像将和起始时就将图像 I 压缩成 A 位而解压出来的图像完全一样。

下面的列表给出了这个新的标准在哪些相应的区域要对已有标准做出改进与提过。

（1）高压缩效率。

（2）能够处理较大尺寸的图像，从 JPEG 可以处理的 216×216 个像素的图像上升到可以处理 232×232 个像素的图像。

（3）逐级图像传输。在所提出的新标准中，可以根据信噪比、分辨率、颜色分量或者感兴趣的区域来逐级解压一幅图像。

（4）方便、快速地切入被压缩数据流的不同点。

（5）解码器可以在解压某一图像的部分时就平移/聚焦这幅图像。

（6）解码器还可以在解压某一图像的同时旋转、剪切这幅图像。

（7）错误抵制，纠错码可以包括在压缩后的码流中，用以提高在噪声环境下的传输可靠度。

JPEG 2000 的主要信息来源是【JPEG 00】和【Taubman 和 Marcellin 02】。然而这里讨论的内容是基于 ISO/IEC 00 的，它是委员会的最终草案（FCD），发表于 2000 年 3 月。这份文件定义了参照为位流的压缩后的码字流和解码器的工作过程。它包含编码器的信息，但任何可以产生有效位流的编码器都可以考虑为是一个有效的 JPEG2000 编码器。

一般来说，JPEG 2000 解码器可以用相对于原图像来说较低的质量或较低的分辨率来解压整个图像。它还可以对感兴趣的图像的某一部分进行解压缩，在进行这种部分解压时，它可以用最高的图像质量和分辨率，也可以用较低的图像质量和分辨率。更进一步，解码器还可以在不需要对原压缩码流做任何解码工作的时候从压缩后的码流中抽取部分码流而新建另外一个压缩码流。这样一来，解码器在不需要对原压缩码流做任何解压的情况下，就可以建立一个低分辨率、低质量的图像。这种逼近的好处有如下两点：

（1）节省时间和空间；

（2）防止了图像噪声的堆积，这种噪声在对一幅图像做有损压缩、解压缩来回几次的时候是常出现的。

JPEG 2000 也可以剪切和变换一幅图像。当一幅图像起始被压缩时，可以特定地逃出几个感兴趣的区域。解压器可以进入被压缩数据流的任一区域的数据而将它写成一个新的压缩数据流。这样做的结果使得对于感兴趣区域的边缘需要做一些特殊处理，但不需要将整个图像的数据进行解压，然后再对感兴趣的区域进行重新压缩了。更进一步的操作，如找出图像的镜像，或将图像旋转 90°，180°，或 270°都几乎可以在不解压的情况下只对压缩数据流操作而完全做的。

下面简单看一下 JPEG 2000 是如何工作的。

JPEG 2000 与传统 JPEG 最大的不同，在于它放弃了 JPEG 所采用的以离散余弦变换（Discrete Cosine Transform，DCT）为主的数据块编码方式，而改用以小波变换（Wavelet Transform，WT）为主的多解析编码方式。小波变换的主要目的是要将影像的频率成分抽取出来。

JPEG 2000 文件采用的扩展名为 ".j2k"。

下面几段简单地叙述了 JPEG 算法的计算过程。

如果被压缩的图像是彩色图像，它就先要被分成三种分量。类似于 JPEG 算法，每一个分量划分成互不重叠矩形区域称为数据砖块（Tile，类似于 JPEG 中的数据块），然后对每个数据砖块分别进行压缩。

和 JPEG 算法很相像，对每个数据砖块的压缩大致可分为以下四步。

第一步是对数据砖块进行小波变换而得到次带小波变换系数。

第二步是对小波变换系数进行量化，用户可以指定一个特定的码率而进行量化。指定的码率越低，对小波变换系数进行量化可以越粗糙。

第三步是使用 MQ 编码器对量化后的小波变换系数进行算术性的编码，在这里用上了

EBCOT 算法，EBCOT 算法的基本原理是把每一个次带再分成块，称为码块，然后对每一个码块进行编码。对几个码块进行编码后的码位组成的一个或几个包就是最后输出码流的组件。

最后一步是组建输出码流。这一步中，是把包以及标志码组合在输出码流中。这些标志码可以被解码器用来很快地跳过编码码流中一定的区域而到达一些特定的点。例如，通过使用标志码，解码器可以在解码其他一些码块之前而先解码一些特定的码块，从而在显示图像其他区域之前而显示一些特定的区域。标志码的另一个用途是使得解码器可以以几种方式中的任一种去逐步的解码图像。编码码流是以层为组织结构的，每一层包含着更高分辨率的图像信息，因此，一层接一层的解压图像是一种很自然的方式去达到逐步进行图像传输和解压的目的。

5.5.7　JPEG 2000

The data compression area is very active, with new approaches, ideas, and technique being developed and implemented all the time. Nothing in the world is perfect same as the JPEG. It is widely used for image compression but is not perfect. The use of the DCT on 8×8 blocks of pixels results sometimes in a reconstructed image that has a blocky appearance, especially when the JPEG parameters are set for much more loss of information. This is why the JPEG committee has decided, as early as 1995, to develop a new, wavelet-based standard for the compression of still images, to be known as the JPEG 2000.

Perhaps the most important milestone in the development of JPEG 2000 occurred in December 1999, when the JPEG committee met in Maui, Hawaii and approved the first committee draft of Part 1 of the JPEG 2000 standard. At its Rochester meeting in August 2000, the JPEG committee approved the final draft of this International Standard. In December 2000 this draft was finally accepted as a full International Standard by the ISO and ITU-T.

One of the new, important approaches to compression introduced by JPEG 2000 is the "compress once, decompress many ways" paradigm. This is also actually shown below in the expected improving lists. The JPEG 2000 encoder selects a maximum image quality Q and maximum resolution R, and it compresses an image using these parameters. The decoder can decompress the image at any image quality up to and including Q and at any resolution less than or equal to R. Suppose that an image I was compressed into B bits. The decoder can extract A bits from the compressed stream, where $A < B$ and produce a lossy decompressed image that will be identical to the image obtained if I was originally compressed lossily to A bits.

Following is a list of areas where this new standard is expected to improve on existing methods:

(1) High compression efficiency.

(2) The ability to handle larger images, up to 232×232 pixels instead of the original JPEG handling images of up to 216×216.

(3) Progressive image transmission. The proposed new standard can decompress an image progressively by SNR, resolution, color component, or region of interest.

(4) Easy, fast access to various points in the compressed stream.

(5) The decoder can pan/zoom the image while decompressing only parts of it.

(6) The decoder can rotate and crop the image while decompressing it.

(7) Error resilience. Error-correcting codes can be included in the compressed stream, to improve transmission reliability in noisy environments.

The main sources of information on JPEG 2000 are [JPEG 00] and [Taubman and Marcellin 02]. However, what we talked here is based on ISO/IEC 00, the final committee draft (FCD), released in March 2000. This document defines the compressed stream referred to as the bitstream and the operations of the decoder. It contains informative sections about the encoder, but any encoder that produces a valid bitstream is considered a valid JPEG 2000 encoder.

As we talked, in general, the JPEG 2000 decoder can decompress the entire image in lower quality and/or lower resolution. It can also decompress parts of the image, where we are interested at, at either maximum or lower quality or resolution. Furthermore, the decoder can extract parts of the compressed stream and assemble them to create a new compressed stream without having to do any decompression. Thus, a lower-resolution and/or lower-quality image can be created without the decoder having to decompress anything of the original image. The advantages of this approach are:

(1) it saves time and space;

(2) it prevents the buildup of image noise, common in cases where an image is lossily compressed and decompressed several times.

JPEG 2000 also makes it possible to crop and transform the image. When an image is originally compressed, several regions of interest may be specified. The decoder can access the compressed data of any region and write it as a new compressed stream. This necessitates some special processing around the region's borders, but there is no need to decompress the entire image and recompress the desired region. In addition, mirroring (flipping) the image or rotating it by 90°, 180°, and 270° can be carried out almost entirely in the compressed stream without decompression.

Next let's have a briefly look at how the JPEG 2000 works.

Comparing with the JPEG, the biggest difference is that the JPEG 2000 gives up the dicrete cosine trasform, DCT to the 8×8 data unit and uses the wavelet transform, WT instead. The major purpose of the wavelet transform is to abstract the frequence components from the imgae.

The extention name of the JPEG 2000 files is ".j2k".

The following paragraphs are a short summary of the algorithm.

If the image being compressed is a color one, it will be divided into three components. Similar to the JPEG algorithm, each component is partitioned into rectangular, nonoverlapping regions called tiles (similar to data unit in JPEG), which are compressed individually.

Very similar as in JPEG, a tile is compressed in four main steps.

The first step is to compute a wavelet transform that results in subbands of wavelet coefficients.

In the second step, the wavelet coefficients are quantified. This is done when the user specifies a target bitrate. The lower the bitrate, the coarser the wavelet coefficients have to be quantified.

The step three uses the MQ coder to arithmetically encode the quantified wavelet coefficients. The EBCOT algorithm has been adopted for the encoding step. The principle of EBCOT is to divide each subband into blocks, termed code-blocks that are coded individually. The bits resulting from coding several code-blocks become a packet and the packets are the components of the bitstream.

The last step is to construct the bitstream. This step places the packets, as well as many markers, in the bitstream. The markers can be used by the decoder to skip certain areas of the bitstream and to reach certain points quickly. Using markers, the decoder can, e.g., decode certain code-blocks before others, thereby displaying certain regions of the image before other regions. Another use of the markers is for the decoder to progressively decode the image in one of several ways. The bitstream is organized in layers, where each layer contains higher-resolution image information. Thus, decoding the image layer by layer is a natural way to achieve progressive image transmission and decompression.

早在 1800 年初期，法国数学家 Joseph Fourier 就发现了任何一个周期函数都可以表示为一个正弦和余弦函数的和。这个和可能包含无限个正弦和余弦函数。这个有趣的事实就是现在所知道的傅里叶变换，它在工程领域里应用很广，主要体现在信号分析中。它能够隔离包含在信号里的各种频率从而使得用户可以通过删除和增加一些频率分量来研究和编辑信号。傅里叶变换的缺点在于它没有告诉在一个给定的信号中，每一个频率分量在什么时候或说沿着时间轴的哪一点上是起作用的，所以说傅里叶变换只提供了频率分辨而没有提供时间分辨。

小波分析或说小波变换对于从时间上和频率上都能对信号进行分析的问题是一个成功的逼近。给定一个随时间变化的信号，选择一个时间段，然后用小波逼近去识别和隔离在这个时间段上组成这个信号的频率成分。这个时间段可以是一个较长的时间间隔，这时是在一个大尺度下分析这个信号，随着时间间隔选的越来越小，分析的尺度越来越小。在大尺度分析中，示出了信号的全局表现，而在小尺度分析中，则示出了信号在一个小的时间区域内的具体表现，这有点像把信号在一个时间域内放大缩小而不是在一个空间域放大缩小。这样，小波背后的基本设想是基于一个尺度来分析一个函数或信号。

从数学上来说，小波就是满足一定要求的函数。这些要求中最重要的一条就是小波函数的积分要为零。如我们在微积分中所学到的，这意味着对于小波函数的每一块位于 x 轴上方的面积，必有一块等同的位于 x 轴下方的面积与之对应。这意味着小波函数必须沿 x 轴上下波动，这调整了它的名字'波'。其他的一些要求导致了这些函数是在有限的空间里的，所以建议使用'小波'而不是'波'来命名这类函数。

限于篇幅关系，这里不再详细叙述每个具体可用小波函数的详细情况，而只通过一个简单的小波变换的例子来给出一个小波变换的基本概念。

例：假没有一幅分辨率只有 4 个像素 p_0, p_1, p_2, p_3 的一维图像，对应的像素值或者称为图像位置的系数分别为

$$[9 \qquad 7 \qquad 3 \qquad 5]$$

计算它的哈尔小波变换系数。

计算步骤如下：

步骤 1：求均值（Averaging）。计算每一对相邻像素对的平均值，得到一幅分辨率比较低的新图像，本例中，像素数目由原来的 4 个变成了 2 个，即新的图像的分辨率是原来图像分辨率的 1/2，相应的像素值为

$$[8 \qquad 4]$$

步骤 2：求差值（Differencing）。很明显，用 2 个像素表示这幅图像时，图像的信息已

经部分丢失。为了能够从由 2 个像素组成的图像重构出原来由 4 个像素组成的原始图像，就需要存储一些图像的细节系数（Detail Coefficient），以便在重构时找回丢失的信息。方法是把每个像素对的第一个像素值减去这个像素对的平均值，或者使用这个像素对的差值除以 2（结果是一样的）。在这个例子中，第一个细节系数是(9−8)＝1（或（9−7)/2=1)。我们可以看到，计算得到的平均值是 8，它比第一个像素 9 小 1 而比第二个像素 7 大 1（8+1=9, 8−1=7），存储这个细节系数 1 就可以恢复原始图像的前两个像素值。使用同样的方法，第二个细节系数是(3−4)＝−1(或（3−5)/2=−1)，存储这个细节系数就可以恢复后 2 个像素值（4+（−1））=3，（4−（−1））=5)。因此，原始图像就可以用下面的两个平均值和两个细节系数表示为

$$[8 \qquad 4 \qquad 1 \qquad -1]$$

步骤 3：重复步骤 1 和 2。把由第一步和第二步分解得到的图像进一步分解成分辨率更低的图像和细节系数。在这个例子中，分解到最后，就用一个像素平均值 6 和三个细节系数 2，1 和−1 表示原来的整幅图像，即

$$[6 \qquad 2 \qquad 1 \qquad -1]$$

所述的整个过程示于表 5-7 中。

表 5-7　哈尔变换过程

分辨率	平均值	细节系数
4	[9　7　3　5]	
2	[8　4]	[1　−1]
1	[6]	[2]

Back in the early 1800s, the French mathematician Joseph Fourier discovered that any periodic fucntion can be expressed as a sum of sines and cosines. The sum may have infinite sines and cosines. This interesting fact is now known as Fourier transform and it has wide applications in engineering, mainly in the analysis of signals. It can isolate the various frequencies that underlie a signal and thereby enable the user to study the signal and also edit it by deleting or adding certain frequencies. The downside of Fourier expression is that it does not tell us when each frequency, at which point or points in time line, is active in a given signal. We therefore say that Fourier expansion offers frequency resolution but no time resolution.

Wavelet analysis or the wavelet transform on another hand is a successful approach to the problem of analyzing a signal both in time and frequency. Given a signal that varies with time, we select a time interval, and use the wavelet approach to identify and isolate the frequencies that constitute the signal in that interval. The interval can be wide, in which case we say that the signal is analysed on a large scale. As the time interval gets narrower, the scale of analysis is said to become smaller and smaller. A large scale analysis illustrates the global behavior of the signal, while each small scale analysis illuminates the way the signal behaves at a short interval of time; it is like zooming in the signal in time, instead of in space. This way, the fundamental idea behind wavelets is to analyze a function or a signal according to scale.

Mathematically, wavelets are any functions that satisfy certain requirements. Among these is the most important requirement that a wavelet integrates to zero. As we learned in integer calculation, this implies that for each area of the wavelet function above the x axis, there must be an equal area below that axis. This means that the wavelet function has to wave above and below the x axis, which justifies the name "wave." Other requirements result in functions that are localized in space, thereby suggesting the use of the diminutive "wavelet" instead of "wave."

To save the space, we won't talk about the details of each available wavelet function. We just give a basic idea of the wavelet transform calculations via the following simple example.

Example: Suppose there is a one dimension image with the resulution of 4 pixels, p_0, p_1, p_2, p_3. The corespndent pixels' value or say the image's position coeffecients is as follows

$$[9 \quad 7 \quad 3 \quad 5]$$

Asking its Haar transform coefficients.

The calculation steps are as follows.

Step1: Get the average value, Averaging. Calculate the average value of each pair of neighbouring pixels and get another lower resolution new image. In this example, the quantity of the pixels changed from 4 to 2. This means that new image's resolution is the half of the original one's. The correspondent pixels are as follows

$$[8 \quad 4]$$

Step2: Find out the difference, differencing. Obviously, if we only use the new 2 pixels to repesent the image, the information has been partialy lost. To reconstruct the original image containing the 4 pixels from the new one containing the 2 pixels, we need to store some original image's detail coefficients to find back the lost information while rebuiding the image. The way to get the detail coefficient is to use the first pixel of each pair of pixels minus this pair's average or to use the difference of this pair of pixels divieded by 2. (The result will be the same.) In this example, the frist detail coefficient is $(9-8=1)$ (or $(9-7)/2=1$). As we can see, their average is 8 which is 1 less than the first pixel 9 and 1 more than the second pixel 7 $(8+1=9, 8-1=7)$. As soon as we keep this detail coefficient 1, we could recover the first pair of pixels in the original image. Same thing, the second detail coefficient could be found as $(3-4) =-1$ (or $(3-5)/2= -1$). Keeping this detail coefficient, we could recover the second pair of pixels in this example, $(4+ (-1) =3, 4-(-1) =5)$. For this reason, the original image could be represented by the following two averages and two detail coefficients

$$[8 \quad 4 \quad 1 \quad -1]$$

Step3: Repeat the step 1 and 2. Seperate the image got from the step1 and 2 into lower resolution image and more detail coefficients. In this example, at last, we could use one pixel average value 6 and three detail coefficients 2, 1, and -1 to represent the original image as follows

$$[6 \quad 2 \quad 1 \quad -1]$$

The whole process is illustrated in the following Table 5-7.

Table 5-7　Haar Transform Process

Resolution	Averages	Detail Coefficients
4	[9　7　3　5]	
2	[8　4]	[1　−1]
1	[6]	[2]

第六章 视频压缩编码

视频又称运动图像，是由相继拍摄并存储的一幅幅单独的画面（称为帧）序列组成的。这些画面以一定的时间间隔或速率（单位为帧率，即每秒钟显示的帧数目）连续地投射在屏幕上播放出来，由于人眼的视觉暂留效应，使观察者产生平滑和连续的动态画面的感觉。典型的帧率为 24～30f/s，这样的视频图像看起来是光顺和连续的。通常，伴随着视频图像还有一个或多个音频轨，以提供声音。常见的视频有电影、电视等。本章将讨论如下内容。

视频压缩基本概念；

运动图像压缩标准；

MPEG 压缩标准（Motion Picture Experts Group）；

H.264 编码技术。

6.1 视频压缩基本概念

6.1.1 运动图像简介

我们这里谈到的运动图像即指数字视频，又称为电视或影像。运动图像是简单地由一幅接一幅的静止图像形成的。当它们以不小于 24 帧/s 的数率连续显示时，由于人眼的视觉暂留特性，使人产生了连续的感觉。

运动图像可以实现更高压缩度，因为即使在运动图像里有许多的动作，但与单个静止图像里拥有的巨大的信息相比，在两幅相邻的静止图像间的差别一般是很小的。这给了我们一个启示，与相邻静止图像的压缩算法相比，运动图像编码具有更为广阔的压缩空间。

6.1.2 视频制式

模拟电视信号的标准也称为视频的制式，世界各地使用的视频制式标准不完全相同，不同的制式，对视频信号的解码方式、色彩处理的方式以及屏幕扫描频率的要求都有所不同。目前世界上彩色电视的制式主要有 PAL（Phase Alternate Line）、SECAM（Sequential Color Memory System）和 NTSC（National Television System Committee）三种制式。

PAL 制式：是前联邦德国制定的彩色电视广播标准，采用逐行列相正交平衡调幅的技术调制电视信号，德国、英国、新加坡、中国等国家采用这种制式。

SECAM 制式：是法国制定的一种新的彩色电视制式，是顺序传送彩色电视信号与存储和恢复彩色电视信号。法国、东欧和中东等国家采用这种制式。

NTSC 制式：是由美国国家电视标准委员会指定的彩色电视广播标准，由于采用正交平衡调幅的技术调制电视信号，故也称正交平衡调幅制。美国、加拿大、日本、韩国等均采用这种制式。

其中，PAL、SECAM 制式播放速度 25 帧/s，而 NTSC 制式的播放速度为 30 帧/s。

6.1.3　视频的基本参数

（1）帧和帧速，视频中的一幅画面称之为帧。每秒播放的帧数称为帧速。

（2）帧频、场频和行频。

① 帧频：定义每秒扫描多少帧为帧频。NTSC 制式为 29.97 帧，PAL 和 SECAM 制式为 25 帧。

② 场频：定义每秒扫多少场为场频。电视画面一般采用隔行扫描的方式把一帧画面分成奇、偶两场。所以 NTSC 制式的场频为 59.94、PAL 和 SECAM 制式的场频为 50。

③ 行频：定义每秒扫多少行为行频。它在数值上等于帧频乘以每帧的行数。每帧 525 行的 NTSC 制式的行频为 15734，而 625 行的 PAL 和 SECAM 制式行频为 15625。

（3）分辨率。电视的清晰度一般用垂直方向和水平方向的分辨率来表示。垂直分辨率与扫描行数密切相关。扫描行数越多、分辨率越高。我国电视图像的垂直分辨率为 575 行（线）。但电视接收机实际垂直分辨率约 400 行（线）。

6.1.4　视频分类

按视频信号的组成和存储方式可分为模拟视频和数字视频。

1．模拟视频

模拟视频是由连续的模拟信号组成的视频图像，通过在电磁信号上建立变化来支持图像和声音信息的传播和显示。电影、电视、VHS 录像带上的画面通常都是以模拟视频的形式出现的，传统的摄像机、录像机、电视机等视频设备所涉及的视频信号都是模拟视频信号。

模拟视频中的电视信号分为全电视信号、复合视频信号、S-Video 分量信号、色差信号或分量信号。

（1）全电视信号：一帧电视画面的信号一般就是一个全电视信号，由奇数场信号和偶数场信号构成。彩色全电视信号定义为包括亮度（Y）、色度（C）、复合同步信号（H/V）和伴音信号的模拟电视信号。

（2）复合视频信号：是从全电视信号中分离出伴音信号后的视频信号，由亮度信号和色度信号间插在一起。为了便于同步传输伴音，复合视频输入/输出端口都配有音频输入/输出端口（AV Audio Video 口），视频卡可直接从这些端口采集视频信号。

（3）S-Video 分量信号：是把复合视频信号中的亮度和色度信号分两路记录在模拟磁带的一种分量视频信号。S-Video 把亮度和色度分开传输，比复合视频信号能更好地重现色彩。高档摄像机、高档录像机、激光视盘 LD 机的输出均支持分量视频格式。

（4）色差信号或分量信号：视频信号主要由 Y 和 C 构成，C 信号可解调出 Cr 和 Cb 两路信号（Cr、Cb 是 RGB 输入信号中红色、蓝色信号与其亮度值之间的差异），称为色差信号（、Cr、Cb）或分量信号（Y、R—Y、B—Y），NTSC 表示为 YIQ，PAL 和 SECAM 则表示为 YUV。

2．数字视频

数字视频是以二进制数字方式记录的视频信号，是用计算机数字技术把图像中的每一个点（称为像素）都用二进制数字组成的编码来表示，这种信号是离散的数字视频信号。

将原来的模拟视频经过采样量化变为计算机能处理的数字信号的过程称为视频信号的数

字化。模拟视频的数字化不像声音、图像那样简单，由于视频信号既是空间函数，又是时间函数，视频信号的数字化过程远比静态图像的数字化过程复杂。首先模拟视频信号采用复合 YUV 的方式记录，而计算机则将视频分解为像素点以 RGB 形式记录；其次电视机采用隔行扫描方式，而显示器目前基本都采用逐行扫描。因此，模拟视频的数字化就非常复杂，计算机系统必须具备连接不同类型的模拟视频信号的能力，可将录像机、摄像头（机）、VCD 机、DVD 机等提供的不同视频源接入多媒体计算机系统，然后再进行具体的数字化处理。模拟视频信号采样时先把复合视频信号中的 Y 和 C 分离，得到 YUV 分量，然后用模/数转换器分别对三个分量进行数字化，最后再转换成对应的 RGB 形式进行存储。

数字视频与模拟视频相比具有很多优点：一是采用二进制数字编码，信号精确可靠且不易受到干扰；二是数字化的视频信号通过索引表处理，无论复制多少次画面质量几乎都不会下降；三是可以将视频编辑融入计算机的制作环境；四是视频数字信号可以被大比例的压缩，在网络上可以流畅的双向传输。

Chapter 6　Video Compression Coding

Video is also called moving images. It is formed by a series of consecutive single images which are also known as frames taken and saved independently. Those images or frames are consecutively projected and shown in the screen in certain time intervals or rates which have the unit of frame rate, which are the frames displayed per second. Since the human being's eyes have visual temporary staying effect, the observer will have the feeling of watching continuous and smooth dynamic images. The typical frame rates are from 24 to 30 frames per second, fps, under which, the visual images look smooth and continuous. Usually along with the video images, there will be one or more audio tracks to provide the correspondent sounds. The most common videos include movie, TV, and etc. In this chapter, we're going to talk about the following contents.

The basic video compression ideas;

Moving imaged compression standards;

MPEG compression standards (Motion Picture Experts Group);

H.264 encoding technology.

6.1　The Basic Video Compression Ideas

6.1.1　Moving Image Introduction

The moving images we talked here mean the digital videos, which are also known as TV or movie images. The moving images are simply formed by the one by one displayed still images. When they are consecutively displayed in a not lower than 24 frames/second rate, people will see the continuous images because of the human being eyes' visual temporary staying effect characteristics.

The moving images could reach even higher compression rate. The reason for this is that even there are many moving acts in the moving images, compared with the huge information residing in the still images, the difference between the two neighbouring images is usually small. This gives us an idea that compared with still images' compression algorithms used in the neighbouring images, the moving images' encoding will have much more space to go to compress the data.

6.1.2　Video Signal's Modes

The analog TV signal standards are also called video signal's modes. There are different video signal's modes are applied in the different areas all over the world. There are different requirements to the video signals' decoding formats, color processing formats, and screen scanning frequencies for the different video signal's modes. Currently, the major video signal's modes used in the world include the following three types, PAL, phase alternate line, SECAM, sequential color memory system, and NTSC, National Television System Committee.

PAL mode: It is a color TV standard proposed by the former Federal Germany. It applies line by line and column by column vertically cross balance amplitude modulation technology to model the TV signals. It is currently applied in Germany, England, Singapore, China, and some other countries.

SECAM mode: It is a new color TV mode proposed by France. It sequentially transmits color TV signals and then saves and recovers color TV signals. It is applied in France, Easton Europe, Middle East, and some other countries.

NTSC mode: It is the color TV broadcasting standard proposed by American National TV standards Committee. It is also called perpendicular balanced amplitude modulation system since it uses perpendicular balanced amplitude modulation technology to modulate the TV signals. It is applied in US, Canada, Japan, Korea, and some other countries.

Among those, the broadcasting speed of PAL and SECAM modes is 25 frames/s and the broadcasting speed of NTSC mode is 30 frames/s.

6.1.3　The Basic Coefficients of Video

(1) Frame and Frame Rate. Each image within a sequential video is called a frame. The frames broadcasted every second is called frame rate.

(2) Frame rate, screen rate, and line rate.

① Frame rate. It defines how many frames will be scanned per second. NTSC's is 29.97 frames and PAL and SECAM's are 25 frames.

② Screen rate. It defines how many screens will be scanned per second. The TV system typically uses scanning every other line format to divide one frame image into odd and even two separate screens. The screen rate the NTSC mode used is 59.94 and PAL and SECAM modes used is 50.

③ Line rate. It defines how many lines will be scanned per second. It could be got by multiplying frame rate and the number of lines within one frame. The line rate of NTSC mode which has 525 lines per frame is 15734. The line rate of PAL and SECAM which have 625 lines per frame is 15625.

(3) Resolution. The clarity of TV is usually claimed as the resolution of vertically and horizontally. The vertical resolution is closely related with the scanning lines. The more scanning lines are, the higher resolution is. In our country, the TV image's vertical resolution is 575 lines. The TV set's actual vertical resolution is 400 lines.

6.1.4 Video Classifications

Based on the video signals' composition format and saving format, the video signals could be classified into analogue video and digital video.

1. Analogue video

The analogue video is composed of the video images of continuous analogue signals. It supports the video and audio information transmitting and displaying via controlling the variations based on the electromagnetic signals. Usually the video images used in movie, TV, and VHS video tape are analogue videos. The video signals related to the traditional video cameras, VCR, TV, and etc. are usually analogue video signals.

The TV signals within the analogue video are classified as full TV signals, complex TV signals, S-Video component signals, color differential signals, or component signals.

(1) Full TV signals. One frame TV image signal is usually a full TV signal. It is composed of odd field signal and even field signal. It is defined as an analogue TV signal which includes luminance (Y), Chrominance (C), complex synthetic signal (H/V), and accompanying audio signal.

(2) Complex video signal. It is the TV video signal after separating the audio signal from the full TV signal. It is composed of mixed luminance signal and chrominance signal. In order to transmit the synchronized audio signal, the complex video signal's input/output pots provide the audio input/output pots which are called AV audio video pots as well. The video cards could directly collect the video signals from those pots.

(3) S-Video component signal. In this mode, the luminance signal and chrominance signal contained inside the complex video signal are separately recorded in the analogue tapes. S-Video transmits the luminance and chrominance signals separately. This way, it could better recover the colors than complex video signal does. The outputs of higher end video cameras, VCR, and laser video CD (LD) all support this S-Video component signal mode.

(4) Color differential signal or component signal. In this mode, the video signal is mainly composed of luminance (Y) and chrominance (C). From the C signal, we could get the Cr and Cb two single signals (Cr and Cb are obtained from the red minus luminance and blue minus luminance within the RGB color system). They are called color differential signals (Y, Cr, Cb) or component signals (Y, R-Y, B-Y). In NTSC system, it is represented as YIQ. In PAL and SECAM systems, it is represented as YUV.

2. Digital video

Digital video is the video signal recorded with binary numbers. It applies the computer digital technology to represent each dot of an image (called pixel) with the encoded codes consisted of binary numbers. This kind of signal is a discrete digital video signal.

The process to convert the original analogue video into the digital video could be handled by computer via the sampling and quantification is called video signal's digitalization. The analogue video signal's digitalization is not as simple as audio and image's. Since the video signal is not only a space zone function, but also a time zone function, its digitalizing process is much more complicated than the still image's. First of all, the analogue video signal is recorded in a complex YUV format. But the computer decomposes the digital video into pixels and records them in RGB formats. Secondly, the TV applies every other line scanning format and the current computer monitor usually applies line by line scanning. This makes the analogue video signal's digitalizing complicated. The computer system must has the ability to connect the different type of analogue video signals. It should be able to input the different video resources into the computer ssytem from VCR, video camera, VCD, DVD, and etc. and then process the detailed digitalization. When sampling the analogue video signal, we separate the *Y* and *C* signals first from the complex video signal and get the YUV components. After this, we digitalize the three components respectively via the analogue/digital converter. Finally, we transform the color space from YUV to RGB and save them.

Comparing to the analogue video, the digital video has many advantages. One of them is that it uses the binary number system to encode, which promises the signal is accurate, reliable, and not easy to be distorted. Another one of them is that the digitalized video signal is processed via the index form, which makes its quality would be the same even after times of duplications. Even more, we can bring the video editing into the computer working environment. Furthermore, the digital video signal could be compressed in a big scale, which makes it much easier to bidirectional transmit them in the internet.

6.1.5　数字视频文件格式

首先看一下一些常用的数字视频文件格式。

1. AVI 格式

AVI（Audio Video Interleaved），即音频视频交错格式。是微软公司开发的将语音和影像同步组合在一起的文件格式。它对视频文件采用了一种有损压缩方式，但压缩比比较高，因此尽管画面质量不是太好，但其应用范围仍然非常广泛。

2. MOV 格式

MOV 格式即 QuickTime 影片格式，它是 Apple 公司开发的一种视频文件格式，具有跨平台、存储空间要求小等技术特点，被包括 Apple Mac OS、Microsoft Windows 95/98/NT 在内的所有主流计算机平台支持。

3. MPEG/MPG/DAT 格式

MPEG（Moving Pictures Experts Group），是运动图像压缩算法的国际标准，现已被几乎所有的计算机平台支持。DAT 格式是基于 MPEG 压缩/解压缩技术的数字视频格式，被广泛地应用在 VCD 的制作中。

4．RM 格式

RM 格式是 Real Networks 公司开发的目前主流网络视频格式。可以通过其 Real Server 服务器将其他格式的视频转换成 RM 视频并由 Real Server 服务器负责对外发布和播放。RM 视频文件的图像质量会比 MPEG-2 差些。

5．ASF 格式

ASF（Advanced Streaming Format）格式，是 Microsoft 公司推出的在 Internet 上实时传播多媒体的技术标准，能依靠多种协议在多种网络环境下支持数据的传送。

6．WMV 格式

WMV（Window Media Video）格式，也是 Microsoft 公司推出的一种采用独立编码方式并且可以直接在网上实时观看视频节目的文件压缩格式。WMV 格式的主要优点包括：本地或网络回放、可伸缩的媒体类型、多语言支持、环境独立性、丰富的流间关系以及扩展性等。一般要使用 Windows Media Player 8．0 以上的版本才能播放。

7．RMVB 格式

是由 RM 视频格式升级的新视频格式，RMVB 视频格式打破了原先 RM 格式那种平均压缩采样的方式，在保证平均压缩比的基础上合理利用比特率资源，在静止和动作场面少的画面场景采用较低的编码速率，以留出更多的带宽空间在出现快速运动的画面场景时被利用。这样在保证了静止画面质量的前提下，大幅地提高了运动图像的画面质量，图像质量和文件大小之间就达到了微妙的平衡。

6.1.5　Digital Video File Formats

Let's have a look at the normal digital video file formats that we often used.

1. AVI format

AVI (Audio Video Interleaved) as it's named; it is an interleaved audio video format. It is a format synchronized the audio and video together developed by Microsoft. It applies a lossy compression method to the video file with a higher compression ratio. With a result of poor video image quality, but it's still widely accepted because of its higher compression ratio.

2. MOV format

The MOV format is QuickTime movie format. It is developed by Apple. It has the technical characteristics of crossing user platforms, smaller storage space requirement, and etc. It is supported by Apple Mac OS, Microsoft Windows 95/98/NT, and etc main computer working platforms.

3. MPEG/MPG/DAT format

MPEG (Moving Pictures Experts Group) is the international standard of moving pictures compression algorithms. It has been supported by almost all computer working platforms. The DAT format is kind of data video format based on the MPEG data compression/decompression technology. It is widely used in the VCD producing.

4. RM format

RM format is developed by Real Networks and is one of the current main network video formats. One could convert other video formats into RM format via the Real Server. The Real server could further distribute and broadcast the video. The RM video files' image quality is a little bit lower than the MPEG-2's.

5. ASF format

ASF, Advanced Streaming Format is the real time multimedia transmitting technical standard proposed by Microsoft. It could apply variable protocols to transmit the data within different network environments.

6. WMV format

WMV, Windows Media Video is a file compression format that takes an independent encoding format and could directly watch real time video programs from the internet. It is proposed by Microsoft as well. The advantages of the WMV include local or internet playing back, raising and shrinking different multimedia types, multi languages supporting, environmental independent, reach streaming relationships, expandable, and etc. The files need the Windows Media Player 8.0 and higher version to play.

7. RMVB format

It is the new video format upgraded from RM video format. The RMVB video format breaks up the average compression sampling format adopted in the original RM video format. It reasonably arranges the bit rate resource with guaranteed average compression rate. It applies a lower encoding rate at the pictures where there are more still scenes and less moving scenes to save the bandwidth for the fast moving scenes. This way, while maintaining the still images' quality, it greatly increases the moving scene images' quality and skilfully balances the file size and the image quality.

6.1.6 运动图像的数据率

运动图像数据率的计算和它们的采样格式有关。运动图像的采样格式有三种，它们分别是 YCbCr 4：4：4，YCbCr 4：2：2 和 YCbCr 4：2：0（4：1：1），分别示于图 6-1 到图 6-3 中。

假设一个标清数字电视（PAL 制）的图像有大约 720×576（D1 格式）的分辨率，使用 4：2：2 的采样格式，采样深度 10b/像素，每秒 25 帧的质量传输，求其数据率。其数据率可由下面的计算得到：

亮度（Y）：

720 样本/行×576 行/帧×25 帧/s×10b/样本 =104Mb/s（PAL）

色差（Cr，Cb）：

360 样本/行×576 行/帧×25 帧/s×10b/样本×2=104Mb/s（PAL）

总计：约 207 Mb/s

如果每个样本的采样精度由 10b 降为 8b，使用 4：2：0 的采样格式,彩色数字电视信号的数据传输率就降为 124 Mb/s。

注意一下计算过程，色度的计算式中所用的每行的样本值是变化的。这实际上是直接和所采用的采样模式相关联的。在上述例子中，前一部分所使用的采样模式是 4：2：2，看一下相应的图 6-2，图中示出了 4：2：2 的采样模式。从图中我们可以看到，在这种模式中，对应于每 4 个亮度样本值，会有各 2 个色度样本值与之对应，Cr 和 Cb 各两个。这就意味着，在计算中，色度 Cr 和 Cb 的样本值，对应每一个扫描行来说，是亮度样本值的一半。这就是为什么在此例中当计算色度数据率时，用了每行 360 个样本值去取代亮度数据率计算中所有的每行 720 个样本值。在计算中，应该注意到了在色度计算式的最后，乘了一个 2，这实际上是简化了计算 Cr 和 Cb 的过程,用一个计算式一次计算出 Cr 和 Cb 总的数据率来简化分别计算 CrCb 的过程。

在例子中,减小了采样精度的同时改变了采样模式,由原来的 4：2：2 变成了 4：2：0。可以参照图 6-3 来理解这种采样模式。从图中可以看到,对应于每 4 个亮度采样值,各有一个 Cr 和 Cb 的采样值。这种模式下,计算色度数据率时,每行所用的样本数值应该是亮度样本数值的四分之一,在本例中就是 720 个样本数值的四分之一,即 180 个样本数值。另外值得一提的是,这种模式的计算和 4：1：1 模式的计算是一样的,所对应示意图也是一样的。本例中色度的详细计算过程如下所示。

180 样本/行×576 行/帧×25 帧/s×8b/样本×2=41.47Mb/s(PAL)

其相应的亮度数据率为

720 样本/行×576 行/帧×25 帧/s×8b/样本 =82.94Mb/s(PAL)

所以其总的传输数据率为

41.47+82.94 =124.41Mbits/s

近似为 124Mbits/s。

让我们来看一下另外一个例子。

例：CIF 格式（Common Intermediate Format）数字视频的分辨率为 352×240（NTSC 制, 30 帧/s）, 352×288（PAL 制, 25 帧/s）, 采样精度为 8b, 采样格式为 4：2：0, 求其数据传输率。

亮度（Y）： 352×240×30×8 = 20.28 Mb/s (NTSC)

352×288×25×8 = 20.28 Mb/s (PAL)

色差（Cr, Cb）：88×240×30×8×2= 10.14 Mb/s (NTSC)

88×288×25×8×2=10.14 Mb/s (PAL)

总计：20.28+10.14=30.42Mb/s

这就是我们如何根据不同的采样模式去计算其相应的数据传输率的过程。

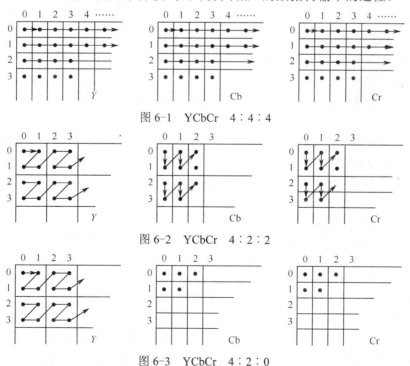

图 6-1　YCbCr　4：4：4

图 6-2　YCbCr　4：2：2

图 6-3　YCbCr　4：2：0

从原理上来讲，数字视频是一系列我们称之为帧的图像，这些图像以一定的帧速率（每秒放送多少帧，或说 f/s）来产生一种仿真的幻觉。这种速率同时包括图像的尺寸和像素深度很大程度上取决于具体的应用场合。例如监控摄像机使用很低的帧速，只有 5f/s，而 HDTV 则使用 25f/s 的帧速。表 6-1 给出了一些典型的视频应用和相应的视频参数。

表 6-1　典型应用中的视频参数

应用	帧率	分辨率	像素深度
监控	5	640×480	12
视频电话	10	320×240	12
多媒体	15	320×240	16
模拟 TV	25	640×480	16
HDTV (720p)	60	1280×720	24
HDTV (1080i)	60	1920×1080	24
HDTV (1080p)	30	1920×1080	24

此表本身示出了数据压缩的必要性。从表中我们可以看到，即使是最便宜应用中的保安摄像机，也会每秒都产生 5×640×480×12=18432000bit 的数据，也就是大约每秒 230 万个字节的数据。这种数据通常是需要保存至少几天的，典型的保安摄像机数据需要保存大约 30 天。此外，大部分视频应用是有声音的，这种声音数据是整个视频数据的一个有机部分，需要和视频图像一起进行压缩。

6.1.6　Moving Images' Data Rates

The calculation of moving image's data rates is related to the moving image's sampling formats. There are three sampling formats existing and they are YCbCr 4 : 4 : 4, YCbCr 4 : 2 : 2 and YCbCr 4 : 2 : 0 (4 : 1 : 1) as shown in Fig. 6-1 through Fig. 6-3.

Suppose a standard high definition digital TV (PAL)'s image has the resolution of 720×576 (D1 format). It uses 4 : 2 : 2 sampling format. Its sampling depth is 10bits/pixel. It transmits 25 frames per second. What are its data rates?

We could get its data rates via the calculation showing below.

Luminance (Y):

720samples/line×576lines/frame×25frames/second×10bits/sample=104Mb/second (PAL)

Chrominance (Cr, Cb):

360samples/line×576lines/frame×25frames/second×10bits/sample×2=104Mb/second (PAL)

Total: About 207Mb/s

If each sample's sampling accuracy reduced from 10bits to 8bits, use 4:2:0 sampling format, the color digital TV signal's data transmission rate will be reduced to 124Mb/s.

If we pay attention to the calculation, we could notice that the samples/line used in the chrominance calculations vary. It is actually directly related to the sampling format. The first part of this example used the sampling format of 4:2:2. Please refer to the Fig. 6-2. It shows the sampling format of 4:2:2. From the figure we can see that in this sampling format, every 4 luminance samples, there will be 2 chrominance samples of Cb and Cr respectively. This actually means that in the calculation, for chrominance, Cr and Cb samples, each scanning line will have half of luminance samples. That's why in the calculation, we took 360samples/line instead of 720samples/line. In the calculation, we timed 2 at the end of chrominance calculation. This is a simplified calculation for both Cr and Cb. Otherwise, we need to write the same calculation process for both Cr and Cb.

In the example, when we reduced the sampling accuracy, we used sampling format of 4:2:0. We could refer to the Fig. 6-3 for this kind of sampling format. From the figure we can see that for every 4 luminance samples, we have 1 Cr and Cb chrominance samples respectively. When we calculate the data rate in this format, the samples/line used should be quarter of the 720, 180samples/line for both Cr and Cb. One more thing worth to mention is that this sampling format's calculation is the same as the format 4:1:1's. The illustration figure is also the same. The detailed chrominance calculation shall be as follows.

180samples/line×576lines/frame×25frames/s×8b/sample×2 = 41.47Mb/s (PAL)

Its luminance is as follows.

720samples/line×576lines/frame×25frames/s×8b/sample = 82.94Mb/s (PAL)

This makes the total transmission data rate showing below.

41.47 + 82.94 = 124.41Mb/s

It is roughly 124Mb/s.

Let's have a look at another example.

Example: CIF format (Common Intermediate Format) digital video's resolution is 352×240 (NTSC system, 30frames/second) and 352×288 (PAL system, 25 frames/second). Its sampling accuracy is 8bits. It uses 4:2:0 sampling format. What is its data transmission rate?

Solution:

Luminance (Y):   352×240×30×8 = 20.28Mb/s (NTSC)

                         352×288×25×8 = 20.28Mb/s (PAL)

Chrominance (Cr, Cb):   88×240×30×8×2 = 10.14Mb/s (NTSC)

                         88×288×25×8×2 = 10.14Mb/s (PAL)

The total transmission rate for both NTSC and PAL is 20.28+10.14=30.42 Mb/s.

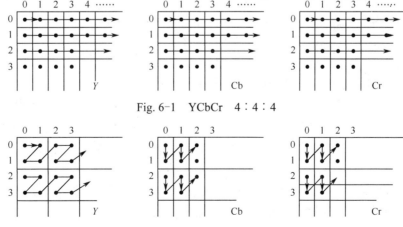

Fig. 6-1   YCbCr 4 : 4 : 4

Fig. 6-2   YCbCr 4 : 2 : 2

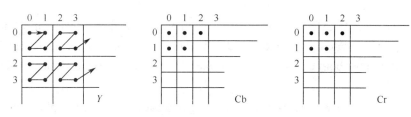

Fig. 6-3 YCbCr 4 : 2 : 0

This is how we calculate the video signal's transmission data rate with different sampling formats.

Digital video is, in principle, a sequence of images, called frames, displayed at a certain frame rate (so many frames per second, or fps) to create the illusion of animation. This rate, as well as the image size and pixel depth, depend heavily on the applications. Surveillance cameras, for example, use the very low frame rate of five fps, while HDTV uses 25 fps. Table 6-1 shows some typical video applications and their video parameters.

Table 6-1　Video Parameters for Typical Applications

Application	Frame rate	Resolution	Pixel depth
Surveillance	5	640×480	12
Video telephony	10	320×240	12
Multimedia	15	320×240	16
Analog TV	25	640×480	16
HDTV (720p)	60	1280×720	24
HDTV (1080i)	60	1920×1080	24
HDTV (1080p)	30	1920×1080	24

The table illustrates the need for compression. Even the most economic application, a surveillance camera, generates 5×640×480×12 = 18,432,000 bits per second! This is equivalent to more than 2.3 million bytes per second, and this information has to be saved for at least a few days before it can be deleted. The typical data saved for security surveillance is about 30 days. Most video applications also involve audio. It is part of the overall video data and has to be compressed with the video image.

6.1.7 视频压缩基本原理

从以上的叙述已经得知视频数据的压缩是必须的。

从前面的例子中我们也看到，207 Mb/s 的数据量，换算成字节约为 26MB/s，那么 20/s 的未压缩视频图像将占用 520MB 的存储空间，此外，如我们上边所指出的，PAL 声音信号还要使比特率再增加一些。这相当于一张 CD-ROM 光盘只能储存 20s 的未压缩电视节目。显然这样的要求是难以接受的，所以视频信号必须压缩。

实际上视频图像数据有极强的相关性，也就是说有大量的冗余信息。其中冗余信息可分为空域冗余信息和时域冗余信息。压缩技术就是将数据中的冗余信息去掉。图 6-4 显示了去掉视频图像帧间冗余概念图，详细概念将在后面介绍。

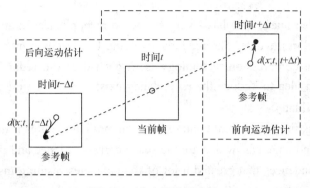

图 6-4　运动估计概念图

　　概括地讲，运动图像压缩的原理就是利用了图像中的两种特性：空间相关性和时间相关性。这两种相关性使得图像中存在大量的冗余信息。如果我们能找到一种算法，将这些冗余信息去除，就可以大大节省传输频带。而接收机按照同样的解码算法，可以在保证一定的图像质量的前提下恢复原始图像。

　　小百科：什么是算法（一）

　　什么叫算法？

　　算法（Algorithm）是解题的步骤。可以把算法定义成解决某一确定类问题的任意一种特殊的方法，是对解题方案的准确与完整的描述。

　　小百科：什么是算法（二）

　　人们的生产活动和日常生活离不开算法——例如人们到商店购买物品，会首先确定购买哪些物品，准备好所需的钱，然后确定到哪些商场选购、怎样去商场、行走的路线，若物品的质量好如何处理，对物品不满意又怎样处理，购买物品后做什么等。只不过平常不叫算法，但如果通过计算机实现就要称为算法。

6.1.7　Video Compression Basic Principles

From the discussion above, we have known that the video data compression is a must.

From the examples we had, the data quantity of 207Mb/s is equivalent to roughly 26MB/s (Mega Bytes per second). With this data rate, for a 20 seconds uncompressed video, it will occupy 520MB storage space. Besides, as we indicated, the accompanied audio will add some bits to the overall video data quantity. This is to say that one CD-ROM disk only can save about 20 seconds uncompressed TV program. Obviously, this is not acceptable. This tells us again that the video data compression is a must.

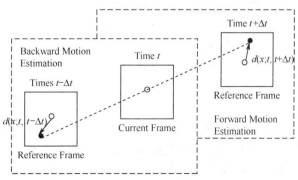

Fig. 6-4　Motion Estimation Sketch

Actually, the video image data has a very strong correspondence. In another words, there is a lot of redundancy within the data. The redundancy could be classified into space redundancy and time redundancy. The compression technology is to get rid of the redundancy within the data. Fig.6-4 gives a rough idea to Yeduce the redundancg existing between the frames. Were going to talk about this in detail later.

General speaking, the principle of compressing the moving pictures is to take advantage of the two characters that the image has, which are the space correspondence and time correspondence. It is those two correspondences that created a lot of redundancies residing inside the image data. If we could find an encoding algorithm to take out those redundancies, we could save a lot of transmission bandwidth. Based on the same decoding algorithm, the receiver could recover the original images with keeping certain image qualities.

Common sense knowledge: What is algorithm? (1)

What is called algorithm?

Algorithm is the steps to solve a problem. We could define the algorithm as a special method to solve any one kind of certain problems. It is a precise and complete explanation of a problem solving solution.

Common sense knowledge: What is algorithm? (2)

People's manufacturing activities and daily life heavily depend on the algorithms. For example, when people go to the store shopping, will first of all decide what to buy and prepare certain amount of money. Then will need to decide which shop one's heading to, how to get there, and the routes one will take. If the product's quality is good, what will one do? If the product's quality is poor, then what to do? What will one do after the purchasing? This is a typical daily life algorithm. The only difference is that in people's daily life, it is not called a algorithm. But if the steps realized through the computer, it will be called an algorithm.

6.2　MPEG 压缩标准

首先看一下什么是国际标准化组织 ISO。

ISO，the International Organization for Standardization 是一个非政府结构组织。因此，它和联合国不一样，ISO 的成员国们不是那些国家的政府代表，而是来自于公众和私人团体。美国是由 ANSI（American National Standards Institute）代表的。

MPEG，Moving Picture Experts Group 实际上是 the International Organization for Standardization, ISO 的一个工作团体的别名。注意 ISO 缩写的顺序与正常英文缩写的顺序是不同的。在 ISO 组织内部，MPEG 更正式的是指"运动图像和声音编码"。

6.2.1　MPEG 历史

MPEG 在这个组织中分配到了 WG11，这样的分配不仅仅是对 MPEG 单独做的，比如 WG1 分配给了静止图像编码组，WG2 分配给了多媒体及媒体链接，在实际中，这些工作组更著名于它们相应的别名，比如 WG1 的别名 JPEG, Joint Photographic Experts Group 和 WG2 的别名 MHEG, Multimedia and Hypermedia Experts Group。

MPEG 的主要工作范围可以解释为：开发运动图像，声音及二者结合的数据的压缩、解压缩、处理和编码表示的国际标准以满足宽范围应用的需求。换句话说，MPEG 为所有类型的语音视频数据的压缩、解压缩制定技术标准。

深入到了 ISO 组织工作组层面上，MPEG 被分在 WG11 组。实际的工作组包含有几百个工作人员，分布在世界各地，对 MPEG 标准的产生积极地贡献着自己的力量。

简而言之，MPEG（Moving Picture Expert Group）是在 1988 年由国际标准化组织（ISO）和国际电工委员会（IEC）联合成立的专家组，负责开发电视图像数据和声音数据的编码、解码和它们的同步等标准。这个专家组开发的标准称为 MPEG 标准。

MPEG 技术是近 10 年国际图像压缩编码技术的结晶，也是伴随信息时代应运而生的热门技术。（Three layers, the audio, the video, and the system data）MPEG 主要包括 MPEG 视频、MPEG 音频和 MPEG 系统（视音频同步）三个部分。

MPEG 压缩标准是针对运动图像而设计的，其平均压缩比可达 50∶1，压缩率比较高，且又有统一的格式，兼容性好。

MPEG 是一个大家族。

最初 MPEG 专家组的工作项目是 3 个，即在 1.5Mb/s，10Mb/s，40Mb/s 传输速率下对图像编码，分别命名为 MPEG-1，MPEG-2，MPEG-3，MPEG-3 后被取消。

为了满足不同的应用要求，MPEG 又陆续增加其他一些标准 MPEG-4，MPEG-7，MPEG-21。

6.2.2　MPEG-1 系统

标准名称为"信息技术"——用于数据速率高达大约 1.5 Mb/s 的数字存储媒体的电视图像和伴音编码（Information Technology——Coding of Moving Pictures and Associated Audio for Digital Storage Media at up to About 1.5 Mb/s）。

技术特点：MPEG－1 可适用于不同带宽的设备，如 CD-ROM、Video-CD 等。它可针对 CIF 标准分辨率(对于 NTSC 制为 352×240；对于 PAL 制为 352×288)的图像进行压缩，传输速率为 1.15Mb/s，每秒播放 30 帧，具有 CD 音质，质量级别基本与 VHS（广播级录像带）相当 。

Video-CD（VCD）使用的是 MPEG-1 标准。这种格式主要用在中国，很成功，主要用于电影的传播。流行的这种模式不包括工作忠诚、识别及版权保护系。但尽管有这些瑕疵，解码设备的厂家们对这个行业的生意还是满意的。MPEG-1 的早期也被用于小段网络传输视频格式。仅在中国就已经卖掉了七千多万的 VCD 机。

MP3（MPEG-1 声音层 3）使用的也是 MPEG-1 标准，它已经变成了一个很响亮的名字，代表着网络上高质量的音频文件，可用来在点对点网络上下载、分享音乐。它永久性的改变了人们欣赏音乐的方式。MP3 的文件已经足够小使得人们可以从网上下载它或把它存储到小的移动设备上，在此同时又可提供一定质量的立体声音频。正因为如此，MP3 音频编码技术为内容使用开辟了一个新的市场和模式。音乐的版权及乐曲的非法复制和传播案子仍然没有结束并在热议中，一旦消费者的权利和版权所有者找到一个平衡点，MP3 编码的数字音乐的大好前程是不言而喻的。

其典型的图像数据如下：

$$352×240×30×8×1.5 =30 \text{ Mb/s (NTSC)}$$

$$352×288×25×8×1.5 =30 \text{ Mb/s (PAL)}$$

压缩比为 $\qquad$ 30/1. = 30:1

应用范围：应用 MPEG－1 技术最成功的产品非 VCD 莫属了，VCD 作为价格低廉的影像播放设备，得到广泛的应用和普及。MPEG－1 也被用于数字电话网络上的视频传输，如非对称数字用户线路（ADSL），视频点播（VOD）以及教育网络等。

6.2.3 MPEG–2 系统

标准名称为"信息技术——电视图像和伴音信息的通用编码（Information Technology—Generic Coding of Moving Pictures and Associated Audio Information）"。

技术特点：MPEG－2 制定于 1994 年 11 月，在 MPEG-2 发布两年以后。它的设计动力是提供一种压缩技术，以支持向数字电视服务的转换。设计目标是高级工业标准的图像质量以及更高的传输率。MPEG－2 所能提供的传输率在 3～10MB/s 间，分辨率可达 720×486（NTSC），720×562（PAL），MPEG－2 能够提供广播级的视像和 CD 级的音质。MPEG－2 的音频编码可提供 5.1 伴音声道。MPEG－2 的另一特点是，可提供一个较广范围的可变压缩比，以适应不同的画面质量、存储容量以及带宽的要求。

应用范围：今天，整个世界已经有八千多万使用 MPEG-2 标准的盒子应用在有线、卫星以及区域电视广播上了。另外，DVD 也已经显示出了极大的成功，关于它的飞速发展，我们不必说太多，这已经把模拟视频 VHS 逼向绝路。DVD 使用 MPEG-2 视频编码，MPEG-2 结合杜比数字（原来的 AC-3）音频和 MPEG-2 系统。MPEG-2 音频最起始面对的是欧洲市场的，而 AC-3 则考虑作为北美的选择。尽管如此，迄今为止大部分的 DVD 支持两种音频模式，但我们实际上已经有一段时间没见到过使用 MPEG-2 音频的 DVD 了，但这只能说明 MPEG-2 标准需要进一步的改进，应用的发展可以选择它最喜欢的技术。如今，全世界有无数软、硬件的 DVD 播放机已被卖掉，目前已超出了成千万上亿个单元，更有甚者是这个数字还在继续增加着。

综上所述，我们可以说 MPEG-2 技术就是实现 DVD 的标准技术，现在 DVD 播放器也开始在家庭中普及起来了。除了作为 DVD 的指定标准外，MPEG－2 还可用于为广播、有线电视网、电缆网络以及卫星直播提供广播级的数字视频。

6.2.4 MPEG–4 系统

标准名为其低速率视听编码"Very-low bitrate audio-visual coding"，数据速率 64kb/s。

MPEG－4 从 1994 年开始工作，它是为视听（Audio-Visual）数据的编码和交互播放开发算法和工具，是一个数据速率很低的多媒体通信标准。MPEG -4 的目标是要在异构网络环境下能够高度可靠地工作，并且具有很强的交互功能。

6.2.5 MPEG–7 系统

1996 年启动,标准名称为多媒体内容描述接口（Multimedia Content Description Interface）。

准确说来，MPEG－7 并不是一种压缩编码方法，而是一个多媒体内容描述接口。继 MPEG－4 之后，要解决的矛盾就是对日渐庞大的图像、声音信息的管理和迅速搜索。MPEG－7 就是针对这个矛盾的解决方案。MPEG－7 力求能够快速且有效地搜索出用户所需的不同类型的多媒体影像资料，比如在影像资料中搜索有长江三峡镜头的片段。

6.2.6 MPEG-21 系统

MPEG—21 将由 MPEG—7 发展而来，刚刚才开始启动。据透露，MPEG—21 主要规定数字节目的网上实时交换协议。

6.2 MPEG Compression Standards

Let's have a look first at the ISO.

ISO is a non-governmental organization (NGO). Therefore, unlike the United Nations, the national members of ISO are not delegations of the governments of those countries but come from public and private sectors. The U.S. is represented by ANSI (American National Standards Institute)

MPEG, "Moving Picture Experts Group." is actually a nickname for a working group of the International Organization for Standardization, abbreviated as ISO (notice the flipping of the letters in the abbreviation).

Within the ISO organization, MPEG is more formally referred to as "Coding of Moving Pictures and Audio,"

6.2.1 MPEG History

MPEG is assigned to WG11. MPEG is not alone; sibling working groups exist, such as WG1, coding of still pictures and WG12, multimedia and hypermedia. In practice, these working groups are better known by their corresponding nicknames, such as JPEG, Joint Photographic Experts Group or MHEG, Multimedia and Hypermedia Experts Group respectively.

MPEG's major "Area of Work," could be read as follows:

Development of international standards for compression, decompression, processing, and coded representation of moving pictures, audio, and their combination, in order to satisfy a wide variety of applications. In another words, MPEG creates a technical specification for the (de)compression of all sorts of audio-visual data.

We have arrived at the Work Group level within the ISO org chart, and MPEG has been identified as WG11. The actual group counts several hundred people worldwide, actively contributing to the creation of MPEG standards.

Simply speaking, MPEG, Moving Picture Experts Group is founded by ISO, the International Organization for Standardization and IEC, International Electrotechnical Commission in 1988. It is responsible to develop the international standards for compression, decompression, processing, and coded representation of moving pictures, audio, and their combination, in order to satisfy a wide variety of applications. The standards developed by this experts group are called MPEG standards.

MPEG technology is like a concentrated result of international image compression technologies for the past 10 years. It is also a new technology coming along with the information era. MPEG mainly includes three parts, MPEG video, MPEG audio, and MPEG system (video and audio synchronizing) or say it has three layers, the audio, the video, and the system data.)

The MPEG compression standard is designed for moving images. Its average compression ratio could reach 50:1 which is pretty high. It has formulated format and has a better compatibility.

MPEG is actually a big famly.

At the beginning, the MPEG working group was concentrate on three projects which are under three different transmission rates, 1.5Mbps, 10Mbps, and 40Mbps, to encode the moving images respectively. They are named MPEG-1, MPEG-2, and MPEG-3 respectively. The MPEG-3 was abandoned later on.

To meet the different applications' requirements, MPEG has added MPEG-4, MPEG-7, and MPEG-21 since then.

6.2.2　MPEG-1

Its formal name is Information technology — coding of moving pictures and associated audio for digital storage media at up to about 1.5 Mb/s.

Its typical technical characteristics include the follows. It's suitable for different bandwidth equipments, such as CD-ROM, Video-CD, and etc. It could compress the images with various CIF standard resolutions (in NTSC is 352×240 and in PAL is 352×288). The transmission rate is 1.15Mbit/s. It broadcasts 30 frames per second and has CD musical quality. Its quality level is basically equivalent to the VHS's.

The Video CD (VCD) uses the MPEG-1 standard, which has turned out to be a success mainly in China, where the format is often used to distribute movie material. The distribution model didn't include a working royalty system and the recognition and protection of copyrights. However, in spite of this deficiency, decoder hardware manufacturers didn't complain about the business. MPEG-1 has also been used as a format for video clips on the Web in its early days. So far in China alone, more than 70 million Video CD players have been sold.

MP3 (MPEG-1 Audio Layer III) uses the MPEG-1 standard as well. It has turned into a buzz word to denote high-quality audio on the Web for downloading music, sharing music over peer-to-peer networks. It has changed the way people experience music forever. MP3 files are small enough to be downloaded from the Internet or stored on small mobile devices, while offering stereo audio of acceptable quality. In this regard, MP3 audio coding technology has created a new market and a new paradigm for content use. The case of copyrights on music and illegal copying and distribution of music titles is still open and widely discussed. Once the rights of consumers and copyright holders find a point of equilibrium, the future of MP3-coded digital music will be ensured.

Its typical data is as follows

$$352×240×30×8×1.5 =30 \text{ Mb/s (NTSC)}$$
$$352×288×25×8×1.5 =30 \text{ Mb/s (PAL)}$$

The compression ratio is 30/1, 30:1

Its major application range is as follows. One of the most successful applications is VCD. As a low cost video playing back equipment, VCD has been widely accepted. MPEG-1 is also applied in transmitting videos over digital telephone networks, such as ADSL, asymmetric digital subscriber line, VOD, video on demand, and education networks.

6.2.3 MPEG-2

Its formal name is Information technology — Generic coding of moving pictures and associated audio information.

Its major technical characters are as follows. MPEG-2 was published in November, 1994, two years after the MPEG-1 released. The driving force behind MPEG-2 was to offer compression technology to support the migration to digital television services. It aimed the high level industry standard and higher transmission rate. It could provide a 3 to 10MB/s transmission rate and could reach the resolutions of 720×486 for NTSC and 720×562 for PAL. MPEG-2 could provide the broadcasting level video images and CD quality audio. MPEG-2's audio encoding could provide 5.1 associated audio channels. Furthermore, MPEG-2 could provide a wide range of variable compression ratio to adjust for the requirements of different image qualities, storage capacities, and bandwidth.

Its application range is as follows. Today, the world has more than 80 million set-top boxes used the MPEG-2 standard serving the cable, satellite, and terrestrial TV broadcasting businesses. In addition, the DVD has turned out to be the legitimate successor of the celebrated Audio Compact Disc. Not much more needs to be said about the overwhelming rise of the DVD, which is killing the business of analog VHS video. The DVD uses MPEG-2 video coding, and MPEG-2 audio, along with Dolby Digital (formerly AC-3) and MPEG-2 systems. MPEG-2 audio was chosen initially for the European markets and AC-3 was considered an option for the United States. However, by now most DVD players support both audio formats, whereas we haven't seen a DVD using MPEG-2 audio in a long time. But that only proves the point of the modularity of the MPEG standard. Application development can choose its favourite technology. Today, endless numbers of hardware and software DVD players are sold and distributed worldwide, now in excess of tens of millions of units, and we are still counting.

In conclusion, we could say that MPEG-2 technology is the standard technology to produce DVD. The DVD player has been popular in families. Besides to be the specified standard of DVD, MPEG-2 could provide broadcasting level digital videos for broadcasting, wired TV network, cable network, and satellite directly broadcasting.

6.2.4 MPEG-4

Its formal name is Very-low bit rate audio-visual coding, 64kb/s.

MPEG has been working on MPEG-4 since 1994. The purpose is to develop the algorithms and tools for audio-visual data coding and interchange broadcasting. It is a very low data bit rate multimedia communication standard. It aims at the working in different infrastructure networks with high reliability and strong interchange ability.

6.2.5 MPEG-7

It has been started since 1996 and formally called Multimedia Content Description Interface.

Precisely speaking, MPEG-7 is not a compression coding method. But more like a multimedia content description interface. After MPEG-4, the major conflict needs to be solved is that for the bigger and bigger video image data quantities, the audio information's management and quick searching. MPEG-7 is the solution for this issue. It aims to quickly and efficiently find the user required different types of multimedia video audio files, such as the scenes of Sanxia, Changjiang River within a quantity of video files.

6.2.6 MPEG-21

MPEG-21 will be evolved from MPEG-7. It's has been just started. As known information, MPEG-21 majorly set the digital programs' internet real time exchanging protocols.

6.3　MPEG 压缩算法

MPEG-Video 图像压缩技术基本思想和方法可以归纳成两个要点。

（1）在空间方向上，图像数据压缩采用 JPEG（Joint Photographic Experts Group）压缩算法来去掉冗余信息。

（2）在时间方向上，图像数据压缩采用运动补偿（Motion Compensation MC)算法来去掉冗余信息，在前面的章节已经讨论过 JPEG 算法，这里就不再赘述了，主要来看看第二点。

6.3.1　运动补偿

运动补偿地基本原理是，当编码器对图像序列中地第 N 帧进行处理时，利用运动补偿中的核心技术－运动估值 ME（Motion Estimation），得到第 N 帧的预测帧 N'。在实际编码传输时，并不传输第 N 帧，而是第 N 帧和其预测帧 N' 得差值 Δ。如果运动估计十分有效，Δ 中的概率基本上分布在零附近，从而导致 Δ 比原始图像第 N 帧得能量小得多，编码传输 Δ 所需得比特数也就少得多。这就是运动补偿技术能够去除信源中时间冗余度得本质所在。

运动补偿是当前视频图像压缩技术中最关键的技术之一。

如上所述，为了消除运动图像的空间冗余，人们普遍采用的办法是从当前帧中减去参考（前一帧），从而得到通常含有较少能量（或者成为信息）的"残差"，从而可以用较低的码率进行编码。解码器可以通过简单的加法完全恢复编码帧。

这个概念稍微难接受一些，为了压缩两帧之间的活动图像，把新一帧和上一帧相减，只把变化部分送过去，第二帧编码的内容就很少了。举一个最简单的例子，假定说背景里面都是静止的，就有一个乒乓球是白的，上一帧在这个地方我打过去时，乒乓球跑到这个地方，两帧相减会出现什么问题呢？上一帧的地方有乒乓球，你要编码，下一帧乒乓球移动了以后再一减，中间又出现了一个相反的东西。因为乒乓球在移动，在这种情况下，编码两帧的时候，绞尽脑汁觉得效率仍然不是最高的，编码新的一帧，能不能把上一帧估算一下，乒乓球无非是移动吗，移动当然可能会旋转，你打弧旋球，但是短时间内，可以认为它是直线运动，编码新一帧的时候，把上一帧乒乓球的位置，根据直线的估算，把乒乓球位置移动到估算的位置，然后再把这两个相减。相减以后，得到变化的部分就更少。这就叫图像的运动估算的运动补偿，这个名词稍微专业一点。这样压缩以后，活动图像就得到了更进一步的压缩。

运动补偿是一种描述相邻帧（相邻在这里表示在编码关系上相邻，在播放顺序上两帧未必相邻）差别的方法，具体来说是描述前面一帧（相邻在这里表示在编码关系上的前面，在播放顺序上未必在当前帧前面）的每个小块怎样移动到当前帧中的某个位置去。这种方法经常被视频压缩/视频编解码器用来减少视频序列中的时间冗余。

一个稍微复杂一点的设计是估计一下整帧场景的移动和场景中物体的移动，并将这些运动通过一定的参数编码到码流中去。这样预测帧上的像素值就是由参考帧上具有一定位移的相应像素值而生成的。这样的方法比简单的相减可以获得能量更小的残差，从而获得更好的压缩比。当然，用来描述运动的参数不能在码流中占据太大的部分，否则就会抵消复杂的运动估计带来的好处。

通常，图像帧是一组一组进行处理的。每组的第一帧（通常是第一帧）在编码的时候不使用运动估计的办法，这种帧称为帧内编码帧（Intra Frame）或者 I 帧。该组中的其他帧称为帧间编码帧，简称 P 帧，是我们所使用的预测帧（Prediction Frame）。这种编码方式通常被称为 IPPPP，表示编码的时候第一帧是 I 帧，其他后续帧都是 P 帧。

另外，在进行预测的时候，不仅仅可以从过去的帧来预测当前帧，还可以使用未来的帧来预测当前帧。当然在编码的时候，未来的帧必须比当前帧更早的编码，也就是说，编码的顺序和播放的顺序是不同的。通常这样的当前帧是使用过去和未来的 I 帧或者 P 帧同时进行预测，被称为双向预测帧，即 B 帧。这种编码方式的编码顺序的一个例子为 IBBPBBPBBPBB。

总体来说有两种运动补偿方法，全局运动补偿和分块运动补偿。

在全局运动补偿中：

运动模型基本上就是反映摄像机的各种运动，包括平移，旋转，变焦等等。这种模型特别适合对没有运动物体的静止场景的编码。全局运动补偿有下面的一些优点：

该模型仅仅使用少数的参数对全局的运行进行描述，参数所占用的码率基本上可以忽略不计。

该方法不对帧进行分区编码，这避免了分区造成的块效应。

在时间方向的一条直线的点如果在空间方向具有相等的间隔，就对应了在实际空间中连续移动的点。其他的运动估计算法通常会在时间方向引入非连续性。

但是，缺点是，如果场景中有运动物体的话，全局运动补偿就不足以表示了。这时候应该选用其他的方法。

在分块运动补偿中：

每帧被分为若干像素块（在大多数视频编码标准，如 MPEG 中，是分为 16×16 的像素块）。从参考帧的某个位置的等大小的块对当前块进行预测，预测的过程中只有平移，平移的大小和方向称为运动矢量。

对分块运动补偿来说，运动矢量是模型的必要参数，必须一起编码加入码流中。由于运动矢量之间并不是独立的（例如属于同一个运动物体的相邻两块通常运动的相关性很大），通常使用差分编码来降低码率。这意味着在相邻的运动矢量编码之前对它们作差，只对差分的部分进行编码。使用熵编码对运动矢量的成分进行编码可以进一步消除运动矢量的统计冗余（通常运动矢量的差分集中于 0 矢量附近）。

运动矢量的值可以是非整数的，此时的运动补偿被称为亚像素精度的运动补偿。这是通过对参考帧像素值进行亚像素级插值，而后进行运动补偿做到的。最简单的亚像素精度运动补偿使用半像素精度，也有使用 1/4 像素和 1/8 像素精度的运动补偿算法。更高的亚像素精度可以提高运动补偿的精确度，但是大量的插值操作大大增加了计算复杂度。

图 6-5～图 6-10 的几组插图可以帮助理解运动估计和运动补偿的基本概念。

图 6-5　宏预测与运动补偿示意图

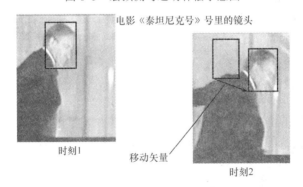

图 6-6　分块运动估计

图 6-7　视频序列：乒乓球移动第 0 帧

图 6-8　视频序列：乒乓球移动第 1 帧

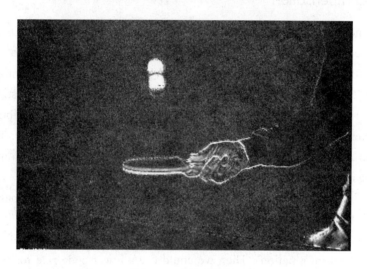

图 6-9　两帧的差值

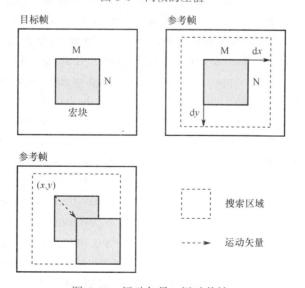

图 6-10　运动矢量—运动估计

6.3 MPEG Compression Algorithm

MPEG-video image compression technology's basic idea and algorithms could be concluded into two major points.

(1) In space zone, image data compression applies JPEG, joint photographic experts group compression algorithm to get rid off the redundancy.

(2) In time zone, image data compression applies motion compensation to get rid off the redundancy. We have talked about the JPEG algorithm and won't bother anymore. We just have a look at the part 2.

6.3.1 Motion Compensation

The basic principle of motion compensation is that when encoder deals with the N's frame of the image sequence, it applies the core technology of the motion compensation, ME, motion estimation to get the N's frame's estimated frame N'. In actual codes transmission, instead of transmitting the N'th frame, transmitting the difference of the frame N and its estimation frame N', Δ. If the motion estimation is efficient enough, the probability of Δ will be distributed around zero. This makes the energy of Δ being much smaller than the energy of Nth frame. The encoded bits needed to transmit Δ will be much less. This is the key point why the motion compensation could get rid off the time redundancy resided in the symbol resources.

The motion compensation is one of the key technologies of current video images compression.

As indicated, to get off the space redundancy of the moving images, the popular way is to deduct the reference frame (previous frame) from current frame and get the difference with much less energy (or evolved information). Then we could use a lower code rate for the encoding. The decoder could completely recover the original frames by the simple addition.

This idea may not be that easy to be accepted. As we just talked, in order to compress the active part of two consecutive images, deduct the previous frame form current frame, only transmit the changed part, and reduce the current frame's content dramatically. We could explain the situation via a simple example. Suppose all the background is still with dark color and only one pingpong ball is white. In last frame when I hit the ball and the ball moved over. When we directly deduct the previous frame from current frame, we'll have some problems. In previous frame, there was a pingpong ball here. We'll encode the frame. In current frame, the pingpong ball has been moved. After the direct deduction, there will be something else appearing at this position. Since the pingpong ball is moving, in this situation, while encoding the consecutive two frames, whatever you think, the efficiency is still not that high. Then, we may think while encoding a new frame, could we estimate the previous one? The pingpong ball is just moving. Of course, the ball may be spinning when you gave it a special hit. But within a short time, we can ignore the spinning and suppose it is just doing a straight line movement. This way, when encoding the new frame, we estimate the ball's new position from its old position in the previous frame based on a straight line movement. We actually move the ball to the estimated location. Then within those two frames, deduct one from the other. After the deduction, the changing part will be much smaller. This is called the moving images' motion compensation with motion estimation. It's two of the terminologies. After this kind of compression, the moving images got further compression.

The motion compensation is a kind of describing the difference of neighbouring frames. (The neighbouring here means they are neighbours in encoding. But it doesn't mean that they'll be neighbours while broadcasting.) It's actually to say that how each small block in the previous frame (Here means the one frame before the current one when encoding. It's not necessary to be in front of the current one when broadcasting.) moves to a certain position within the current frame. This method is often adopted by video compressor/video encoder decoder to reduce the time redundancy residing inside the video sequential images.

A little bit complicated design is to estimate the movement of whole frame's background and the aimed object in this background. Put the movement into some related coefficients encoded into the output bit stream. This way, the pixels within the predicted frame are the original pixels of the referent frame plus correspondent shifts. With this method, we could get the nearly difference of smaller energy then the directly deduction's to get a better compression ratio. On the other hand, the coefficients describing the movements shouldn't be too much in the output bit stream. Otherwise, we won't get any benefits from the complicated motion estimation.

The image frames are usually processed in groups. We don't apply the motion estimation method to the first frame of each group (usually is the sequential first frame in the group). Those kinds of frames are called intra frames or I frames. The others in the groups are called inter frames, simply called P frames. They are the prediction frames we talked. This kind of encoding format is called IPPPP format, which means the first frame is I frame and all the others are P frames during the encoding.

Besides this, when predicting, we could not only predict the current frame from the previous frame, but also predict the current frame from the next frame as soon as this next frame had been encoded before this current frame. In another words, the encoding sequence is different from the broadcasting sequence. Usually this kind of current frame is predicted from both the either previous or next I frames or P frames and is called bidirectional predicting frames, B frames. One of the examples of this kind of sequence of encoding is as follows IBBPBBPBBPBB.

Basically there are two types of motion compensation, the whole image scene motion compensation and the block motion compensation.

In whole image scene motion compensation:

The motion model basically reflects the various movements of the video camera including line moving, turning, zoom in and out, and etc. This type of model is mostly suitable to the encoding of still scenes without moving object/s. The whole image scene motion compensation has the following advantages.

The model only uses limited coefficients to describe the whole image scene's operating. The bits those coefficients occupied could be ignored because of its quantity rate.

This method is not processing the block encoding which avoids the block effects got from the blocking.

If the dots on one line along with the time axel have the same intervals as along with the space axel, those dots will be actually correspondent to the continuous moving dots in the space. Using other motion estimation methods would introduce discontinuity along with the time axel.

The major disadvantage of this method is that as soon as there is moving object/s in the scene, it can't efficiently represent it/them anymore. Some other compensation ways need to be applied.

In block motion compensation:

Each frame would be separated as some pixel blocks. In most video encoding standards, such as MPEG, each frame would be separated as 16×16 pixel blocks. We start from a position within the reference frame with an equivalent block to predict the current block. There are only line movements during the prediction process. The direction and size of the movement is called moving vector.

To block motion compensation, the moving vector is the model's necessary coefficient and need to be encoded into bit stream. Since the moving vectors are not independent from each other, (For example, same object's neighbouring two blocks' movements are closely correlated each other.) we usually apply the differential encoding to reduce the bit rate. This means that we get the difference of the moving vectors of neighbouring blocks before encoding them. After this is done, we only encode the differential part. We further use entropy encoding to encode the moving vector part to furthermore reduce the moving vector's statistical redundancy. (The differential energy of the moving vector is usually concentrated near the zero vector.)

The value of the moving vector is not necessarily always integer. There is a particular name for this kind of moving compensation, inferior pixel accuracy motion compensation. It is done by processing the inferior pixel level inserting to the reference frame's pixels first and then going ahead with the motion compensation. The simplest inferior pixel accuracy motion compensation is to use the half pixel accuracy. Some are using the quarter or 1/8 pixel accuracy motion compensation algorithms. The even higher inferior pixel accuracy could improve the accuracy of motion compensation. But the much more insertion operating greatly increases the calculation complexity.

The Fig. 6-5 to Fig. 6-10 gives out a basic idea of the motion estimation and motion compensation.

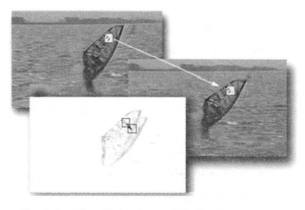

Fig. 6-5   Micro Estimation and Motion Compensation

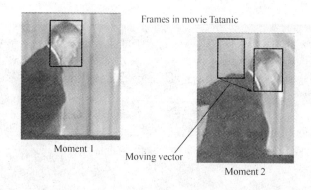

Frames in movie Tatanic

Moment 1

Moving vector

Moment 2

Fig. 6-6   Block Motion Estimation

Fig. 6-7   Video sequence: Tennis frame 0

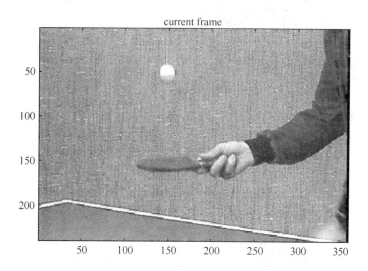

Fig. 6-8   Video sequence: Tennis frame 1

315

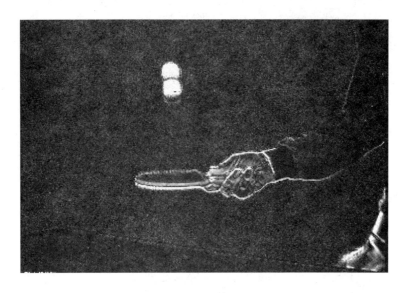

Fig. 6-9　Frame difference：frame 0 and 1

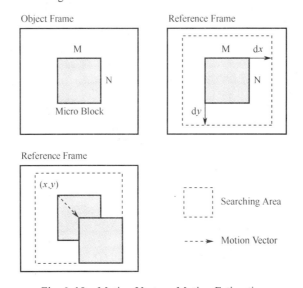

Fig. 6-10　Motion Vector - Motion Estimation

6.3.2　MPEG-1 视频的表示

如前所述，MPEG 有不同的数据处理方式，下面具体的看看它们的数据结构。图 6-11 给出了 MPEG-1 具体的数据结构，图中相应的技术术语如下：

（1）运动图像序列（Sequence）——视频；

（2）图片组（GOP，Group Of Pictures）——运动图像序列中的一个连续的图像集；

（3）图片切片（Slice）——重新同步单元；

（4）宏块（Microblock）（16×16）——运动补偿单元；

（5）块（Block）（8×8）——变换单元；

每一个宏块由四个亮度信号 Y 块和两个色度信号 U、V 块组成。

图 6-12 给出了 MPEG-1 色度和亮度的位置关系。

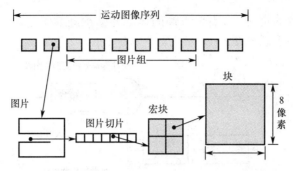

图 6-11　MPEG-1 数据体系结构

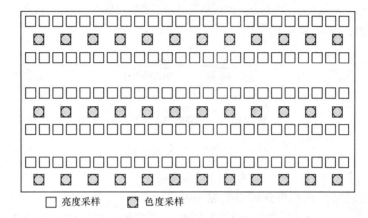

□ 亮度采样　　◻ 色度采样

图 6-12　MPEG-1 色度和亮度的位置关系

亮度信号 Y 有偶数个行和偶数个列，色度信号 U、V 分别取 Y 信号在水平、垂直方向的 1/2。如图 6-12 所示，黑点代表色度 U、V 位置，亮度 Y 位置用白圈表示。

在 MPEG-1 中将图像分为三种类型（这一点我们在前面已经谈到过）：

（1）I（Intra Picture）图像。

利用图像自身的相关性压缩，提供压缩数据流中的随机存取的点。

（2）P（Predicted Picture）图像。

用最近的前一个 I 图像（或 P 图像）预测编码得到（前向预测）。

（3）B（Bidirectionally Interpolated Picture）图像。

B 图像在预测时，既可使用前一个图像作参照，也可使用下一个图像做参照或同时使用前后两个图像作为参照图像（双向预测）。

图 6-13 给出了 MPEG-1 定义的三种图像。

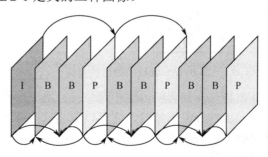

图 6-13　MPEG-1 定义的三种图像

有了上述三种图像的定义，来看一下帧间运动补偿预测编码技术，图6-14给出了帧间运动补偿预测编码技术的示意框图。

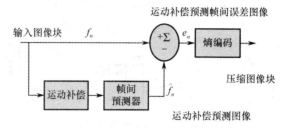

图6-14　帧间运动补偿预测编码技术

帧间预测编码，参照图6-15所示。

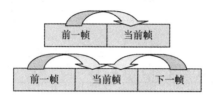

图6-15　帧间预测编码

（1）I帧：不进行预测、进行帧内编码（参考帧）；

（2）P帧：通过向前预测得到的误差编码帧；

（3）B帧：通过双向预测得到的误差编码帧，因图像序列存放在存储器中，可以使用下一帧来预测当前帧；

（4）P帧和B帧中的向前预测；

（5）B帧中的双向预测。

前文已经讲到过运动补偿。

物体在空间上的位移，用有限的运动参数（如运动矢量）加以描述，并和预测误差一同参与编码。

实现时，如前所述，画面一般划分成一些不连接的像素块（在MPEGl和MPEG2标准中一个像素块为16×16像素），对于每一个这样的像素块，只估算一个运动矢量，如图6-16所示。

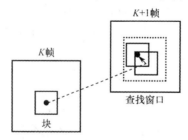

图6-16　宏块中的运动矢量

对于运动的物体，估计出物体在相邻帧内的相对位移，用上一帧中物体的图像对当前帧的物体进行预测，将预测的差值部分编码传输，就可以压缩这部分图像也减少它的码率。这种考虑了对应区域的位移或运动的预测方式就称为运动补偿预测编码。帧间预测是运动补偿预测在运动矢量为零时的特殊情况。

图6-17示出了I图像的压缩编码算法。

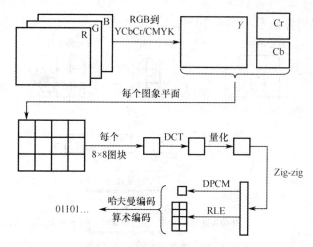

图 6-17　Ⅰ图像的压缩编码算法

P 图像的压缩编码算法示于图 6-18 中。

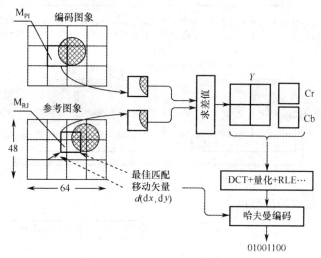

图 6-18　P 图像的压缩编码算法

移动矢量的算法框图如图 6-19 所示。从图中可以看出，移动矢量有水平分量和垂直分量，分别代表了移动矢量在水平和垂直两个分量的移动距离。

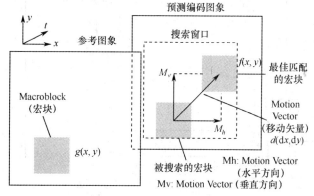

图 6-19　移动矢量的算法框图

B 图像的压缩编码算法示于图 6-20 中。

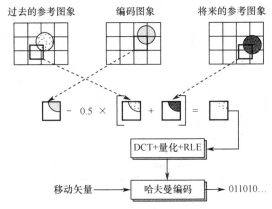

图 6-20 B 图像的压缩编码算法

图 6-21 示出了运动序列流的组成。

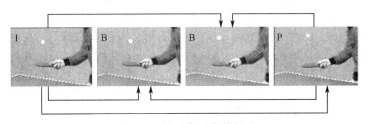

图 6-21 运动序列流的组成

图 6-22 示出了相应的计算框图。

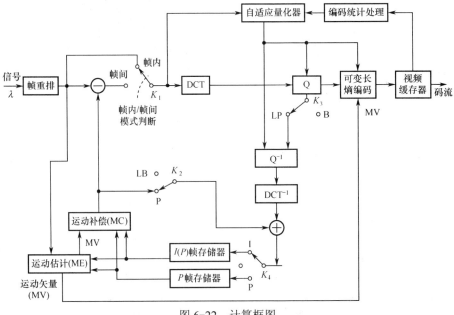

图 6-22 计算框图

表 6-2 示出了 MPEG 三种图像压缩后的典型值。

I、P 和 B 图像压缩后的大小示于表 6-2 中，单位为比特。从表中可以看到，I 帧图像的数据量最大，而 B 帧图像的数据量最小。

表 6-2　MPEG 三种图像压缩后的典型值

图像类型	I	P	B	平均数据/帧
MPEG-1 CIF 格式（1.15 Mb/s）	150 000	50 000	20 000	38 000
MPEG-2 601 格式（4.00 Mb/s）	400 000	200 000	80 000	130 000

6.3.2　MPEG-1 Video Expression

As we talked, MPEG has different data processing format. Let's have a look at its data structures. Fig. 6-11 shows its detailed data structure. The terminologies used in the figure are listed below.

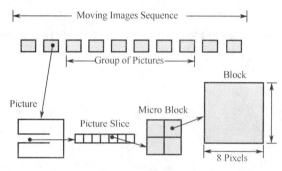

Fig. 6-11　MPEG-1 Data System Structure

(1) Moving images sequence, sequence - video;

(2) Group of pictures, GOP - a consecutive image collection within the sequence;

(3) Picture slice, slice - re-synchronizing unit;

(4) Macroblock (16×16) - the 16×16 motion compensation unit;

(5) Block (8×8) - the 8×8 transform unit.

Each macroblock consists of 4 luminance signal Y blocks and 2 chrominance signal U and V blocks.

The Fig. 6-12 shows the position relationship of luminance and chrominance within MPEG-1.

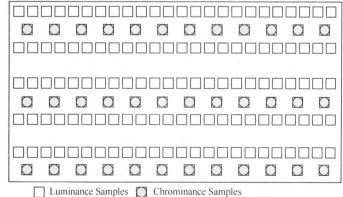

Fig. 6-12　MPEG-1 Luminance and Chrominance Position Relationship

The luminance Y signals have even lines and columns. The chrominance U and V signals reside between and take half of the Y signals horizontally and vertically as shown in Fig. 6-12. The color ones represent the chrominance U, V positions and white ones represent the luminance Y positions.

In MPEG-1, the images are classified into three types (as we mentioned).

(1) I (Intra Picture) Picture.

Compressing based on the image's own correspondence provide the random access point within the compressed data stream.

(2) P (Predica ted Picture) Picture.

Got based on the closest previous I frame (or P frame) to predict and encode.

(3) B (Bidirectional Interpolated Picture) Picture.

While predicting the B frame, could either use the previous frame as a reference or use the next following frame as a reference to encode the current frame. Even more, could use both previous and next following frames at the same time as reference to encode the current frame.

Fig. 6-13 shows the three types of pictures defined in MPEG-1.

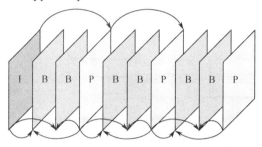

Fig. 6-13　Three Types of Frames Defined in MPEG-1

As soon as we have the definitions of those three types of pictures, we could have a look at the inter frame motion compensation prediction encoding technology. The Fig. 6-14 is a sketch block diagram of the inter frame motion compensation prediction encoding technology.

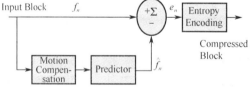

Fig. 6-14　Inter Frame Motion Compensation Prediction Encoding Technology

Please refer to the Fig. 6-15 for the inter frame encoding explanation.

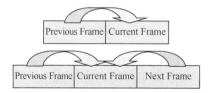

Fig. 6-15　Inter Frame Prediction Encoding

(1) I frame: There is no prediction and directly encoding within the frame (reference frame);

(2) P frame: The difference encoding frame predicted from the previous frame;

(3) B frame: The difference frame predicted bidirectional from both the previous and the next frames. Since the frame sequence is stored in the memory, we could use the next frame to predict the current frame.

(4) The forward prediction used in P frame and B frame:

(5) The bidirectional prediction used in B frame:

We have talked about motion compensation.

We describe an object's shifts in space with limited motion characters such as motion vectors and put them into the encoding with the predicted differences.

In reality, as we mentioned, a picture usually divided into not connected macroblocks (in MPEG-1 and MPEG-2, the size is 16×16 picels.). We only predict one motion vector for each macroblock as shown in Fig. 6-16.

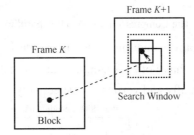

Fig. 6-16   The Motion Vector in Macroblock

For the moving object, estimate the object's relative shifting in the neighboring frames. We use the object's position in the previous frame to predict its new position in the current frame. We then encode and transmit the difference got from the prediction. This way, we could compress this portion image and reduce the code rate. As we have known, this kind of prediction considered the relative shift or movement in the correspondent area is called motion compensation prediction encoding. Inter frame prediction is the motion compensation prediction's special situation when the motion vector equals zero.

Fig. 6-17 shows I image's compression encoding algorithm.

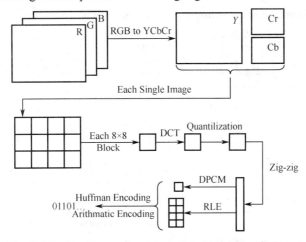

Fig. 6-17   The Compression Encoding Algorithm of I Image

The P image's compression encoding algorithm is shown in Fig. 6-18.

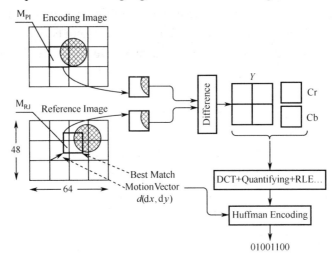

Fig. 6-18　The Compression Encoding Algorithm of P Image

The calculation block diagram of motion vector is shown in Fig. 6-19. From the figure, we could see that the motion vector has both horizontal and vertical components that represent the vector's moving distance horizontally and vertically respectively.

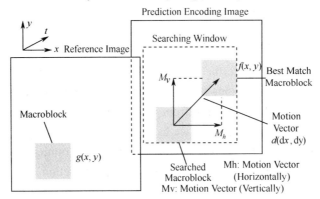

Fig. 6-19　Motion Vector Calculation Block Diagram

The B image's compression encoding algorithm is shown in Fig. 6-20.

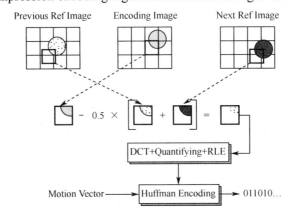

Fig. 6-20　The Compression Encoding Algorithm of B Image

The Fig. 6-21 shows the format of moving frame sequence.

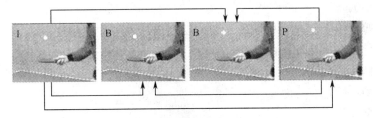

Fig. 6-21   The Creation of Motion Image Sequence Stream

The Fig. 6-22 shows the correspondent calculation block diagram.

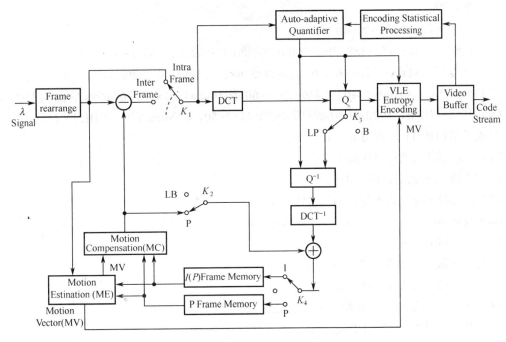

Fig. 6-22   Calculation Block Diagram

The Table 6-2 gives out the three typical compressed image data sizes.

Table 6-2   MPEG Three Typical Compressed Data Quantities

Image Types	I	P	B	Average Data/Frame
MPEG-1 CIF Format（1.15 Mb/s）	150 000	50 000	20 000	38 000
MPEG-2 601Format（4.00 Mb/s）	400 000	200 000	80 000	130 000

The data sizes of the compressed I, P, and B images are shown in the table and the unit is bit. From the table we could see that the compressed I image has the biggest data size and the compressed B image has the smallest data size.

6.3.3　MPEG-2 标准

在MPEG-2中,考虑到要适应不同数据速率设备的应用,引入了"档次(配置)——Profiles"和"等级——Levels"的概念,档次对应一套算法,而等级指定一套参数范围(如图像大小、帧速率和位速率)。

由配置和等级组合起来,构成了 MPEG-2 所支持的各种视频规格。

MPEG 文件的分类有如下的特点。

分为 5 个档次,如下:

(1)简单型;

(2)基本型;

(3)信噪比可调型;

(4)空间可调型;

(5)增强型;

含有 4 个等级(Levels),如下:

(1)低级(Low)352×288×30,4Mb/s,它面向 VCD 并与 MPEG-1 兼容;

(2)基本级(Main)720×460×30 或 720×576×25,15Mb/s,它面向视频广播信号;

(3)高级 1440(High-1440)1440×1080×60 或 1440×1152×50,60Mb/s,它面向 HDTV;

(4)高级(High)1920×1080×60 或 1920×1152×50,80Mb/s,它面向 HDTV。

形成了常用的 11 种规范如下:

(1)高级的基本型,MP@HL;

(2)高级的增强型,HP@HL;

(3)高-1440 级的基本型,MP@H1440;

(4)高-1440 级的空间可调型,SSP@H1440;

(5)高-1440 级的增强型,HP@H1440;

(6)基本级的简单型,SP@ML;

(7)基本级基本型,MP@ML;

(8)基本级的信噪比可调型,SNP@ML;

(9)基本级的增强型,HP@ML;

(10)低级的基本型,MP@LL;

(11)低级的信噪比可调型,SNP@LL。

表 6-3 给出了这些常用规范的主要参数。

表 6-3　MPEG-2 的配置与等级

Level\Profile（等级\配置）	Simple（简化型）	Main（基本型）	SRN Scalability（信噪比可变型）	Spatial Scalability（空间分辨率可变型）	High（高档型）
High（高级）		4:2:0 1920×1152×60 80 Mb/s I, P, B			4:2:0, 4:2:2 1920×1152×60 80 Mb/s I, P, B

Level\Profile（等级\配置）	Simple（简化型）	Main（基本型）	SRN Scalability（信噪比可变型）	Spatial Scalability（空间分辨率可变型）	High（高档型）
High-1440（高级 1440）		4:2:0 1440×1152×60 60 Mb/s I, P, B		4:2:0 1440×1152×60 60 Mb/s I, P, B	4:2:0，4:2:2 1440×1152×60 60 Mb/s I, P, B
Main（基本级）	4:2:0 720×576×30 15 Mb/s I, P	4:2:0 720×576×30 15 Mb/s I, P, B	4:2:0 720×576×30 15 Mb/s I, P, B		4:2:0 720×576×30 20 Mb/s I, P, B
LOW（低级）		4:2:0 352×288×30 4 Mb/s I, P, B	4:2:0 352×288×30 4 Mb/s I, P, B		

例：说明数字视频规格 MP@ML 和 HP@HL 各自的含义。

MP@ML（Main Profile, Main Level），可译成"基本级基本型"，它指的是：帧速率为 25 帧/s（PAL 制），分辨率为 720×576，子采样格式为 4：2：0，位速率达 15 Mb/s 的视频。

HP@HL（High Profile, High Level），可译成"高级的增强型"，它指的是：帧速率为 50 帧/s（PAL 制），分辨率为 1920×1152，子采样格式为 4：2：0 或 4：2：2，位速率达 80Mb/s 的视频。

6.3.3　MPEG-2 Standard

In MPEG-2, to meet the different requirements of various equipments with different data rates, introduced the ideas of profiles and levels. The profiles related to a set of algorithms and the levels specify the range of a set of parameters (such as the image's size, frame rate, and bit rates).

The profiles and the levels together form the various video formats supported by MPEG-2.

The MPEG files classification has following characteristics.

There are 5 profiles as shown below.

(1) Simple;

(2) Main;

(3) SNR (signal noise ratio) Scalable;

(4) Spatial Scalable;

(5) High (Enhanced).

There are 4 levels as shown below.

(1) Low, 352×288×30, 4Mb/s. It is mainly for VCD and is compatible with MPEG-1.

(2) Main, 720×460×30 or 720×576×25, 15Mb/s. It is mainly for video broadcasting signals.

(3) High - 1440, 1440×1080×60 or 1440×1152×50, 60Mb/s. It is mainly for HDTV.

(4) High, 1920×1080×60 or 1920×1152×50, 80Mb/s, It is mainly for HDTV as well.

From above, forms 11 most often used formats as follows.

(1) Main profile at high level. MP@HL;

(2) High profile at high level. HP@HL;

(3) Main profile at high - 1440 level. MP@H1440;

(4) Spatial scalable profile at high - 1440 level. SSP@H1440;

(5) High profile at high - 1440 level. HP@H1440;

(6) Simple profile at main level. SP@ML;

(7) Main profile at main level. MP@ML;

(8) SNR scalable profile at main level. SNR@ML;

(9) High profile at main level. HP@ML;

(10) Main profile at low level. MP@LL;

(11) SNR scalable profile at low level. SNR@LL.

The table 6-3 shows the major characteristics of those formats.

Table 6-3　MPEG-2 Match Profiles and Levels

Profile/Level	Simple	Main	SRN Scalability	Spatial Scalability	High
High		4:2:0 1920×1152×60 80 Mb/s I, P, B			4:2:0，4:2:2 1920×1152×60 80 Mb/s I, P, B
High-1440		4:2:0 1440×1152×60 60 Mb/s I, P, B		4:2:0 1440×1152×60 60 Mb/s I, P, B	4:2:0，4:2:2 1440×1152×60 60 Mb/s I, P, B
Main	4:2:0 720×576×30 15 Mb/s I, P	4:2:0 720×576×30 15 Mb/s I, P, B	4:2:0 720×576×30 15 Mb/s I, P, B		4:2:0 720×576×30 20 Mb/s I, P, B
LOW		4:2:0 352×288×30 4 Mb/s I, P, B	4:2:0 352×288×30 4 Mb/s I, P, B		

Example: Please explain the meanings of digital video formats MP@ML and HP@HL.

Solution: MP@ML is main profile at main level. It means the video which has the following characteristics. The frame rate is 25 frames/second (for PAL), the resolution is 720×576, the sub-sampling format is 4:2:0, and the bit rate can reach 15Mb/s.

HP@HL is high profile at high level. It means the video which has the following characteristics. The frame rate is 50 frames/second (for PAL), the resolution is 1920×1152, the sub-sampling format is 4:2:0 or 4:2:2, and the bit rate can reach 80Mb/s.

6.3.4　MPEG-4 标准

MPEG-4 编码算法支持由 MPEG-1 和 MPEG-2 提供的所有功能，同时引入了 AV 对象

（Video Object Plane—AVO）的概念，使得更多的交互操作成为可能："AV对象"可以是一个孤立的人，也可以是这个人的语音或一段背景音乐等。图6-23到图6-25给出了这种技术的基本概念。

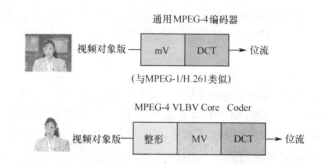

图6-23　普通MPEG-4编码器和MPEG-4 VLBV核心编码器

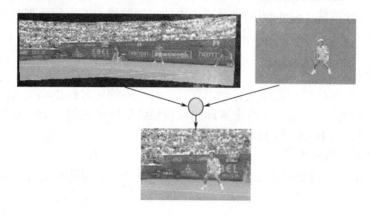

图6-24　MPEG-4应用实例　背景全景图+视频对象（VO）＝合成图像

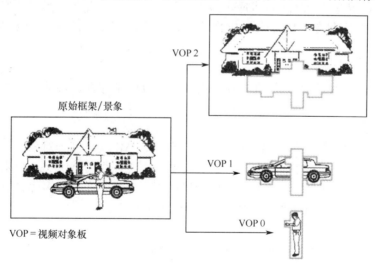

图6-25　MPEG-4应用实例　背景全景图+视频对象（VO）＝合成图像

6.3.4.1　MPEG-4的关键技术

MPEG-4除采用第二代视频编码的核心技术如变换编码、运动估计与运动补偿、量化和第一代熵编码外，还提出了一些新的有创见性的关键技术，并在前二代视频编码技术基础上进行了卓有成效的完善和改进。下面列出了这样的关键技术。

1．视频对象提取技术

MPEG-4 实现基于内容交互的首要任务就是把视频/图像分割成不同对象或者把运动对象从背景中分离出来，然后针对不同对象采用相应编码方法，以实现高效压缩。因此视频对象提取即视频对象分割，是 MPEG-4 视频编码的关键技术，也是新一代视频编码的研究热点和难点。

2．VOP 视频编码技术

视频对象平面（VOP，Video Object Plane）是视频对象（VO）在某一时刻的采样，VOP 是 MPEG-4 视频编码的核心概念。MPEG-4 在编码过程中针对不同 VO 采用不同的编码策略，即对前景 VO 的压缩编码尽可能保留细节和平滑；对背景 VO 则采用高压缩率的编码策略，甚至不予传输而在解码端由其他背景拼接而成。

3．视频编码可分级性技术

视频编码的可分级性（Scalability）是指码率的可调整性，即视频数据只压缩一次，却能以多个帧率、空间分辨率或视频质量进行解码，从而可支持多种类型用户的各种不同应用要求。

4．运动估计与运动补偿技术

在 MPEG-4 视频编码中，运动估计相当耗时，对编码的实时性影响很大。因此这里特别强调快速算法。运动估计方法主要有像素递归法和块匹配法两大类，前者复杂度很高，实际中应用较少，后者则在 H.263 和 MPEG 中广泛采用。

6.3.4.2　MPEG4 的应用领域

凭借着出色的性能，MPEG4 技术目前在多媒体传输、多媒体存储等领域得到了广泛的应用。

1．精彩的视频世界

精彩的视频世界是 MPEG4 技术应用最多也是最为广大朋友所熟悉的的形式。目前它主要以两种形式出现，一种是 DIVX－MPEG4 影碟（国内市面上已出现），另一种是网上 MPEG4 电影。

2．低比特率下的多媒体通信

MPEG4 技术已经广泛地应用在如视频电话、视频电子邮件、移动通信、电子新闻等多媒体通信领域。

3．实时多媒体监控。

多媒体监控领域原来一直是 MPEG1 技术担当重任，MPEG4 基于对象的调整压缩方法可以获得比 MPEG1 更大的压缩比，使压缩码流更低。因此，尽管 MPEG4 技术一开始并不是专为视频监控压缩领域而开发的，但它高清晰度的视频压缩，在实时多媒体监控上，无论是存储量，传输的速率，清晰度都比 MPEG1 具有更大的优势。

4．硬件产品上面的应用。

MPEG-4 已经用在了下列产品上。

日本夏普公司的数字摄像机 VN－EZ1。

飞利浦公司于 2004 年八月份推出了一款支持 DivX 的 DVD 播放机 DVD737 。

手机：在手机领域，推出了可拍摄 MPEG4 动态视频的手机型号，如西门子 ST55、索尼爱立信 P900/P908、LG 彩屏 G8000 等。

MPEG4 数字硬盘等。

6.3.4 MPEG-4 Standard

MPEG-4 algorithm supports all function abilities provided by MPEG-1 and MPEG-2. At the same time, it introduces the AV object idea (video object plane, AVO) which makes it's possible to provide more interexchange operations. The AV object could be a person, a speech of this person, and a piece of background music. The Fig.6-23 through Fig.6-25 gives a basic idea of this technology.

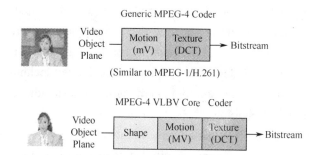

Fig. 6-23   Normal MPEG-4 Encoder and MPEG-4 VLBV Core Coder

Fig. 6-24   MPEG-4 Real Example   Background Full Scene+Video Object=Synthesis Image

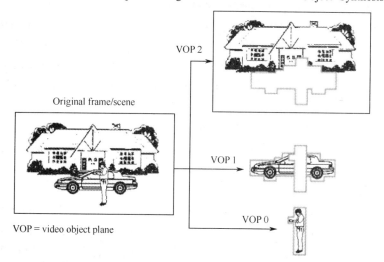

Fig. 6-25   MPEG-4 Real Example   Background Full Scene+Video Object=Synthesis Image

6.3.4.1　MPEG–4 Core Technology

Besides applies the second generation of video encoding core technologies such as transforming encoding, motion estimation motion compensation, quantification, and first generation entropy encoding, the MPEG-4 proposed some new creative core technologies. Furthermore, based on the previous two generation video encoding technologies, MPEG-4 efficiently modified and improved those technologies. The following are some of them.

1. Video Object Pick-up Technology

The priority thing the MPEG-4 needs to do to reach the goal of content exchange is to separate the video/image into different objects or separate the moving object from the background. Then encode the different objects with various methods to get the high efficiency compression. That's why the video object pick-up or video object separation is the key technology in MPEG-4 video encoding and also a research hot and tough topic in modern video encoding.

2. VOP Video Encoding Technology

Video Object Plane, VOP is VO's certain time point sample. VOP is the core idea of MPEG-4 video encoding. MPEG-4 applies different encoding strategies for various VO during the encoding process. For the front view video object, VO, the compressing encoding goes to as detail as possible to keep the details and smooth. On the other hand, for the background VO, the compressing encoding goes to as high as possible compression rate and even more doesn't transmit it at all where reconstruct the background at the receiver end by composing other backgrounds together.

3. Video Encoding Scalability Technology

The video encoding scalability usually means the adjustability of the encoded bit rates. That is to say that one could compress encoding the video data once and decode it in many ways. One could decode it in different frame rates, different space resolutions, and different video qualities. This way, it could support various application requirements from different types of users.

4. Motion Estimation and Motion Compensation Technology

In MPEG-4 video encoding, the motion estimation costs quite much time which greatly affect the real time encoding. That's why we significantly emphasise the faster calculation algorithms. The motion estimation algorithms are roughly classified into two categories. One of them is the pixel deliver return method and the other is block match method. The formers calculation process is too complicated to actually apply and seldom used. The later one is widely applied in H.263 and MPEG.

6.3.4.2　The Major Application Areas of MPEG–4

Because of its outstanding attributes, the MPEG-4 technology has been widely applied in multimedia transmission, multimedia storage, and etc.

1. Colourful Video World

The colourful video world is the most popular field the MPEG-4 applied for. This is also where people are most familiar with. Currently, it has majorly two formats. One of them is DIVX - MPEG-4 video disc. (It has been seen in our country.) The other is the MPEG-4 movies on the internet.

2. Low Bit Rate Multimedia Communications

MPEG-4 technology has been widely applied in many multimedia communication areas such as video telephone, video emails, mobile communications, electronic news etc.

3. Real Time Multimedia Video Surveillance

The MPEG-1 has been applied in real time multimedia video surveillance for a while. MPEG-4 object based adjusting compression methods could get a higher compression ratio then MPEG-1's to get a shorter compressed bit stream. The MPEG-4 is not particularly developed for video surveillance application at the very beginning. But its high resolution video compression makes it the number one in real time multimedia surveillance application. Its storage occupancy, transmission rate, and video image clearance are all better than MPEG-1's.

4. Hardware Products Application

The MPEG-4 has been applied on the following products.

Japan Sharp Company's digital video camera VN - EZ1

Philip proposed one DVD player, DVD737 in August, 2004, which supporting DivX.

Cellphones. In this field, cellphones taking MPEG-4 dynamic video have been seen in the market, such as Simons ST55, Sony Ellice P900/P908, LC color screen C8000 etc.

MPEG-4 digital hard disk etc.

6.4 其他形式的视频压缩

MPEG 不是仅有的视频压缩标准。

视频压缩国际标准除了 MPEG 系列外，主要还有由 ITU-T 制定的 <u>H.261、H.262、H.263、H.264</u>。图 6-26 表示了他们之间的基本关系。

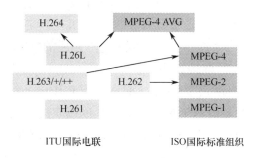

图 6-26　其他形式的视频压缩

其中 H.262/MPEG-2 由 ITU-T 与 MPEG 联合制定。

它们主要应用于实时视频通信领域，如会议电视（又叫视频会议，VC——video conference，它是通过网络通信技术来实现的虚拟会议，使在地理上分散的用户可以共聚一处，支持人们远距离进行实时信息交流与共享、开展协同工作的应用系统）、远程教学、远程医疗等。

运动图像编解码标准示于图 6-27 中。

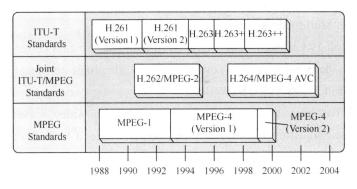

图 6-27　运动图像编解码标准

6.4.1　H.264 编码技术

（1）为什么叫 H.264；

（2）H.264 的特点；

（3）H.264 的应用。

什么是 H.264？H.264 是一种最新的视频高压缩技术，又叫 MPEG-4 AVC（活动图像专家组-4 的高等视频编码），或称为 MPEG-4 Part10。它是由国际电信标准化部门 ITU-T 和规定 MPEG 的国际标准化组织 ISO/国际电工协会 IEC 共同制订的一种活动图像编码方式的国际标准格式。

在制订完最初的 H.263 标准之后，ITU-T 的视频编码专家组（VCEG——Video Coding Experts Group）将开发工作分为两部分：一部分称为"短期（short-term）"计划，目的是给 H.263 增加一些新的特性（这一计划开发出了 H.263+ 和 H.263++）；另一部分被称为"长期（long-term）"计划，其最初的目标就是要制定出一个比当时其他的视频编码标准效率提高一倍的新标准。这一计划在 1997 年开始，其成果就是作为 H.264 前身的 H.26L（起初称为 H.263L）。在 2001 年年底，由于 H.26L 优越的性能，ISO/IEC 的 MPEG 专家组加入到 VCEG 中来，共同成立了联合视频小组（JVT——Joint Video Team），接管了 H.26L 的开发工作。这个组织的目标是："研究新的视频编码算法，其目标是在性能上要比以往制定的最好的标准提高很多。"

这一标准正式成为国际标准是 2003 年 3 月在泰国 Pattaya（芭堤雅）举行的 JVT 第 7 次会议上通过的。由于该标准是由两个不同的组织共同制定的，因此有两个不同的名称：在 ITU-T 中，它的名字叫 H.264；而在 ISO/IEC 中，它被称为 MPEG-4 的第 10 部分，即高级视频编码（AVC——Advanced Video Coding)）。

JVT 是由 ISO/IEC MPEG 和 ITU-T VCEG 成立的联合视频工作组（Joint Video Team），致力于新一代数字视频压缩标准的制定。

由此可得出如下结论：JVT 标准。

（1）在 ISO/IEC 中的正式名称为：MPEG-4 AVC 标准；

（2）在 ITU-T 中的名称为：H.264。

其基本情况如图 6-28 所示，表 6-4 给出了主要视频编码技术的许可花费。

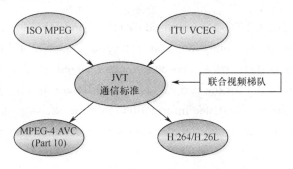

图 6-28　H.264 编码技术结构

表 6-4　主要视频编码技术许可费一览表

编码技术名称	MPEG-2	MPEG-4 visual	Windows Media video9
免费版权运用	未见叙述	一年 5 万个编码器和解码器，5 万人使用以下的产品和服务	2003 年 1 月 1 和～12 月 31 日出厂的产品和在 Windows 上工作的产品
解码器	2.5 美元/个	0.25 美元/个	0.1 美元/个
解码器许可费上限（每年）	无限制	100 万美元	40 万美元
编码器	2.5 美元/个	0.25 美元/个	0.2 美元/个
编码器许可费上限（每年）	无限制	100 万美元	80 万美元
编码器/解码器	未见叙述	未见叙述	0.25 美元/个
编码器/解码器许可费上限（每年）	未见叙述	未见叙述	100 万美元
DVD 碟片（二小时电影场合）	0.03 美元/层（两层以上 0.02 美元/层）	0.04 美元（5 年内摄制的电影），0.02 美元（5 年前摄制的电影）	无
内容配送者许可费	无	0. 000333 美元/分钟	无
内容配送者许可费上限	无	100 万美元/年	无
有效期限	2010 年 12 月 31 日	2008 年 12 月 31 日	2012 年 12 月 31 日

6.4　Other Formats Video Compression

The MPEG is not the only video compression standard.

Besides the MPEG series, the international video compression standards also have the standards H.261, H.263, H.263, and H.264 set up by ITU-T, Fig.6-26 Shows their basic Ye lations-hips.

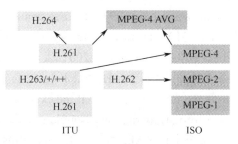

Fig. 6-26　Other Formats Video Compressions

Within those, the H.262/MPEG-2 is proposed by ITU-T and MPEG together.

They majorly applied in real time video communications, such as TV conference, (It is called video conference, VC as well. It is a virtual conference realized via the network communication technology. It makes it possible for the people stayed in different location to stay together having the meeting. It supports the long distance information real time exchanging and sharing. Based on this people can develop the cooperation application systems.) remote education, remote medical treatment etc.

Moving pictures encoding decoding standards are shown in Fig. 6-27.

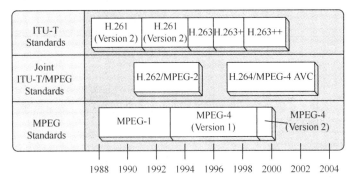

Fig. 6-27   Moving Pictures Encoding Decoding Standards

6.4.1   H.264 Coding Technology

(1) Why named H.264;

(2) The attributes of the H.264;

(3) H.264's applications.

What is H.264? The H.264 is a new video high compression technology. It is named MPEG-4 AVC (The advanced video coding of moving picture experts group) or MPEG-4 Part 10 as well. It is proposed by Telecommunication Standardization Sector, ITU-T and International Organization for Standardization, ISO/International Electrotechnical Commission, IEC together and is a international standard format for moving pictures coding.

After finished the H.263 standard, ITU-T VCEG, Video Coding Experts Group separated the standardization job into two parts. One of them is the short term plan that aimed to add some new attributes for H.263, which created the H.263+ and H.263++. The other part is the long term plan that initially aimed to develop a new standard doubling the coding efficiency of any other existed video coding standards. This plan started in 1997. Its first achievement is the H.264's former format H.26L (initially called H.263L). Because of the clear advantages of H.26L, ISO/IEC's MPEG experts group joined the VCEG at the end of 2001. A new group, Joint Video Team, JVT was created. It took over the development job of H.26L. Its goal is to research new video coding algorithms and to greatly improve the functionalities that the existed standards had.

This standard has become a formal international standard since it was passed in the seventh JVT meeting held in Pattaya, Thailand in March, 2003. Since this standard is created by two different organizations, it got two different names. In ITU-T, it is called H.264 and in ISO/IEC, it is

336

called MPEG-4 part 10 that is advanced video coding, AVC.

JVT, joint video team is a new team formed by ISO/IEC MPEG and ItU-T and it works on the new generations of digital video compression standards.

We got the following conclusion.

JVT standards:

(1) In ISO/IEC, it is formally named MPEG-4 AVC standard;

(2) In ITU-T, it is named H.264.

The idea is shown in Fig. 6-28.

The Table 6-4 lists the major video coding technology permission cost.

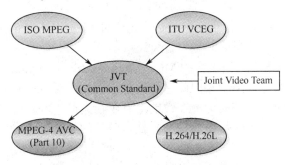

Fig. 6-28　H.264 Coding Technical Structure

Table 6-4　Major Video Coding Technology Permission Cost List

Coding Technology Title	MPEG-2	MPEG-4 Visual	Windows Media Video 9
Free to use	N/A	50000 encoder and decoder per year; Equal to or less than 50000 users.	The products manufactured between Jan 1st and Dec 31st, 2003 and based on Windows
Decoder	$2.5/each	$0.25/each	$0.1/each
Maximum decoder permission fee per year	Unlimited	$1000000	400000
Encoder	$2.5/each	$0.25/each	$0.2/each
Maximum encoder permission fee per year	Unlimited	$1000000	800000
Encoder/Decoder	N/A	N/A	$0.25/each
Maximum encoder/decoder permission fee per year	N/A	N/A	$1000000
DVD disk (2 hours movie)	$0.03/layer　($0.02/layer if more than 2 layers)	$0.04 (Movies within 5 years), $0.02 (Movies older than 5 years)	None
Content attach permission fee	None	$0.000333/minute	None
Maximum Content attach permission fee	None	$1000000/year	None
Expiring date	December 312010	December 312008	December 312012

6.4.2 H.264 技术的优点

图 6-29 给出了 H.264 编码的基本原理。

以下是 H.264 一些技术优点。

（1）低码流（Low Bit Rate）：最多节省 50％位速率：相较于 H.263v2（H.263+）或是 MPEG-4 Simple Profile，在相同编码器最佳化的条件下，H.264 编码最多可节省 50％的位速率。

（2）高质量的视频图像：H.264 能提供连续、流畅的高质量视频图像（DVD 质量）。

（3）容错能力强：H.264 提供了解决在不稳定网络环境下容易发生的丢包等错误的必要工具。

（4）网络适应性强：H.264 提供了网络适应层（Network Adaptation Layer），使得 H.264 的文件能容易地在不同网络上传输。

H.264 在编码框架上还是沿用以往的 MC（Motion estimation/compensation）-DCT 结构，即运动补偿加变换编码的混合结构，因此它保留了一些先前标准的特点。然而，以下介绍的技术使得 H.264 比之前的视频编码标准在性能上有了很大的提高。应当指出的是，这个提高不是单靠某一项技术实现的，而是由各种不同技术带来的小的性能改进而共同产生的。

H.264 采用了许多新技术以提高压缩效率，其主要技术特点包括：

（1）用 4×4 块的整数变换代替离散余弦变换；

（2）多参考帧、7 种宏块预测模式：16×16，16×8，8×16，8×8，8×4，4×8，4×4，运动估计和补偿更加精确；

（3）帧内预测；

（4）改进的去块效应滤波器（Deblocking filter）；

（5）增强的熵编码方法 UVLC（Universal VLC）、CAVLC（Context adaptive VLC）和 CABAC。

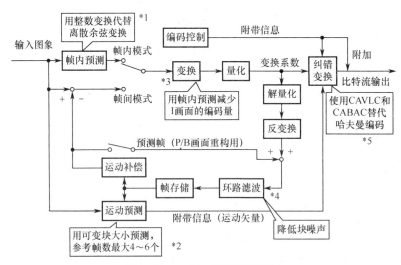

图 6-29　H.264 编码原理框图

图中：*1：H.264/AVC 中的 DCT 变换与以前的情况不同，它采用的变换单位不是 8×8 块，而是 4×4 块，且变换是整数操作，而不是实数操作。其优点为:运算速度快、精度高并占用较少内存。整数操作，编译码有严格的反变换，避免了截取误差，减少了运动边缘块的编码噪声。同时，4×4 变换比 8×8 变换产生的方块效应亦要小。

*2：采取多参考帧模式，进行多于一帧的先前帧（最多 5 帧）动目标估值，可以改善动目标估值性能，提高译码器的误码恢复能力，取得更好的预测效果。

变尺寸块运动补偿：例如平坦区可取 16×16 尺寸块，而细节区可采用 8×8 甚至 4×4 等更精细尺寸块，此时 MC 预测精度更高、灵活性更好，更符合视频内容不断变化的实际情况。

*3：简单地说，帧内预测编码就是用周围邻近的像素值来预测当前的像素值，然后对预测误差进行编码。

由于单个图像帧内有较高空间冗余度，在空间域上进行帧内方向空间预测可获得更高压缩效率。

*4：又称为抗块效应滤波器（Deblocking Filter），它的的作用是消除经反量化和反变换后重建图像中由于预测误差产生的块效应，即块边缘处的像素值跳变，从而一来改善图像的主观质量，二来减少预测误差。

*5：H.264 标准采用的熵编码有两种：一种是基于内容的自适应变长编码（CAVLC——Context-based Adaptive Variable Length Coding）；另一种是基于内容的自适应二进制算术编码（CABAC——Context-based Adaptive Binary Arithmetic Coding）。CAVLC 与 CABAC 根据相邻块的情况进行当前块的编码，以达到更好的编码效率。CABAC 比 CAVLC 压缩效率高，但要更复杂一些。

表 6-5 中列出了 H.264 与 MPEG-2 及 MPEG-4 一些主要技术比较。图 6-30 列出了 MPEG-4 及 H.264 对同一图像的编码结果。

表 6-5　H.264 与 MPEG-2/MPEG-4 主要技术比较

主要技术	H.264	MPEG-2	MPEG-4
帧内预测	4×4 像素以 9 种，16×16 像素块 4 种预测模式	无	无
帧间预测	分 16×16 像素块 7 种模式，SDTV 图像最大预测 5 帧（预测 1/4 像素）	以 16×16 像素块为单位从前面帧预测（预测 1/2 像素）	以 16×16 和 8×8 像素块为单位从前面帧预测（预测 1/4 像素）
变换	4×4 像素单位的整数变换	8×8 像素单位的离散余弦变换	
纠错编码	CAVLC 和 CABAC	哈夫曼编码	
环路滤波器	有	无	无

与 MPEG-4（Part 2）效果比较（资料来源：MPEG/IEC VCEG-N18）

（a）　　　　　　　　　　　　　（b）

图 6-30　（a）图为 MPEG-4 @ 1 Mb/s，（b）图为 H.264 @ 512 kbps，H.264 序列在一半比特率的情况下更让人偏爱一些。

6.4.2　H.264 Technology Benefits

Fig. 6-29 shown the basic idea of H.264 coding.

Some of the H.264 technology's advantages are listed below.

(1) Low bit rate. It could save up to 50% bit rate comparing to H.263v2 (H.263+) or MPEG-4 Simple Profile under the same optimum encoder condition.

(2) High quality video image. H.264 could provide continuous smooth high quality video image (DVD quality).

(3) Strong error capacity. H.264 provided solutions to deal with the missing package error under the unstable network condition.

(4) Strong network adaptation. H.264 provided network adaptation layer to make it easier to transmit H.264 files via different types of networks.

In coding frame, H.264 still use ME/MC, motion estimation/motion compensation - DCT structure, which is motion compensation plus transform coding hybrid structure. So, it keeps some characteristics of previous standards'. But the technologies it used has made it greatly improved the functionalities of previous video coding standards. What is worth to be mentioned is that the improvement is not achieved by simply applying one technology only. The result is actually coming from the integration of small functionality improvements got from different technologies.

H.264 applied many new technologies to improve the compression efficiency. Some of the major ones are listed below.

(1) Replace the discrete cosine transform, DCT with 4×4 block integration transform;

(2) Multiple reference frames. There are seven macroblock prediction formats have been used and they are 16×16, 16×8, 8×16, 8×8, 8×4, 4×8, and 4×4. This makes the motion estimation and motion compensation more accurate;

(3) Intra frame prediction;

(4) Modified deblocking filter;

(5) Enhanced entropy coding algorithms including universal VLC, UVLC, context adaptive VLC, CAVLC, and CABAC.

Abstracts in the Fig. 6-29:

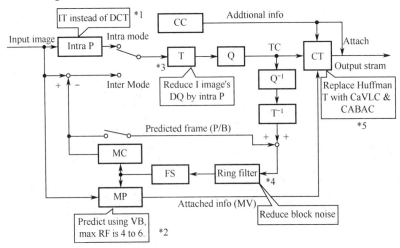

Fig. 6-29　H.264 Coding Block Diagram

IT——integration transform;

DCT——discrete cosine transform;

P——prediction;

CC——coding control;

T——transform;

Q——quantification;

CT——correction transform;

Q^{-1}——inverse quantification;

T^{-1}——inverse transform;

DQ——data quantity;

MC——motion compensation;

MP——motion prediction;

MV——motion vector;

FS——frame storage;

VB——variable block;

Notes in the figure:

*1. The DCT transform in H.264/AVC is different from before. The transform unit it used is not 8×8, but 4×4 blocks. Furthermore, the transform is integer operation instead of real number operation. The advantage of this is that the calculation speed is faster, high accuracy, and smaller memory occupancy. With the integration operation, both encoding and decoding have strict inverse transform, which avoids the cutting off errors and reduces the motion block edge coding noises. Meanwhile, the 4×4 block transform produces less block noises than the 8×8 block transform.

*2. Apply the multiple reference frames mode. It could use more than one previous frame (up to 5 frames) to do the motion estimation to improve the motion estimation accuracy, increase the decoder's error correction ability, and achieve a better prediction effect.

Apply variable block size motion compensation, such as taking 16×16 blocks in the smooth (not change that much) area and 8×8, or 4×4, or even smaller blocks in the detailed area. This way, the motion compensation could achieve higher prediction accuracy and more flexibility, which is closer to the real video continuous changing situation.

*3. Simply speaking, the intra prediction is to use the close surrounding pixels to predict the current pixel and then encode the predicted difference.

Since the single image frame has higher space redundancy, in space, intra prediction within the frame could gain higher compression ratio.

*4. It is also called deblocking filter. It is used to get rid off the block effect produced from the prediction error appearing in the images after the inverse quantification and inverse transform. The block effect means the pixel jump at the edge of the blocks. This way could change the image's watching quality and reduce the prediction error.

*5. H.264 standard applies two kinds of entropy encoding. One of them is the context based adaptive variable length coding, CAVLC and the other is the context based adaptive binary arithmetic coding, CABAC. The CAVLC and CABAC encode the current block based on the situation of neighbouring blocks to reach a better coding efficiency. CABAC has higher compression efficiency and is more complicated.

The Table 6-5 shows the major technologies compression between H.264 and MPEG-2/MPEG-4. The Fig. 6-30 shows the encoding results of H.264 and MPEG-4 to same video image.

Table 6-5   Major Technologies Compression between H.264 and MPEG-2/MPEG-4

Major Technology	H.264	MPEG-2	MPEG-4
Intra frame prediction	4×4 pixels with 9 kinds of predictions and 16x16 with 4 kinds of predictions	None	None
Inter frame prediction	16×16 pixels block, 7 modes. SDTV image 5 prediction frames max (predict 1/4 pixel).	16x16 pixels block. Predict from previous one frame (predict 1/2 pixel).	16×16 and 8×8 pixels block. Predict from previous one frame (predict 1/4 pixel).
Transform	The integrate numbers of 4×4 pixels transform	8x8 pixels block discrete cosine transform, DCT	
Error correction coding	CAVLC and CABAC	Huffman coding	
Round circuit filter	Has	None	

（a）   （b）

Fig. 6-30   The Fig. (a) is MPEG-4 @ 1Mb/s and the Fig. (b)is H.264 @ 512kbps.

We obviously prefer the H.264 with half of the bit rate.

6.4.3   H.264 的应用

H.264 的应用场合相当广泛，包括可视电话（固定或移动）、实时视频会议系统、视频监控系统、互联网视频传输以及多媒体信息存储等，涉及电视广播（家电）、计算机网络和通信三大行业，如图 6-31 所示。

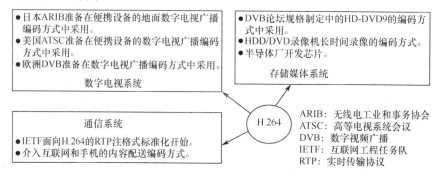

图 6-31   H.264 广泛的应用领域

其中的可视电话是利用电话线路实时传送人的语音和图像（用户的半身像、照片物品等）的一种通信方式。如果说普通电话是"顺风耳"的话，可视电话就"千里眼"了。可视电话设备是由电话机、摄像设备、电视接收显示设备及控制器组成的。

除此之外，H.264 在移动电话、视频监控、多媒体（超媒体）视频、IPTV、手机电视、宽带电话以及视频信息存储等方面也正在得到越来越广泛的应用。

IPTV（Internet Protocol Television）也称为网络电视，是指基于 IP 协议的电视广播服务。该业务将电视机或个人计算机融为一体，通过宽带网络向用户提供数字广播电视、视频服务、信息服务、互动社区、互动休闲娱乐、电子商务等宽带业务。IPTV 的主要特点是交互性和实时性。

6.4.4　关于 H.264 发展的策略思考

（1）H.264 比 H.263、MPEG4（Part 2）可节省一倍带宽而维持同等视频图像质量，并且提高了网络适应性和解码器的差错恢复能力，因而确有其巨大魅力。

但这是以增加复杂性为代价的，特别是其动目标估值补偿改进方面最为明显。据估计，H.264 的编码计算复杂度约为 H.263 的三倍，其译码复杂度约为 H.263 的两倍。

因此，一定时期内及一定应用对象与应用场合（如移动通信手机），其高复杂度往往成为实际市场应用的瓶颈。

当然，随着芯片技术的进展，这一问题会逐步得到解决，因此，适时切入的决策便非常重要。

（2）H.264 的设计初衷是期望在低带宽与速率情况下获得更良好的图像质量，因此在高码率情况下运作，H.264 的图像质量和 H.263 相比并无明显改进。由此在什么情况下选用 H.264，其网络带宽状况是用户必须考虑的重要因素。

（3）H.264 标准仅推出两年左右，其产品成熟性与互操作性还需要市场的进一步检验。

我国 AVS 与 H.264 的关系。

AVS（Audio Video coding Standard）是中国具备自主知识产权的第二代信源编码标准，它解决的重点问题是数字音视频海量数据的编码压缩问题，也称数字音视频编解码标准技术。它是数字信息传输、存储、播放等环节的前提。

我国的 AVS 以 H.264 为基础的。

6.4.3　The Applications of H.264

H.264 has wide range of applications that include video telephone (fixed and mobile), real time video conference system, video surveillance system, network video transmission, and multimedia information storage etc. Those applications distributed in three major industries, TV broadcasting (home use electronics), computer network, and communication as shown in Fig. 6-31.

Among those, the video telephone is a communication way to transmit the speaker's voice and the video images (user's above part photo, pictures, and other objects etc) real time via the telephone lines. If we say the normal phone is "ear along the wind", the video phone will be "eyes thousands miles away". The video telephone equipments include telephone, video equipment, TV receiving and displaying equipment, and controlling equipment.

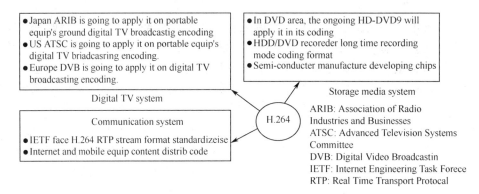

• Japan ARIB is going to apply it on portable equip's ground digital TV broadcastig encoding • US ATSC is going to apply it on portable equip's digital TV briadcasring encoding. • Europe DVB is going to apply it on digital TV broadcasting encoding.	• In DVD area, the ongoing HD-DVD9 will apply it in its coding • HDD/DVD recoreder long time recording mode coding format • Semi-conducter manufacture developing chips

Fig. 6-31   The H.264's Broad Application Areas

Besides the above, H.264 has become more and more popular in mobile phone, video surveillance, multimedia (super media) video, IPTV, cell phone TV, wide bandwidth phone, and video information storage etc.

IPTV - IPTV, internet protocol television which is also referred to as network television means the TV broadcasting based on internet protocols. This service integrates the TV and personal computer together and provides the user digital TV broadcasting, video service, information service, interacting community, interacting leisure entertainment, and electronic business etc wide bandwidth services via wide bandwidth networks. Its major characteristics are interactivity and real time.

6.4.4   The H.264's Developing Strategies

(1) Compared with H.263 and MPEG4 (Part 2), the H.264 could save half of the band width and keep the same video image quality at the same time. Meanwhile, it increases the network adaptability and the decoder's error recovery ability. Those make it really attractive.

But those advantages are not free and they are the trade off of complexity especially in motion estimation and motion compensation. Roughly, the H.264's encoding calculation is three times more complicated than H.263's. The H.264's decoding calculation is roughly two times more complicated than H.263's.

For this reason, for a certain time frame and certain applications such as mobile phone application, the calculation complexity is easy to create a bottle neck in the real application market.

On the other hand, along with the development of chip technology, this issue should be gradually solved. So, get into the field at a suitable time point is appeared very important.

(2) The initial design purpose of H.264 is to get a better video image quality under the narrower band width and lower bit rates. For this reason, if it's operating under the high bit rates environment, it doesn't have a obviously advantage comparing with H.263. So, when user chooses to apply H.264, the network bandwidth is a very important factor to take into the consideration.

(3) H.264 standard has been just proposed for about two years. Its maturation and inter action characteristics still need market evaluations.

The relationship between our country's AVS and H.264.

AVS, audio video coding standard, is the second generation of source coding standard solely owned by China. It emphasizes on digital audio video huge quantity of data encoding compression. It is also referred to digital audio video encoding decoding standard technology. It is the base of digital information transmission, storage, and broadcasting.

Our country's AVS takes H.264 as the base.

6.4.5 复合视频和分量视频

通常看到的家用电视接收机从发射机接收到的是一个复合信号，其中的亮度分量和色度分量是复合在一起的。这种信号模式是在 20 世纪 50 年代早期当颜色加到电视信号的传输中时设计的。基本的黑白电视信号变成了由亮度分量和另外两个加进来的色度分量 C1，C2 组成的复合信号。其中的两个色度信号可以是 U 和 V，也可以是 Cb 和 Cr，或者是 I 和 Q，或是其他任意两个色度分量。在这种模式中，最根本的要点是要传输的是一个信号。如果这个信号是无线传输的，那么只需要一个传输频率，如果它是有线传输的，那么只需要一条传输线。

NTSC 制中使用 YIQ 分量模式，它的定义如下：

$$Y = 0.299R' + 0.587G' + 0.114B'$$

$$I = 0.596R' - 0.274G' - 0.322B'$$
$$= -(\sin 33°)U + (\cos 33°)V$$

$$Q = 0.211R' - 0.523G' + 0.311B'$$
$$= (\cos 33°)U + (\sin 33°)V$$

在接收端，伽马修正的 $R\_G\_B\_$ 分量由下面的反变换式恢复出来。

$$R' = Y + 0.956I + 0.621Q$$

$$G' = Y - 0.272I - 0.649Q$$

$$B' = Y - 1.106I + 1.703Q$$

在 PAL 制中，使用基本的 YUV 彩色空间，其定义如下：

$$Y = 0.299R' + 0.587G' + 0.114B'$$

$$U = -0.147R' - 0.289G' + 0.436B' = 0.492(B' - Y)$$

$$V = 0.615R' - 0.515G' - 0.100B' = 0.877(R' - Y)$$

其反变换为

$$R' = Y + 1.140V$$

$$G' = Y - 0.394U - 0.580V$$

$$B' = Y - 2.030U$$

(YUV 彩色空间的最初引进是为了保证黑白电视标准和彩色电视标准的兼容)。

SECAM 制使用复合彩色空间 *YDrDb*，它的定义如下：

$$Y = 0.299R' + 0.587G' + 0.114B'$$

$$Db = -0.450R' - 0.833G' + 1.333B' = 3.059U$$

$$Dr = -1.333R' + 1.116G' - 0.217B' = -2.169V$$

其反变换为

$$R' = Y - 0.526Dr$$

$$G' = Y - 0.129Db + 0.268Dr$$

$$B' = Y + 0.665Db$$

复合视频便宜但存在一些问题，比如在显示的图像中出现的亮度交错和色度交错的人工迹象。高质量的视频中常常使用分量视频，这时要么使用三种不同的频率无线传输，要么使用三条不同的电线有线传输三个不同的彩色分量。常用的分量视频标准是 ITU-R 601 建议，它采用了 *YCbCr* 彩色空间。在这个标准中，亮度 *Y* 的取值范围为[16, 235],两个色度分量各自的取值范围为[16，240]，色度的中心是 128，表示零色度。

6.4.5 Composite and Components Video

The common television receiver found in many homes receives from the transmitter a composite signal, where the luminance and chrominance components [Salomon 99] are multiplexed. This type of signal was designed in the early 1950s, when color was added to television transmissions. The basic black-and-white signal becomes the luminance (Y) component, and another two chrominance components C1 and C2 added. Those can be U and V, Cb and Cr, I and Q, or any other chrominance components. In this format, the main point is that only one signal is needed. If the signal is sent on the air, only one transmission frequency is needed. If it is sent on a cable, only one cable is used.

NTSC uses the YIQ components, which are defined by

$$Y = 0.299R' + 0.587G' + 0.114B'$$

$$I = 0.596R' - 0.274G' - 0.322B'$$
$$= -(\sin 33°)U + (\cos 33°)V$$

$$Q = 0.211R' - 0.523G' + 0.311B'$$
$$= (\cos 33°)U + (\sin 33°)V$$

At the receiver, the gamma-corrected $R\_G\_B\_$ components are extracted using the inverse transformation

$$R' = Y + 0.956I + 0.621Q$$

$$G' = Y - 0.272I - 0.649Q$$

$$B' = Y - 1.106I + 1.703Q$$

PAL uses the basic *YUV* color space, defined by

$$Y = 0.299R' + 0.587G' + 0.114B'$$

$$U = -0.147R' - 0.289G' + 0.436B' = 0.492(B' - Y)$$

$$V = 0.615R' - 0.515G' - 0.100B' = 0.877(R' - Y)$$

whose inverse transform is

$$R' = Y + 1.140V$$

$$G' = Y - 0.394U - 0.580V$$

$$B' = Y - 2.030U$$

(The *YUV* color space was originally introduced to preserve compatibility between black-and-white and color television standards.)

SECAM uses the composite color space *YDrDb*, defined by

$$Y = 0.299R' + 0.587G' + 0.114B'$$

$$Db = -0.450R' - 0.833G' + 1.333B' = 3.059U$$

$$Dr = -1.333R' + 1.116G' - 0.217B' = -2.169V$$

The inverse transformation is

$$R' = Y - 0.526Dr$$

$$G' = Y - 0.129Db + 0.268Dr$$

$$B' = Y + 0.665Db$$

Composite video is cheap but has problems such as cross-luminance and cross-chrominance artifacts in the displayed image. High-quality video systems often use component video, where three cables or three frequencies carry the individual color components. A common component video standard is the ITU-R recommendation 601, which adopts the *YCbCr* color space. In this standard, the luminance *Y* has values in the range [16, 235], whereas each of the two chrominance components has values in the range [16, 240] centered at 128, which indicates zero chrominance.

练　习

1. 单选题

（1）视频会议系统最著名的标准是（　　）。

    A．H.261 和 H.263   B．H.320 和 T.120

    C．G.723 和 G.728   D．G.722 和 T.127

（2）下列格式中不属于 MPEG 标准中定义的图像形式是（　　）。

    A．*I* 帧内图　　　　　　　　　　B．*P* 预测图

    C．B 双向预测图　　　　　　　　D．A 交流分量图

（3）国际上常用的电视频制式有（　　）。

    （1）PAL 制　　　　　　　　　　（2）NTSC 制

    （3）MPEG 制　　　　　　　　　（4）SECAM 制

    A．（1）　　　　B．（1），（2）　　　C．（1），（2），（4）　　　D．全部

（4）对视频文件进行压缩是为了使（　　）。

    A．图像更清晰　　　　　　　　　B．对比度更高

    C．声音更动听　　　　　　　　　D．存储容量更小

（5）数字视频信息的数据量相当大，对 PC 机的存储、处理和传输都是极大的负担，为此必须对数字视频信息进行压缩编码处理。目前 VCD 光盘上存储的数字视频信息采用的压缩编码标准是（　　）。

    A．MPEG-1　　B．MPEG-2　　C．MPEG-4　　　　D．MPEG-7

（6）下列哪些是图像和视频编码的国际标准？（　　）。

    （1）JPEG　　　　　　　　　　（2）MPEG

    （3）ADPCM　　　　　　　　　（4）H.261

    A、(1),(2),(3)　　B．(1),(2)　　C．(1),(2),(4)　　　　D．全部

（7）MPEG 视频编码中使用了帧内压缩编码和帧间压缩编码。其中帧间压缩码采用的图像序列分内帧（I 图像）、预测帧（P 图像）和双向预测帧（B 图像）三种。其中（　　）计算得到的。

    A．I 图像是由 B 图像　　　　　　B．P 图像是由 I 图

    C．I 图像是由 P 图像　　　　　　D．P 图像是由 B 图像

2．多选题

（1）MPEG 标准中定义的图像形式有(　　)。

    A．*I* 帧内图　　B．P 预测图　　C．G 组内预测图　　D．B 双向预测图

    E．D 直流分量图

（2）彩色可用(　　)来描述。

    A．亮度　　　　B．色度　　　　C．颜色

    D．饱和度　　　E．对比度

3．叙述题

（1）什么是视频图像数据压缩的基础？

（2）当 MPEG 专家组提出 MPEG-1/2 视频压缩标准时，他们共定义了几种不同的图像？哪几种？哪一种图像的压缩率最高？哪一种的最低？

（3）请分别解释 TV 图像模式 MP@ML and HP@HL 的含义。

（4）什么是 TV 图像空间分辨率？什么是 TV 图像时间分辨率？

Exercises

1. Multiple choices (one single answer)

(1) The most famous video conference system standard is (　　).

A. H.261 and H.263   B. H.320 and T.120

C. G.723 and G.728   D. G.722 and T.127

(2) Which (   ) of the following is not one of the MPEG standard image format?

    A. *I* intra frame   B. *P* prediction frame

    C. B bidirectional prediction image   D. A alternate current compartment frame

(3) The common television modes used internationally are (   ).

    (1) PAL mode   (2) NTSC mode

    (3) MPEG mode   (4) SECAM mode

    A. (1)   b. (1), (2)   C. (1), (2), (4)   d. All

(4) The purpose of compressing the video file is (   ).

    A. To get clearer image   B. To get higher contrast

    C. To get more vivid sound   D. To get smaller storage capacity

(5) The quantity of the digital video information data is huge, which is a big burden to PC's storage, processing, and transmission. That's why we need to do the compression encoding process to the digital video information. Currently, the compression coding standard used for the digital video information stored in VCD is (   ).

    A. MPEG-1   B. MPEG-2   C. MPEG-4   D. MPEG-7

(6) Which of the following is/are image and video coding international standard/s?

    (1) JPEG   (2) MPEG

    (3) ADPOM   (4) H.261

    A. (1), (2), (3)   B. (1), (2)   C. (1), (2), (4)   D. All

(7) The intra frame compressing coding and inter frame compressing coding have been applied in MPEG video coding. Between those two, the inter frame compressing coding classified the video images as intra frame (I image), prediction frame (P image), and bidirectional prediction frame (B image). Based on those, (   ).

    A. I image is calculated from B image   B. P image is calculated from I image

    C. I image is calculated from P image   D. P image is calculated from B image

2. Multiple choices (several answers)

(1) The image formats defined in MPEG include (   ).

    A. I intra image   B. P prediction image

    C. G intra group prediction image   D. B bidirectional prediction image

    E. D direct current component image

(2) The color could be described by (   ).

    A. Luminance   B. Chrominance   C. Color

    D. Saturation   E. Contrast

3. Description questions

(1) What is the base of video image data compression?

(2) When MPEG experts group propose the MPEG-1/2 video compression standards, how many different kinds of images have they defined and what are they? Which image's compression rate is the highest and which one's is the lowest?

(3) Please explain the TV modes MP@ML and HP@HL respectively.

(4) What are the meanings of TV image's space resolution and time resolution?

参 考 文 献

[1] 林福宗.多媒体技术基础[M].北京:清华大学出版社, 2002.

[2] 赵子江.多媒体技术应用教程[M].北京:机械工业出版社, 2004.

[3] 张晓艳.多媒体技术基础[M].沈阳：辽宁科学技术出版社, 2012.

[4] 钟玉琢.多媒体技术基础及应用[M].北京:清华大学出版社, 2012.

[5] David Salomon, Giovanni Motta. Handbook of Data Compressio.Springer, 2010.

[6] Mark Nelson. The Data Compression Book, IDG Books Worldwide Inc., 2005.

[7] Saeed Vaseghi. Multimedia Signal Processing,, John Wiley & Sons Ltd, 2007.

[8] William Pratt. Digital Image Processing , John Wiley & Sons Ltd, 2007.